Studies in Big Data

Volume 40

Series editor

Janusz Kacprzyk, Polish Academy of Sciences, Warsaw, Poland
e-mail: kacprzyk@ibspan.waw.pl

The series "Studies in Big Data" (SBD) publishes new developments and advances in the various areas of Big Data- quickly and with a high quality. The intent is to cover the theory, research, development, and applications of Big Data, as embedded in the fields of engineering, computer science, physics, economics and life sciences. The books of the series refer to the analysis and understanding of large, complex, and/or distributed data sets generated from recent digital sources coming from sensors or other physical instruments as well as simulations, crowd sourcing, social networks or other internet transactions, such as emails or video click streams and other. The series contains monographs, lecture notes and edited volumes in Big Data spanning the areas of computational intelligence including neural networks, evolutionary computation, soft computing, fuzzy systems, as well as artificial intelligence, data mining, modern statistics and Operations research, as well as self-organizing systems. Of particular value to both the contributors and the readership are the short publication timeframe and the world-wide distribution, which enable both wide and rapid dissemination of research output.

More information about this series at http://www.springer.com/series/11970

Robert Bembenik · Łukasz Skonieczny
Grzegorz Protaziuk · Marzena Kryszkiewicz
Henryk Rybinski
Editors

Intelligent Methods and Big Data in Industrial Applications

Springer

Editors
Robert Bembenik
Institute of Computer Science
Warsaw University of Technology
Warsaw
Poland

Marzena Kryszkiewicz
Institute of Computer Science
Warsaw University of Technology
Warsaw
Poland

Łukasz Skonieczny
Institute of Computer Science
Warsaw University of Technology
Warsaw
Poland

Henryk Rybinski
Institute of Computer Science
Warsaw University of Technology
Warsaw
Poland

Grzegorz Protaziuk
Institute of Computer Science
Warsaw University of Technology
Warsaw
Poland

ISSN 2197-6503 ISSN 2197-6511 (electronic)
Studies in Big Data
ISBN 978-3-030-08492-9 ISBN 978-3-319-77604-0 (eBook)
https://doi.org/10.1007/978-3-319-77604-0

Printed on acid-free paper

This Springer imprint is published by the registered company Springer International Publishing AG part of Springer Nature
The registered company address is: Gewerbestrasse 11, 6330 Cham, Switzerland

Preface

This book presents valuable contributions devoted to practical applications of Intelligent Methods and Big Data in various branches of the industry. The contents of the volume are based on submissions to the Industrial Session of the 23rd International Symposium on Methodologies for Intelligent Systems (ISMIS 2017), which was held in Warsaw, Poland.

All the papers included in the book successfully passed the reviewing process. They cover topics of diverse character, which is reflected in the arrangement of the volume. The book consists of the following parts: Artificial Intelligence Applications, Complex Systems, Data Mining, Medical Applications and Bioinformatics, Multimedia Processing and Text Processing. We will now outline the contents of the chapters.

Part I, "Artificial Intelligence Applications", deals with applications of AI in the areas of computer games, finding the fastest route, recommender systems and community detection as well as with forecasting of energy futures. It also discusses the dilemma of innovation—AI trade-off.

- **Germán G. Creamer** ("Nonlinear Forecasting of Energy Futures") proposes the use of the Brownian distance correlation for feature selection and for conducting a lead-lag analysis of energy time series. Brownian distance correlation determines relationships similar to those identified by the linear Granger causality test, and it also uncovers additional nonlinear relationships among the log return of oil, coal and natural gas. When these linear and nonlinear relationships are used to forecast the direction of energy futures log return with a nonlinear classification method such as support vector machine, the forecast of energy futures log return improves when compared to a forecast based only on Granger causality.
- **Mateusz Modrzejewski and Przemysław Rokita** ("Implementation of Generic Steering Algorithms for AI Agents in Computer Games") propose a set of generic steering algorithms for autonomous AI agents along with the structure of the implementation of a movement layer designed to work with these algorithms. The algorithms are meant for further use in computer animation in

computer games, provide a smooth and realistic base for the animation of the agent's movement and are designed to work with any graphic environment and physics engine, thus providing a solid, versatile layer of logic for computer game AI engines.

- **Mieczyslaw Muraszkiewicz** ("The Dilemma of Innovation–Artificial Intelligence Trade-Off") makes use of dialectic that confronts pros and cons to discuss some relationships binding innovation, technology and artificial intelligence, and culture. The main message of this contribution is that even sophisticated technologies and advanced innovations, such as those that are equipped with artificial intelligence, are not panacea for the increasing contradictions, problems and challenges contemporary societies are facing. Often, we have to deal with a trade-off dilemma that confronts the gains provided by innovations with downsides they may cause. The author claims that in order to resolve such dilemmas and to work out plausible solutions one has to refer to culture *sensu largo*.
- **Cezary Pawlowski, Anna Gelich and Zbigniew W. Raś** ("Can we Build Recommender System for Artwork Evaluation?") propose a strategy of building a real-life recommender system for assigning a price tag to an artwork. The other goal is to verify a hypothesis about existence of a co-relation between certain attributes used to describe a painting and its price. The authors examine the possibility of using methods of data mining in the field of art marketing and describe the main aspects of the system architecture and performed data mining experiments, as well as processes connected with data collection from the World Wide Web.
- **Grzegorz Protaziuk, Robert Piątkowski and Robert Bembenik** ("Modelling OpenStreetMap Data for Determination of the Fastest Route Under Varying Driving Conditions") propose an approach to creation of a network graph for determining the fastest route under varying driving conditions based on OpenStreetMap data. The introduced solution aims at finding the fastest point-to-point path problem. The authors present a method of transformation of the OpenStreetMap data into a network graph and a few proposals for improving the graph obtained by almost directly mapping the source data into the destination model. For determination of the fastest route, a modified version of Dijkstra's algorithm and a time-dependent model of network graph is used where the flow speed of each edge depends on the time interval.
- **Krista Rizman Žalik** ("Evolution Algorithm for Community Detection in Social Networks Using Node Centrality") uses a multiobjective evolution community detection algorithm which forms centre-based communities in a network exploiting node centrality. Node centrality is easy to use for better partitions and for increasing the convergence of the evolution algorithm. The proposed algorithm reveals the centre-based natural communities with high quality. Experiments on real-world networks demonstrate the efficiency of the proposed approach.

Part II, "Complex Systems", is devoted to innovative systems and solutions that have applications in high-performance computing, distributed systems, monitoring and bus protocol implementation.

- **Nunziato Cassavia, Sergio Flesca, Michele Ianni, Elio Masciari, Giuseppe Papuzzo and Chiara Pulice** ("High Performance Computing by the Crowd") leverage the idling computational resources of users connected to a network to the projects whose complexity could be quite challenging, e.g. biomedical simulations. The authors designed a framework that allows users to share their CPU and memory in a secure and efficient way. Users help each other by asking the network computational resources when they face high computing demanding tasks. As such the approach does not require to power additional resources for solving tasks (unused resources already powered can be exploited instead), the authors hypothesize a remarkable side effect at steady state: energy consumption reduction compared with traditional server farm or cloud-based executions.
- **Jerzy Chrząszcz** ("Zero-Overhead Monitoring of Remote Terminal Devices") presents a method of delivering diagnostic information from data acquisition terminals via legacy low-throughput transmission system with no overhead. The solution was successfully implemented in an intrinsically safe RFID system for contactless identification of people and objects, developed for coal mines in the end of the 1990s. The contribution presents the goals, and main characteristics of the application system are presented, with references to underlying technologies and transmission system and the idea of diagnostic solution.
- **Wiktor B. Daszczuk** ("Asynchronous Specification of Production Cell Benchmark in Integrated Model of Distributed Systems") proposes the application of fully asynchronous IMDS (Integrated Model of Distributed Systems) formalism. In the model, the sub-controllers do not use any common variables or intermediate states. Distributed negotiations between sub-controllers using a simple protocol are applied. The verification is based on CTL (Computation Tree Logic) model checking, integrated with IMDS.
- **Julia Kosowska and Grzegorz Mazur** ("Implementing the Bus Protocol of a Microprocessor in a Software-Defined Computer") presents a concept of software-defined computer implemented using a classic 8-bit microprocessor and a modern microcontroller with ARM Cortex-M core for didactic and experimental purposes. The device being a proof-of-concept demonstrates the software-defined computer idea and shows the possibility of implementing time-critical logic functions using a microcontroller. The project is also a complex exercise in real-time embedded system design, pushing the microcontroller to its operational limits by exploiting advanced capabilities of selected hardware peripherals and carefully crafted firmware. To achieve the required response times, the project uses advanced capabilities of microcontroller peripherals—timers and DMA controller. Event response times achieved with the microcontroller operating at 80 MHz clock frequency are below 200 ns, and the interrupt frequency during the computer's operation exceeds 500 kHz.

Part III, "Data Mining", deals with the problems of stock prediction, sequential patterns in spatial and non-spatial data, as well as classification of facies.

- **Katarzyna Baraniak** ("ISMIS 2017 Data Mining Competition: Trading Based on Recommendations—XGBoost approach with feature Engineering") presents an approach to predict trading based on recommendations of experts using an XGBoost model, created during ISMIS17 Data Mining Competition: Trading Based on Recommendations. A method to manually engineer features from sequential data and how to evaluate its relevance is presented. A summary of feature engineering, feature selection and evaluation based on experts' recommendations of stock return is provided.
- **Marzena Kryszkiewicz and Łukasz Skonieczny** ("Fast Discovery of Generalized Sequential Patterns") propose an optimization of the GSP algorithm, which discovers generalized sequential patterns. Their optimization consists in more selective identification of nodes to be visited while traversing a hash tree with candidates for generalized sequential patterns. It is based on the fact that elements of candidate sequences are stored as ordered sets of items. In order to reduce the number of visited nodes in the hash tree, the authors also propose to use not only parameters windowSize and maxGap as in original GSP, but also parameter minGap. As a result of their optimization, the number of candidates that require final time-consuming verification may be considerably decreased. In the experiments they have carried out, their optimized variant of GSP was several times faster than standard GSP.
- **Marcin Lewandowski and Łukasz Słonka** ("Seismic Attributes Similarity in Facies Classification") identify key seismic attributes (also the weak ones) that help the most with machine learning seismic attribute analysis and test the selection with Random Forest algorithm. The initial tests have shown some regularities in the correlations between seismic attributes. Some attributes are unique and potentially very helpful for information retrieval, while others form non-diverse groups. These encouraging results have the potential for transferring the work to practical geological interpretation.
- **Piotr S. Maciąg** ("Efficient Discovery of Sequential Patterns from Event-Based Spatio-Temporal Data by Applying Microclustering Approach") considers spatiotemporal data represented in the form of events, each associated with location, type and occurrence time. In the contribution, the author adapts a microclustering approach and uses it to effectively and efficiently discover sequential patterns and to reduce the size of a data set of instances. An appropriate indexing structure has been proposed, and notions already defined in the literature have been reformulated. Related algorithms already presented in the literature have been modified, and an algorithm called Micro-ST-Miner for discovering sequential patterns in event-based spatiotemporal data has been proposed.

Part IV, "Medical Applications and Bioinformatics", focuses on presenting efficient algorithms and techniques for analysis of biomedical images, medical evaluation and computer-assisted diagnosis and treatment.

- **Konrad Ciecierski and Tomasz Mandat** ("Unsupervised Machine Learning in Classification of Neurobiological Data") show comparison of results obtained from supervised—random forest-based—method with those obtained from unsupervised approaches, namely K-means and hierarchical clustering approaches. They discuss how inclusion of certain types of attributes influences the clustering based results.
- **Bożena Małysiak-Mrozek, Hanna Mazurkiewicz and Dariusz Mrozek** ("Incorporating Fuzzy Logic in Object-Relational Mapping Layer for Flexible Medical Screenings") present the extensions to the Doctrine ORM framework that supply application developers with possibility of fuzzy querying against collections of crisp medical data stored in relational databases. The performance tests prove that these extensions do not introduce a significant slowdown while querying data, and can be successfully used in development of applications that benefit from fuzzy information retrieval.
- **Andrzej W. Przybyszewski, Stanislaw Szlufik, Piotr Habela and Dariusz M. Koziorowski** ("Multimodal Learning Determines Rules of Disease Development in Longitudinal Course with Parkinson's Patients") use data mining and machine learning approach to find rules that describe and predict Parkinson's disease (PD) progression in two groups of patients: 23 BMT patients that are taking only medication; 24 DBS patients that are on medication and on DBS (deep brain stimulation) therapies. In the longitudinal course of PD, there were three visits approximately every 6 months with the first visit for DBS patients before electrode implantation. The authors have estimated disease progression as UPDRS (unified Parkinson's disease rating scale) changes on the basis of patient's disease duration, saccadic eye movement parameters and neuropsychological tests: PDQ39 and Epworth tests.
- **Piotr Szczuko, Michał Lech and Andrzej Czyżewski** ("Comparison of Methods for Real and Imaginary Motion Classification from EEG Signals") propose a method for feature extraction, and then some results of classifying EEG signals that are obtained from performed and imagined motion are presented. A set of 615 features has been obtained to serve for the recognition of type and laterality of motion using various classifications approaches. Comparison of achieved classifiers accuracy is presented, and then, conclusions and discussion are provided.

Part V, "Multimedia Processing", covers topics of procedural generation and classification of visual, musical and biometrical data.

- **Izabella Antoniuk and Przemysław Rokita** ("Procedural Generation of Multilevel Dungeons for Application in Computer Games using Schematic Maps and L-system") present a method for procedural generation of multilevel dungeons, by processing set of schematic input maps and using L-system for the shape generation. A user can define all key properties of generated dungeon, including its layout, while results are represented as easily editable 3D meshes. The final objects generated by the algorithm can be used in some computer games or similar applications.

- **Alfredo Cuzzocrea, Enzo Mumolo and Gianni Vercelli** ("An HMM-Based Framework for Supporting Accurate Classification of Music Datasets") use Hidden Markov Models (HMMs) and Mel-Frequency Cepstral Coefficients (MFCCs) to build statistical models of classical music composers directly from the music data sets. Several musical pieces are divided by instruments (string, piano, chorus, orchestra), and, for each instrument, statistical models of the composers are computed. The most significant results coming from experimental assessment and analysis are reported and discussed in detail.
- **Aleksandra Dorochowicz, Piotr Hoffmann, Agata Majdańczuk and Bożena Kostek** ("Classification of Musical Genres by Means of Listening Tests and Decision Algorithms") compare the results of audio excerpt assignment to a music genre obtained in listening tests and classification by means of decision algorithms. Conclusions contain the results of the comparative analysis of the results obtained in listening tests and automatic genre classification.
- **Michal Lech and Andrzej Czyżewski** ("Handwritten Signature Verification System Employing Wireless Biometric Pen") showcase the handwritten signature verification system, which is a part of the developed multimodal biometric banking stand. The hardware component of the solution is described with a focus on the signature acquisition and verification procedures. The signature is acquired by employing an accelerometer and a gyroscope built in the biometric pen, and pressure sensors for the assessment of the proper pen grip.

Chapter VI, "Text Processing", consists of papers describing problems, solutions and experiments conducted on text-based content, including Web in particular.

- **Katarzyna Baraniak and Marcin Sydow** ("Towards Entity Timeline Analysis in Polish Political News") present a simple method of analysing occurrences of entities in news articles. The authors demonstrate that frequency of named entities in news articles is a reflection of events in real world related to these entities. Occurrences and co-occurrences of entities between portals are compared.
- **María G. Buey, Cristian Roman, Angel Luis Garrido, Carlos Bobed and Eduardo Mena** ("Automatic Legal Document Analysis: Improving the Results of Information Extraction Processes using an Ontology") argue that current software systems for information extraction (IE) from natural language documents are able to extract a large percentage of the required information, but they do not usually focus on the quality of the extracted data. Therefore, they present an approach focused on validating and improving the quality of the results of an IE system. Their proposal is based on the use of ontologies which store domain knowledge, and which we leverage to detect and solve consistency errors in the extracted data.
- **Krystyna Chodorowska, Barbara Rychalska, Katarzyna Pakulska and Piotr Andruszkiewicz** ("To Improve, or Not to Improve; How Changes in Corpora Influence the Results of Machine Learning Tasks on the Example of Datasets Used for Paraphrase Identification") attempt to verify the influence of data quality improvements on results of machine learning tasks. They focus on measuring semantic similarity and use the SemEval 2016 data sets. They

address two fundamental issues: first, how each characteristic of the chosen sets affects performance of similarity detection software, and second, which improvement techniques are most effective for provided sets and which are not.

- **Narges Tabari and Mirsad Hadzikadic** ("Context Sensitive Sentiment Analysis of Financial Tweets: A New Dictionary") describe an application of a lexicon-based domain-specific approach to a set of tweets in order to calculate sentiment analysis of the tweets. Further, they introduce a domain-specific lexicon for the financial domain and compare the results with those reported in other studies. The results show that using a context-sensitive set of positive and negative words, rather than one that includes general keywords, produces better outcomes than those achieved by humans on the same set of tweets.

We would like to thank all the authors for their contributions to the book and express our appreciation for the work of the reviewers. We thank the industrial partners: Samsung, Allegro and mBank for the financial support of the ISMIS 2017 Conference and this publication.

Warsaw, Poland
July 2017

Robert Bembenik
Łukasz Skonieczny
Grzegorz Protaziuk
Marzena Kryszkiewicz
Henryk Rybinski

Contents

Part I
Artificial Intelligence Applications

Nonlinear Forecasting of Energy Futures

Germán G. Creamer

Abstract This paper proposes the use of the Brownian distance correlation for feature selection and for conducting a lead-lag analysis of energy time series. Brownian distance correlation determines relationships similar to those identified by the linear Granger causality test, and it also uncovers additional non-linear relationships among the log return of oil, coal, and natural gas. When these linear and non-linear relationships are used to forecast the direction of energy futures log return with a non-linear classification method such as support vector machine, the forecast of energy futures log return improve when compared to a forecast based only on Granger causality.

Keywords Financial forecasting · Lead-lag relationship · Non-linear correlation Energy finance · Support vector machine · Artificial agents

1 Introduction

The major contaminant effects of coal and the reduction of natural gas prices since 2005 have led to a contraction in the proportion of coal and an increase in the share of natural gas used in the production of electricity in the US since the year 2000 (see Fig. 1). According to the US Energy Information Administration [25], natural gas and coal will account for 43 and 27% of total electricity generation in 2040, respectively. The share of oil on electricity generation has also decreased since 2000 in the US; however, it is still important at the world level, where it accounts for about 5% [24]. These change of inputs by electric power plants, due to environmental, political or market considerations, may indicate that several fossil fuel prices are mutually determined or that one price depends on another one.

Mohammadi [20] finds that in the case of the US, oil and natural gas prices are globally and regionally determined, respectively, and coal prices are defined by long-term contracts. Mohammadi [19], using cointegration analysis, exposes a strong relationship between electricity and coal prices and an insignificant relationship between

G. G. Creamer (✉)
School of Business, Stevens Institute of Technology, Hoboken, NJ 07030, USA
e-mail: gcreamer@stevens.edu

R. Bembenik et al. (eds.), *Intelligent Methods and Big Data in Industrial Applications*, Studies in Big Data 40, https://doi.org/10.1007/978-3-319-77604-0_1

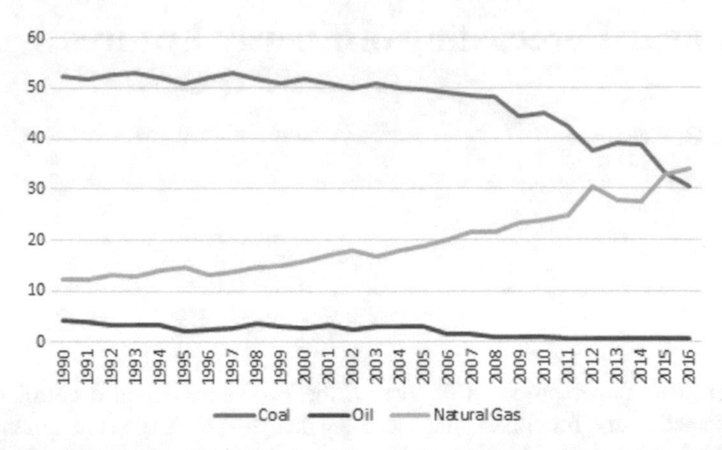

Fig. 1 Evolution of net electricity generation by energy source. *Source*: US Energy Information Administration

electricity and oil and/or natural gas prices. Asche et al. [3] and Bachmeier and Griffin [6] find very weak linkages among oil, coal and natural gas prices using cointegration analysis, while crude oil and several refined product prices are integrated [3]. Hartley et al. [12] notice an indirect relationship between natural gas and oil prices. Furthermore, Aruga and Managi [2] detect a weak market integration among a large group of energy products: WTI oil, Brent oil, gasoline, heating oil, coal, natural gas, and ethanol futures prices.

Mjelde and Bessler [18] observe that oil, coal, natural gas, and uranium markets are not fully cointegrated. Asche et al. [5] indicate that the U. K. energy market between 1995 and 1998 was highly integrated when the demand was for energy rather than for a particular source of energy. Brown and Yucel [8] show that oil and natural gas prices have been independent since 2000; however, when weather and inventories are taken into consideration in an error correction model, crude oil prices have an effect on natural gas prices. Similar results are obtained by Ramberg [21] using cointegration analysis. Amavilah [1] observes that oil prices influence uranium prices.

Causality analysis is also used to evaluate the relationship between spot and future commodity prices. Asche et al. [4]–using a non-linear Granger causality test–shows that neither the futures nor the spot crude oil market leads the relationship.

Most of the studies mentioned are based on cointegration analysis and Granger causality; however, none of these studies have used a non-linear correlation measure to evaluate the lead-lag relationship among the fossil fuels.

In this paper, I propose to use the Brownian distance correlation to conduct a non-linear lead-lag dependence analysis of coal, oil and gas futures log return. I also test if there is any improvement in the forecast of these energy futures using these non-linear dependencies compared to a forecast based only on linear relationships such as those identified by the Granger causality test. Section 2 introduces the different methods explored in this study; Sect. 3 presents the data used; Sect. 4 explains in detail the estimation techniques; Sect. 5 presents the results of the tests; Sect. 6 discusses the results, and Sect. 7 draws some conclusions and final comments.

2 Methods

In this section, I describe the following methods used to evaluate the causality among the fuel futures time series.

2.1 Granger Causality

Granger causality [9–11] is a very popular methodology used in economics, financial econometrics, as well as in many other areas of study, such as neuroscience, to evaluate the linear causal relationship among two or more variables. According to the basic definition of Granger causality, the forecasting of the variable Y_t with an autoregressive process using Y_{t-l} as its lag-l value should be compared with another autoregressive process using Y_{t-l} and the vector X_{t-l} of potential explanatory variables. Thus, X_{t-l} Granger causes Y_t when X_{t-l} happens before Y_t, and X_{t-l} has unique information to forecast Y_t that is not present in other variables.

Typically, Granger causality is tested using an autoregressive model with and without the vector X_{t-1}, such as in the following bivariate example:

$$Y_t = \sum_{l=1}^{L} \alpha_l Y_{t-l} + \epsilon_1 \tag{1}$$

$$Y_t = \sum_{l=1}^{L} \alpha_l Y_{t-l} + \sum_{l=1}^{L} \beta_l X_{t-l} + \epsilon_2 \tag{2}$$

where the residual ϵ_j is a white noise series: $\epsilon_j \sim N(0, \sigma)$, $j = 1,2$.

X_{t-l} Granger causes Y_t if the null hypothesis $H_0 : \beta_l = 0$ is rejected based on the F-test. The order of the autoregressive model is selected according to either the Akaike information criterion or the Bayesian information criterion.

2.2 Brownian Distance

Székely and Rizzo [22] have proposed a multivariate dependence coefficient called distance correlation that can be used with random vectors of multiple dimensions.

Székely and Rizzo [22] also proposed the Brownian distance covariance, which captures the covariance on a stochastic process. Distance covariance between the random vectors X and Y measures the distance between $f_X f_Y$ and $f_{X,Y}$ and is obtained as:

$$\nu(X, Y) = \sqrt{\| f_{X,Y}(t, s) - f_X(t) f_Y(s) \|^2} \tag{3}$$

where $\|.\|$ is the norm, t and s are vectors, f_X and f_Y are the characteristic functions of X and Y respectively, and $f_{X,Y}$ is the joint characteristic function of X and Y.

Empirically, $\nu(X, Y)$ evaluates the null hypothesis of independence $H_0 : f_X f_Y = f_{X,Y}$ versus the alternative hypothesis $H_A : f_X f_Y \neq f_{X,Y}$. In this paper, we refer to this test as the distance covariance test of independence.

Likewise, distance variance is:

$$\nu(X) = \sqrt{\| f_{X,X}(t, s) - f_X(t) f_X(s) \|^2} \tag{4}$$

Once distance covariance is defined, I obtain the distance correlation $R(X, Y)$ from the following expression:

$$R^2 = \begin{cases} \frac{\nu^2(X,Y)}{\sqrt{\nu^2(X)\nu^2(Y))}}, & \nu^2(X)\nu^2(Y) > 0 \\ 0, & \nu^2(X)\nu^2(Y) = 0 \end{cases} \tag{5}$$

Distance correlation takes a value of zero in case of independence and one when there is complete dependence.

In general, this research proposes the evaluation of the non-linear dependence of any financial time series such as the current value of Y (Y_t) on the l lagged value of X (X_{t-l}) with the Brownian distance correlation $R(X_{t-l}, Y_t)$. In particular, I wish to explore the lead-lag relationship among the time series under study. If $R(X_{t-l}, Y_t) \neq 0$ and $l > 0$, then X_{t-l} leads the series Y_t. Additionally, if $R(X_{t-l}, Y_t) \neq 0$, $R(X_t, Y_{t-l}) = 0$ and $l > 0$, then there is an unidirectional relationship from X_{t-l} to Y_t. However, if $R(X_{t-l}, Y_t) \neq 0$, $R(X_t, Y_{t-l}) \neq 0$ and $l > 0$, then there is a feedback relationship between X and Y. On the contrary, if $R(X_{t-l}, Y_t) = 0$ and $R(X_t, Y_{t-l}) = 0$ then there is no lead lag relationship between X and Y [26].

2.3 Support Vector Machine

Support vector machine was proposed by Vapnik [27] as a classification method based on the use of kernels to pre-process data in a higher dimension than the original space. This transformation allows an optimal hyperplane to separate the data X in two categories or values.

A hyperplane is defined as:

$$\{X : F(X) \doteq X^T \beta + \beta_0\}$$

where $F(X) = 0$ and $||\beta|| = 1$
The strong prediction rule learned by a SVM model is the sign of $F(X)$ (see [13]).

3 Data

The dataset allows the exploration of the daily relationship among coal, oil and natural gas futures log returns from January 3, 2006–December 31, 2012. I selected a sample that includes two years before and after the financial crisis period of 2008–2010 to evaluate the causality among log returns during different economic periods.

I used the following daily time series of one month forward futures log prices of the fossil fuel series for the period 2006–2012 to calculate the log returns: West Texas Intermediate oil (WTI), the Central Appalachian [bituminous] coal (Coal) and natural gas (Gas) from the New York Mercantile Exchange (NYMEX).

4 Estimation Techniques

I evaluated the weak stationarity condition of the time series, which implies that both the mean and the variance are time invariant, using the augmented Dickey-Fuller Unit Root (ADF) test. This test indicated that all log price series are non-stationary (Table 1-a) and, as expected after taking the first difference of the log prices, the log return series are stationary with a 99% confidence level for all periods (Table 1-b). So, I used the log returns to conduct the causality tests. These series have some relevant autoregressive effects according to the autocorrelation function (ACF) and the partial ACF; however, the emphasis of this paper is on the lagged cross-correlation.

Table 1 t ratio of the ADF unit-root test by product and period for log prices and log returns. The null hypothesis is the presence of a unit root or that the series is non-stationary. $*$: $p \leq 0.05$, $**$: $p \leq 0.01$

	(a) Log prices			
	2006–12	Pre-crisis	Crisis	Recovery
WTI	–1.70	1.72	–1.21	–2.48
Coal	–2.13	–1.57	–1.14	–2.74
Gas	–2.78	–3.36	–1.43	–1.35
	(b) Log returns			
	2006–12	Pre-crisis	Crisis	Recovery
WTI	–13.25**	–7.07**	–9.24**	–8.13**
Coal	–10.93**	–7.61**	–9.54**	–7.72**
Gas	–11.09**	–8.11**	–8.86**	–7.88**

I applied the Bai-Perron [7] test to detect structural breaks on the time series of the coal/WTI log prices ratio, considering that these are the most dominant products of the causality analysis. The Bai-Perron test is particularly useful when the break date is unknown, and there is more than one break date. For the complete series and each of the periods identified with the Bai-Perron test, I tested the non-linearity of the series using the White [17] and the Terasvirta test [23]. I also conducted a non-linear lead-lag relationship analysis using the Brownian distance correlation between each pair of variables and up to seven lags. I compared these results with the Granger causality test and evaluated the cointegration of the different pairs using the Johansen test [15, 16] to determine the necessity of a VAR error correction model. Two series are cointegrated when each of them are unit-root nonstationary, and a linear combination of both of them is unit-root stationary. In practical terms, it implies that in the short term prices may diverge; however they tend to converge in the long-term. Since none of the log price pairs were cointegrated in the different periods at the 5% significance level according to the Johansen test, I used a vector autoregressive (VAR) model of the log return series to run the Granger causality test with 7 lags instead of using the VAR error correction model. In this paper, $\rightarrow$ denotes relationship. For instance, $X \rightarrow Y$ indicates that X Granger causes Y when Granger causality is used, or Y is dependent on X when the Brownian distance correlation is used.

I evaluated the results of both tests using the relevant variables selected with 70% of the observations of each period, the first lag of the dependent variable, and lags identified by each method to predict the log return direction of the fuel futures using support vector machine with a radial basis kernel. I selected this non-parametric forecasting method because of its flexibility and capacity to model both linear and non-linear relationships. I tested the three periods defined by the structural breaks using a rolling window with 90% of the observations where the training and test datasets are based on 70 and 30% of the observations respectively. The rolling window moved in 5-observation increments until every complete period was tested. The hold-out samples generated by this rolling window are used for cross-validation. Finally, I tested the mean differences of the area under the receiver operating characteristic curve (ROC), and the error rate of the hold-out samples of the models explored based on the Brownian distance correlation and the Granger causality. Whenever the Granger causality test did not show any significant results for a particular product, I used a basic model that included the first lag of the dependent variable.

I calculated the Granger causality, the Brownian distances, the support vector machine models, the structural break points, the Johansen test, and the non-stationarity (Augmented Dickey-Fuller) and nonlinearity (White and Terasvirta) tests using the MSBVAR, Energy, E1071, Strucchange, URCA and Tseries packages for R respectively.[1]

[1] Information about R can be found at http://cran.r-project.org.

Table 2 Descriptive statistics of daily log returns

	2006–12			Pre-crisis		
	Coal	WTI	Gas	Coal	WTI	Gas
Mean (%)	–0.01	–0.02	–0.06	0.06	0.07	–0.04
Standard deviation	1.84	2.39	3.22	1.26	1.88	3.65
Coeff. of variation	–173.39	– 112.42	–53.26	19.99	25.59	–83.04
Min	–28.71	–13.07	–14.89	–5.19	–8.47	–14.89
Max	29.35	16.41	26.87	9.53	7.37	24.96
	Crisis			Recovery		
Mean (%)	–0.02	–0.01	–0.09	–0.03	0.02	–0.05
Standard deviation	2.99	3.25	3.70	1.21	1.91	2.73
Coeff. of variation	–140.25	–223.34	–39.88	–40.91	109.43	–52.72
Min	–28.71	–13.07	–9.78	–5.57	–9.04	–8.25
Max	29.35	16.41	26.87	4.64	8.95	13.27

Table 3 Correlation matrix of log returns

2006–12				Pre-crisis			Crisis			Recovery		
	Coal	WTI	Gas	Coal	WTI	Gas	Coal	WTI	Gas	Coal	WTI	Gas
Coal	1.00	0.27	0.17	1.00	0.07	0.16	1.00	0.35	0.20	1.00	0.30	0.28
WTI	0.27	1.00	0.21	0.07	1.00	0.28	0.35	1.00	0.24	0.30	1.00	0.17
Gas	0.17	0.21	1.00	0.16	0.28	1.00	0.20	0.24	1.00	0.28	0.17	1.00

5 Results

The Bai-Perron test applied to the coal/WTI ratio series split the data into the following periods: January 3, 2006–January 17, 2008 (pre-crisis period), January 18, 2008–November 17, 2010 (financial crisis period), and November 18, 2010–December 31, 2012 (recovery period). I studied these different periods and the complete series 2006–2012.

The most dispersed series measured by the coefficient of variation is gas during the pre-crisis period, and WTI afterwards (see Table 2). During the crisis period, the correlation between WTI and coal log returns increases (Table 3) while in every period the correlation between gas and coal increases and the correlation between gas and WTI decreases. These cross-correlation changes indicate a high interrelationship among the three fossil fuel series; however, the long-term dynamic linkages are better captured by the lead-lag Granger causality and Brownian distance correlation analysis included in Table 4.

Table 4 Brownian distance correlation of log return series. Non-relevant relationships are excluded. Yellow indicates non-linearity according to either the White or Terasvirta test and green means that both tests detect non-linearity with a 5% significance level. Recov. stands for recovery. P-values of null hypothesis that parameters are zero for Brownian distance correlation: ∗: $p \leq 0.05$, ∗∗: $p \leq 0.01$. The table also includes the p-value of the Granger causality test: †: $p \leq 0.05$, ‡: $p \leq 0.01$. For both the Brownian distance correlation and Granger causality test: ±: p-value ≤ 0.01

	Lag→Effect	1	2	3	4	5	6	7
2006-2012	WTI→Coal	0.12±	0.07±	0.09±	0.07±	0.08±	0.08±	0.07±
	Gas→Coal	0.06∗	0.06	0.05	0.04	0.04	0.06	0.07∗
	Coal/WTI	0.09∗∗	0.08∗∗	0.09∗∗	0.10∗∗	0.08∗∗	0.09∗∗	0.07∗
	Coal→Gas	0.05	0.08∗∗	0.07∗	0.05	0.06 ‡	0.05 ‡	0.05 †
	WTI→ Gas	0.09∗∗	0.05	0.04	0.04	0.04	0.05	0.04
Pre-crisis	WTI→ Coal	0.18∗∗	0.09	0.09	0.09	0.07	0.07	0.09
	Gas→ Coal	0.12∗	0.11∗	0.08	0.07	0.08	0.09	0.09
	Gas→ WTI	0.12∗ ‡	0.08 ‡	0.07 ‡	0.08 †	0.07 †	0.07	0.08
Crisis	WTI→ Coal	0.16∗∗ ‡	0.11∗ ‡	0.11∗ ‡	0.10∗ ‡	0.11 ‡	0.11∗ ‡	0.10 ‡
	Gas→ Coal	0.08	0.07	0.09	0.06	0.07	0.08	0.11∗
	Coal→ WTI	0.13∗∗	0.11∗	0.12∗∗	0.14∗∗	0.13∗∗	0.13∗∗	0.09
	Coal→ Gas	0.08	0.11∗	0.12∗	0.07	0.10 †	0.07 †	0.09
	WTI→ Gas	0.11∗	0.07	0.07	0.07	0.07	0.06	0.08
Recov.	Gas→ Coal	0.07	0.09	0.09	0.07	0.08	0.09	0.07
	Coal→ Gas	0.09 ‡	0.13∗ ‡	0.08 ‡	0.07 †	0.08 †	0.07	0.09

During the complete period 2006–2012, WTI and Coal show a two-way feedback relationship according to the Brownian distance, and only the WTI → Coal relationship is supported conforming to the Granger causality test with a 5% significance level (see Table 4). Coal also Granger causes Gas for the lags 5, 6 and 7. Additionally, the Brownian distance recognizes the following dependences (all relevant lags listed between parentheses): Gas (1, 7) → Coal, Coal (2, 3) → Gas, and WTI (1) → Gas. Very similar relationships are observed during the crisis period (2008–2010). So, the crisis years dominate the complete period of analysis. Both tests indicate that the Gas → WTI dependence is relevant during the pre-crisis period, and the Brownian distance also recognizes the importance of the relationship Gas (1,2)→ Coal and WTI (1) → Coal. During the recovery period, only the Coal → Gas relationship is relevant for both tests, especially for the Granger causality tests (first 5 lags). Most of the additional relationships observed using the Brownian distance test, which were not recognized by the Granger causality test, were confirmed to be relevant non-linear relationships according to the White and Terasvirta tests (see Table 4). Hence, the Brownian distance correlation recognizes some important dependencies, some of which are confirmed by the Granger causality test and some, such as the effect of crude oil on natural gas, have been explored before by Brown and Yucel [8], Hartley et al. [12], Huntington [14].

Table 5 Brownian distance correlation and Granger causality of log return series using only the first 70% observations of each period. Non-relevant relationships are excluded. P-values of null hypothesis that parameters are zero for Brownian distance correlation: ∗: $p \leq 0.05$, ∗∗: $p \leq 0.01$, and for Granger causality: †: $p \leq 0.05$, ‡: $p \leq 0.01$

	Lag→Effect	1	2	3	4	5	6	7
2006–12	WTI → Coal	** ‡	** ‡	** ‡	* ‡	** ‡	** ‡	** ‡
	Gas → Coal							*
	Coal → WTI	**	**	**	**	**	**	*
	Gas → WTI				*			
	Coal → Gas		**	*		‡	†	‡
	WTI → Gas	**						
Pre-crisis	Gas → Coal	*						
	Gas → WTI	†	†					
	WTI → Gas	*						
Crisis	WTI → Coal	* ‡	‡	* ‡	‡	‡	‡	‡
	Gas → Coal							*
	Coal → WTI	*		*	**	**	*	
	Coal → Gas		*	*				*
Recovery	WTI → Gas	**						
	Coal → Gas	†	†	†	* †			

Table 6 Mean of test error and area under the ROC curve for prediction of log return direction using Granger causality (GC) and Brownian distance correlation (BD). Means are the result of a cross-validation exercise using samples generated by a rolling window that changes every five days. $*: p \leq 0.05$, $**: p \leq 0.01$ of t-test of mean differences

	Product	Test error		Area under the ROC curve	
		GC	BD	GC	BD
Pre-crisis	Coal	0.56	0.52**	0.50	0.50
	Gas	0.59	0.50**	0.41	0.49**
Crisis	Coal	0.45	0.47	0.56	0.54
	WTI	0.49	0.53	0.51	0.48
	Gas	0.54	0.45**	0.50	0.54**
Recovery	Gas	0.50	0.50	0.51	0.51
All periods		0.52	0.49**	0.50	0.51**

Most of the relationships discussed above still hold using only 70% of the observations of each period as Table 5 indicates. The variables with relevant lags are used to build the models that forecast the direction of the log return for the hold-out sample group (30% of the observations) for each period. According to Table 6, the models built with the variables selected by Brownian distance correlation outperforms the models based on variables selected by the Granger causality when all the products across all periods are evaluated together. In particular, Brownian distance correlation produces the best results for coal and gas during the pre-crisis period, and gas during the crisis as only this method recognized several relevant nonlinear relationships. WTI has the highest coefficient of variation during and after the crisis, so it is very difficult to predict its value during these periods. However, the effect of WTI on coal's return is very strong and was captured by both methods. So, there is not a major difference in its prediction.

The improvement in the forecast of energy futures log return described above validates the relevance of the nonlinear relationships uncovered by the Brownian distance correlation, and its feature selection capability as proposed in this paper.

6 Discussion

The non-linear relationships identified by the Brownian distance correlation complement the view that commodity prices are mostly determined by international market and political forces. Examples of these forces are the decisions of the Organization of the Petroleum Exporting Countries (OPEC) to curtail oil production or a political crisis in the Middle East. These non-linear relationships may have an impact on the selection of inputs used to generate electricity in the US. Electricity generated with coal reached its peak in 2007 and substantially decreased afterwards. In contrast, electricity generated with natural gas has been increasing since 1990, especially since 2009. Between the years 2000 and 2012, the proportion of electricity generated by coal and oil have decreased from 51.7 and 2.9%–37.4 and 0.6% respectively while the proportion of electricity generated by natural gas almost doubled from 15.8–30.4% (see Fig. 1). The increase of the electricity generated with natural gas is equivalent to the contraction of electricity generated with coal. This increase can be partially explained by the decline of natural gas log prices since December 2005–April 2012 and as the non-linear lead-lag analysis indicates both coal and natural gas prices are mutually dependent. The CAA restrictions on SO_2 emissions and the relative reduction of natural gas prices led power plants to partially substitute coal and oil with natural gas as their main input. As more power plants have increased their consumption of natural gas, its price has also increased following similar trends of oil and coal. However, the substitution of coal and oil with natural gas is limited by the fuel-switching capacity of power plants or their ability to invest in new technology. Despite these limitations, Hartley et al. [12] and Huntington [14] demonstrate that the consumption of natural gas is affected by the relative prices of oil and natural gas. This particular case illustrates the non-linear dynamics among the return of the different commodities studied and the major interrelationships that exist among the three fossil fuel series.

7 Conclusions

This paper proposes the use of the Brownian distance correlation to conduct a lead-lag analysis of financial and economic time series. When this methodology is applied to fuel energy log return, the non-linear relationships identified improve the price discovery process of these assets.

Brownian distance correlation determines relationships similar to those identified by the linear Granger causality test and further uncovers additional non-linear relationships among the log return of oil, coal, and natural gas. Therefore, Brownian distance correlation can be used for feature selection, and when combined with a non-linear classification method such as support vector machine, it can improve the forecast of financial time series.

This research can be extended to explore the lead lag relationship between spot and future prices of complex assets such as commodities and foreign currencies applied to different markets.

Acknowledgements The author thanks participants of the Eastern Economics Association meeting 2014, the AAAI 2014 Fall Symposium on Energy Market Predictions, Dror Kennett, Alex Moreno, and three anonymous referees for their comments and suggestions. The author also thanks the Howe School Alliance for Technology Management for financial support provided to conduct this research. The opinions presented are the exclusive responsibility of the author.

References

1. Amavilah, V.: The capitalist world aggregate supply and demand model for natural uranium. Energy Econ. **17**(3), 211–220 (1995)
2. Aruga, K., Managi, S.: Linkage among the U.S. energy futures markets. In: 34th IAEE International Conference Institutions, Efficiency and Evolving Energy Technologies. Stockholm (2011)
3. Asche, F., Gjolberg, O., Volker, T.: Price relationships in the petroleum market: an analysis of crude oil and refined product prices. Energy Econ. **25**, 289–301 (2003)
4. Asche, F., Gjolberg, O., Volker, T.: The relationship between crude oil spot and futures prices: Cointegration, linear and nonlinear causality. Energy Econ. **30**, 2673–2685 (2008)
5. Asche, F., Osmundsen, P., Sandsmark, M.: The UK market for natural gas, oil, and electricity: are the prices decouples? Energy J. **27**, 27–40 (2006)
6. Bachmeier, L., Griffin, J.: Testing for market integration: crude oil, coal, and natural gas. Energy J. **27**, 55–71 (2006)
7. Bai, J., Perron, P.: Estimating and testing linear models with multiple structural changes. Econometrica **66**(1), 47–78 (1998)
8. Brown, S., Yucel, M.: What drives natural gas prices? Energy J. **29**, 45–60 (2008)
9. Granger, C.W.J.: Testing for causality: a personal viewpoint. J. Econ. Dyn. Control **2**, 329–352 (1980)
10. Granger, C.W.J.: Essays in Econometrics: The Collected Papers of Clive W.J. Granger. Cambridge University Press, Cambridge (2001)
11. Granger, C.: Investigating causal relations by econometric models and cross-spectral methods. Econometrica **37**(3), 424–438 (1969)
12. Hartley III, P., Rosthal, K.B.M., K.E, : The relationship of natural gas to oil prices. Energy J. **29**(3), 47–65 (2008)

13. Hastie, T., Tibshirani, R., Friedman, J.: The Elements of Statistical Learning: Data Mining, Inference and Prediction, 2nd edn. Springer, New York (2008)
14. Huntington, H.: Industrial natural gas consumption in the united states: an empirical model for evaluating future trends. Energy Econ. **29**(4), 743–759 (2007)
15. Johansen, S.: Estimation and hypothesis testing of cointegration vectors in Gaussian vector autoregressive models. Econometrica **59**(6), 1551–1580 (1988)
16. Johansen, S.: Statistical analysis of cointegration vectors. J. Econ. Dyn. Control **12**(2–3), 231–254 (1988)
17. Lee, T.H., White, H., Granger, C.W.J.: Testing for neglected nonlinearity in time series models. J. Econ. **56**, 269–290 (1993)
18. Mjelde, J., Bessler, D.: Market integration among electricity markets and their major fuel source markets. Energy Econ. **31**, 482–491 (2009)
19. Mohammadi, H.: Electricity prices and fuel costs: Long-run relations and short-run dynamics. Energy Econ. **31**, 503–509 (2009)
20. Mohammadi, H.: Long-run relations and short-run dynamics among coal, natural gas and oil prices. Appl. Econ. **43**, 129–137 (2011)
21. Ramberg, D.J.: The relationship between crude oil and natural gas spot prices and its stability over time, master of Science thesis, Massachusetts Institute of Technology (2010)
22. Székely, G.J., Rizzo, M.L.: Brownian distance covariance. Ann. Appl. Stat. **3**(4), 1236–1265 (2009)
23. Terasvirta, T., Lin, C.F., Granger, C.W.J.: Power of the neural network linearity test. J. Time Ser. Anal. **14**(2), 209–220 (1993)
24. US Energy Information Administration: Annual Energy Outlook. US Energy Information Administration, Washington D.C (2011)
25. US Energy Information Administration, : Annual Energy Outlook. US Energy Information Administration, Washington D.C. (2013)
26. Tsay, T.: Analysis of Financial Time Series, 3rd edn. Wiley, NJ (2010)
27. Vapnik, V.: The Nature of Statistical Learning Theory. Springer, New York (1995)

Implementation of Generic Steering Algorithms for AI Agents in Computer Games

Mateusz Modrzejewski and Przemysław Rokita

Abstract This paper proposes a set of generic steering algorithms for autonomous AI agents along with the structure of the implementation of a movement layer designed to work with said algorithms. The algorithms are meant for further use in computer animation in computer games - they provide a smooth and realistic base for the animation of the agent's movement and are designed to work with any graphic environment and physics engine, thus providing a solid, versatile layer of logic for computer game AI engines. Basic algorithms (called steering behaviors) based on dynamics have been thoroughly described, as well as some methods of combining the behaviors into more complex ones. Applications of the algorithms are demonstrated along with possible problems in their usage and the solutions to said problems. The paper also presents results of studies upon the behaviors within a closed, single-layered AI module consisting only out of a movement layer, thus removing the bias inflicted by pathfinding and decision making.

Keywords Computer games · Artificial intelligence · Agent systems · Steering Movement algorithms · AI agents · Generic AI

1 Introduction

AI (*artificial intelligence*) in computer games has its beginnings in 1979's *Pac-Man*. Each of the ghosts in *Pac-Man* had a different algorithm for calculating the position of the next goal tile and a set of common, simple movement rules. Although artificial

M. Modrzejewski (✉) · P. Rokita
Division of Computer Graphics, Institute of Computer Science, The Faculty of Electronics and Information Technology, Warsaw University of Technology, Nowowiejska 15/19, 00-665 Warsaw, Poland
e-mail: M.Modrzejewski@ii.pw.edu.pl

P. Rokita
e-mail: P.Rokita@ii.pw.edu.pl

R. Bembenik et al. (eds.), *Intelligent Methods and Big Data in Industrial Applications*, Studies in Big Data 40, https://doi.org/10.1007/978-3-319-77604-0_2

intelligence in computer games has gone a long way since then, the basic principle has remained the same - we want the computer-controlled characters to act as if they were controlled by a human player. This means that, depending on the type of the game, the characters should possess such skills, as smooth, realistic movement, dynamic decision making, autonomous strategy planning, efficient pathfinding and so on [2]. The problem becomes more and more difficult along with the ever-increasing complexity of the games themselves. Satisfying effects, of course, can no longer be achieved using a single algorithm. Although some effects creating the illusion of intelligence may be achieved using only some well-tuned animation or some graphic workarounds, the actual logic of game agent AI has become very complex. Most modern games utilizing some AI require a layered structure of the AI algorithms.

1.1 AI Layers

As stated in [1], we can distinguish *group AI* and *unit AI*, both having multiple internal layers. The former contains algorithms which work on group strategy, formation keeping, cooperative movement and other group behaviors. The latter contains algorithms which determine the behavior of a single artificial intelligence agent, with the three main layers being decision making, pathfinding and movement.

While computer game artificial intelligence may be associated mostly with decision making, the actual basic layer of any AI system is the movement layer, providing realistic motion for the characters. Without good design and implementation of this layer, the agent's movement will be chaotic, unrealistic and very difficult to animate without using workarounds. This layer may be very simple or very complex, depending on the character's movement capabilities and the game itself - an enemy in a platform game may simply patrol one piece of terrain back and forth, while bots in first person shooter games may be able to perform complicated manoeuvres, like smooth path following or predictive pursuit. The decision making layer processes the agent's internal and external knowledge in order to output an abstract decision (e.g. "explore the map", "flee because of injuries" etc.). The pathfinding layer connects decision making with movement, providing the agent with the ability to navigate around the game world in a controlled manner.

Not every layer is needed in every case. The aforementioned enemy in a platform game may not need neither strategy nor pathfinding and have only two simple decisions, to attack or to keep patrolling. A chess simulator, on the other hand, may use only a strategic layer provided with algorithms able to predict the player's moves.

Of course, the AI module has to communicate with the world interface in order to obtain necessary data. The effects of AI algorithms are visible as animations and effects in the game physics.

1.2 Movement Layer

The character's movement has to have a mathematical description. It's very easy to overcomplicate this description, which is a possible major drawback of the movement layer implementation, as it greatly decreases the reusability of implemented AI. Necessary kinematic parameters that are used in our solution include only linear velocity, angular velocity, orientation and the character's position. Orientation is simply a name for the character's angle - this is the most commonly used name in programming computer game physics, in order not to collide with other parameters that may use the name "angle". Linear velocity and position should be represented as vectors, while orientation and angular velocity need only a single floating-point value.

In this document, the algorithms are considered for a two-dimensional space. Full calculations for a three-dimensional space are usually quite complicated, especially in terms of orientation in both of the planes. These calculations, however, are necessary only in specific genres of games, like flight simulators or games with action set in outer space. In most considered cases, the characters will be under the influence of some sort of gravity force - therefore, the orientation in the vertical plane will have an insignificant effect on the character's movement. This way, we can simplify the calculations to $2\frac{1}{2}$ dimensions, meaning that the position of the character is described by a full vector of $[x, y, z]$ coordinates, while the orientation is considered only for the horizontal plane. The characters are treated as material points by the algorithms. This rule has some obvious exceptions: it is impossible to implement a functional obstacle avoiding or collision detecting algorithm without taking the volume of the character into consideration [12].

The movement layer can be demonstrated as in Fig. 1.

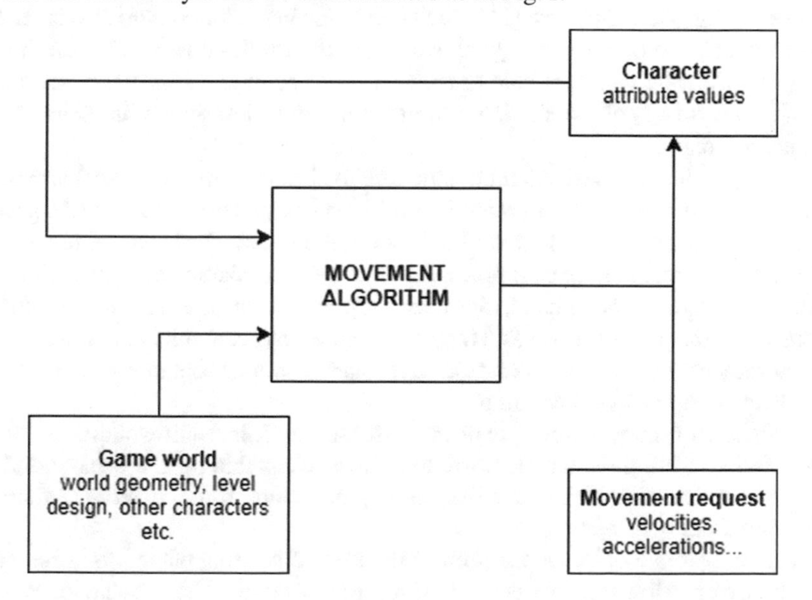

Fig. 1 Movement layer

The movement algorithm communicates with the physics module and modifies the character's attributes either directly, or via a movement request. Such requests may consist of forces, impulses, torques applied to the character or a set of concrete physical parameters, that the character should achieve.

2 Steering

The growing complexity of computer games and bigger arrays of agent movement possibilities result in the needs of using generic solutions [4, 5], that simplify the code from a software engineering point of view and allow to use algorithms without the knowledge (or with partial knowledge) about the actual movement possibilities of the characters. The idea of breaking the character's movement into small building blocks is widely called [1, 3, 10] and in this document, the term steering will be used exclusively in this context.

2.1 Steering Behaviors

One of the ideas of implementing steering is using *steering behaviors*, described in [10]. Steering behaviors describe all possible moves of the character as composites of very simple movements, like "seek the target", "run away from the target" or "face the target". We have implemented each one of said basic behaviors in a separate class using a strategy desing pattern [11]. Each one of the behaviors should also need data needed to achieve only a single goal and return the result allowing to achieve only that goal. We were therefore able to create a set of generic algorithms describing a set of moves for any object sharing a common physical description, implemented in an interface class.

Steering behaviors have a similar structure to simple, commonly used kinematic algorithms. The behaviors are dynamic algorithms - they operate on kinematic parameters of the steered agent and partial data about the target. For instance, in a pursue behavior, the target may be as simple as another moving character. A obstacle avoidance behavior, on the other hand, may need a representation of a piece of the collision geometry of the game's world as a target. Complex targets require simplification, as the movement layer is the lowest AI layer and it is unacceptable to have costly calculations implemented within it.

The behaviors may also operate on multiple targets. This requires additional methods in order to bring the targets down to a form acceptable for the behaviors. This may mean calculating an average of a set of parameters from a given set of targets or sorting them by a known rule.

A good example of the application of the idea of steering behaviors is the basic seek behavior. All it requires is the position of the target. A flee behavior may be implemented in the same way with a minor change, that is the opposite direction of the vector.

Algorithm 1 Seek

calculate the offset vector for the target
2: move with full velocity according to the vector

2.2 *Delegating Behaviors*

Behaviors can delegate the calculations to other behaviors: some more advanced behaviors only need to calculate a new target and call upon a basic behavior to obtain the actual steering. A good example is the pursue algorithm. The pursue will be very effective, if the agent tries to predict the target's position after some time. In this case we assume, that the target will keep moving in the same direction. This allows us to calculate how long would it take the agent to reach the target's current position, and then to predict the target's position after said time. We can now use this position for the seek behavior. The only requirement is introducing an additional parameter T, representing the maximal prediction time for situations, where the agent's speed is very low or the distance to the target is very big.

Algorithm 2 Pursue

calculate the distance from the target d
2: **if** agent's velocity $v < \frac{d}{T}$ **then**
$t = T$
4: **else**
$t = \frac{d}{v}$
6: **end if**
calculate the predicted position of the target after t
8: seek the new target

This way, for research purposes, we have implemented a generic behavior hierarchy, similar to the concrete one proposed by Craig W. Reynolds [9].

- seek and flee,
  - pursue and evade,
  - collision avoidance,
  - obstacle avoidance,
  - follow path,
  - wander,
- align,

- look in front of you,
- face the target,

- match velocity,

  - arrive.

The delegate behaviors are pointed out as bullets beneath their base behavior. While most of the algorithms are self-explanatory, we feel that we need to clear out the arrive and wander behavior.

- wander - makes the agent wander without a purpose. Obtaining calm, realistic movement, has proved to be quite a challenge. We have found that the best results can be achieved with periodically choosing a target within a set radius from the agent and delegating that target to a seek behavior tuned to slow movement.
- arrive - the seek behavior has shown to produce one issue: the agent keeps on seeking the target once he has achieved it, resulting in oscillations around the target. The arrive behavior acts exactly like seek to a certain point, where it starts to slow down the agent, decreasing his velocity to 0 within a certain radius. This is also how the face behaviors works.

The generic collision and obstacle avoidance behaviors, by definition, should not be aware of what an obstacle actually might be. The communication between the algorithms and the world representation of the game must also be generic. We have therefore created interfaces for raycasting algorithms, using an adapter design pattern. These interfaces provide also ready-to-use schemes of raycasting, including single and multiple ray strategies with customizable length, allowing for thorough tests of the behaviors.

3 Behavior Blending

The base principle of implementing steering behaviors is that each behavior allows to achieve only one target. This means that there is no behavior allowing the agent to seek the enemy, avoid obstacles and, in the meantime, collect ammunition. Actually, there is even no behavior allowing the agent to simultaneously seek the enemy and look in the direction that the agent is moving. In order to obtain such results, behavior blending algorithms are necessary.

3.1 Weighted Blending

The most simple algorithm of connecting the behaviors is weighted blending. This can be done by implementing a higher-level steering behavior, using the same exact interface as before. This behavior aggregates other behaviors with weights. The weights don't have to sum up to any certain value: seek and look where you're going behaviors can, and even should, have the same weight of 1. The weight blending algorithm is demonstrated below.

Algorithm 3 Blended steering

```
    for all weighted behaviors do
2:      calculate the steering
        scale the steering according to the weight
4:      add the value to the final steering result
    end for
6:  return the final steering result
```

The obvious problem of this algorithm is blending antagonistic behaviors, which may lead to deadlocks of agents and unwanted equilibria of steering parameters. This may result in the agent standing still or chaotic, shaking animation of the agent. There is also a risk of constraining the agent too much with behaviors like wall avoidance, resulting in unnatural or unexpected movement. A simple example would be a blended behavior, where blending wall avoidance with seeking the target throws the agent in an unwanted direction when the target is running away into a narrow passage between walls - the algorithm may steer the agent into a direction opposite to the wall rather than try to find a way into thc passage. It is therefore necessary to make sure, that a given behavior will never influence the movement of the agent in a way that disorganizes its movement.

3.2 *Priority Arbitrage*

Priority arbitrage is an algorithm that temporarily gives full steering over an agent to one group of steering behaviors (either blended or singular). The aggregating structure holds a list of all groups, sorted by priority. An additional parameter of ε is needed, as it represents the minimal, significant steering values. Whenever the first group returns steering lower than this value, the arbitrage algorithms checks the result returned by the next behavior in terms of priority.

Priority arbitrage allows to solve many problems with deadlocks. The easiest way to get an agent out of a steering equilibrium is to add a wander behavior as the last behavior in terms of priority. When all of the higher-priority behaviors return no steering, even a few turns of the game loop is enough to get the agent back to movement with the wander behavior. Please note that the priorities of the groups are never directly declared. The algorithm checks the behaviors in the order that they were pushed into it. It is, of course, possible to implement priority arbitrage with dynamic weights, but it seems to be much less efficient than the described solution, while not offering better performance.

4 Tests and Results

The following section describes the results of testing our generic class hierarchy. The algorithms were written in C++ with the Box2D library to simulate the physics and provide basic animation. Preliminary unit testing has proved that the algorithms produce correct results of all calculations. Almost all of the quality tests were black-box tests: computer game agent AI algorithms are extremely difficult to test and debug in any other way than by actual observation and experimentation on different cases [8]. We have covered the quality tests by implementing a generic interface to describe the physical possibilities of the agents and implementing some concrete agents with certain physical parameters. We have then placed the agents in a testbed with some obstacles and allowed the player to control one of the characters manually. The rest of the characters were AI-steered. Each character has an active behavior assigned that can be updated at given moment, leaving space for optimization, as not every behavior has to be updated with every turn of the game loop. The simplicity of integrating our algorithms with an actual physical model provided by Box2D was also a key factor in our studies. A view on a part of the test application is shown on Fig. 2. Figure 3 demonstrates creating a basic steering behavior with customizable parameters.

4.1 Algorithm Quality

In a full AI system, a lot of steering work can be done by the higher levels of the AI module. If an agent is unable to reach some point, he can change his decision or find

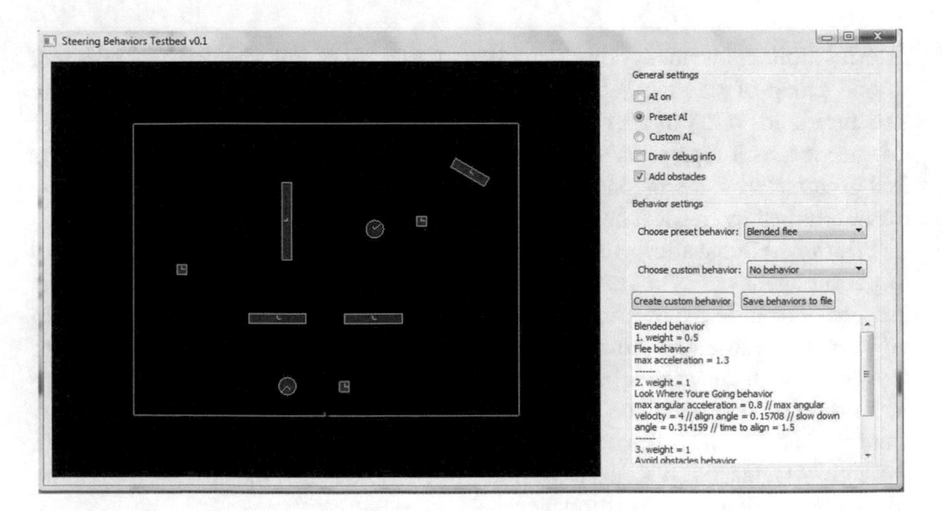

Fig. 2 Test program

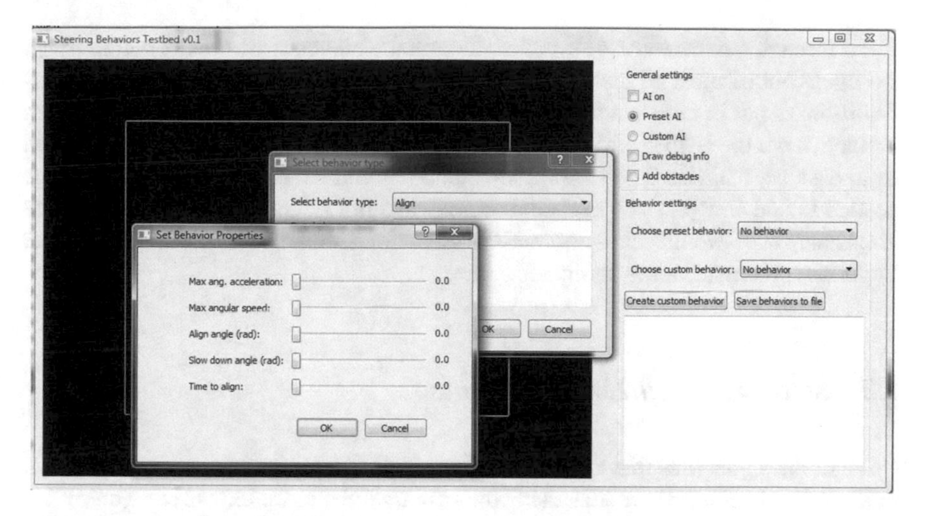

Fig. 3 Test program - creating a behavior

another path to the point. Pathfinding algorithms, on the other hand, in most cases assume unlimited knowledge about the world's geometry, often making it very easy for the agent to navigate through the levels. In order to thoroughly test the possibilities and constraints of steering behaviors, all of our tests are free of any decision-making and pathfinding bias. The agents operate only on the steering behaviors, having no knowledge about possible paths in the testbed and being unable to change their decisions dynamically.

Most of the algorithms compute in constant time ($O(1)$) which meets the requirement of movement algorithms being very fast and as efficient and possible. Any additional calculations are based only on game-specific algorithms included in the steering behaviors, like raycasting and detecting collisions. Although the performance of these calculations may vary, in most cases, they are very well-optimised in actual computer game engines [12].

Kinematic algorithms are still the most commonly used movement algorithms in computer games. They operate on parameters describing data about the statics on the characters and directly apply velocities to the agents. We have implemented kinematic equivalents to the basic proposed dynamic-based algorithms: seek, wander and flee. The kinematic algorithms turned out to provide much rougher animation, especially when the agent is starting or stopping. These situations would need certain workarounds in computer animation of the agents in order to conceal the imperfections of the movement layer itself. With steering behaviors, all of the movement is based on forces and torques, resulting in smooth and realistic acceleration and deceleration.

Aside from aggregating behaviors and combining them into more complex ones, the blending algorithms have proved to produce satisfying results in many nontrivial cases: with some fine-tuning, a priority arbitration algorithm was enough to lead and agent through a narrow corridor while seeking the player. It was also enough to get the agent out of most of the deadlocks that we have designed, including: corners, avoid-boxes put in between the agent and the target, several avoid- and seek-boxes scattered over the testbed and avoid-boxes placed near a moving target. This allowed us to assume, that this type of implementation would be able to cut down the time needed to find new paths or change decisions in a full AI module. This is extremely important, as it would leave more CPU time for animation, rendering and other important processes in the computer game.

4.2 Software Engineering Methods

None of the algorithms that we have implemented depend on any concrete physical model or world geometry model, using only a simple abstract interface (`Kinematic`) for physical parameters. This means that the algorithms may be used with any physical and graphics engine, needing only the implementation of a single, small class to translate the physical representation of the engine to the representation used with the AI algorithms. This structure also allows seamless integration if the game engine adopts the AI engines physical parameter model. For test purposes and integration with the Box2D engine, we have subclassed the `Kinematic` interface into a concrete `Vehicle` class that is the base class for all AI agents. The proposed raycasting interface also does not depend on any concrete solutions and proves to be very flexible and applicable for wide use, as all of the widely used game physics engine implement their own raycasting mechanisms. Any other possible interfaces used in the movement layer that may be needed in further development (ie. area queries) also would share this characteristic. Each of the algorithms is implemented as a separate class, derived from an abstract `SteeringBehavior` interface (including the blending algorithms) with concrete implementations of a `getSteering()` method and necessary helper methods. The usage of the strategy design patterns makes it very simple to dynamically interchange algorithms based on any higher calculations.

Considering a full AI module, decisions combined with pathfinding could easily produce a steering request ready to use with the proposed algorithms. From a software engineering point of view, the class design that we have proposed has proved to be very easy to use, tune and, when needed, further developed by adding new behaviors.

5 Comparison and Usability with Existing Approaches

As we have found, the support for a movement layer in existing engines is very sparse and poorly described. Moreover, many game engines don't implement AI mechanisms at all, utilizing a separate add-on AI engine instead. Below is a quick review of AI support in some of the well-known engines (Fig. 4):

Engine	AI support
Unity - RAIN (add-on engine)	behavior tree editor
Cocos2d	no support
Ogre	no support
JMonkey - Monkey Brains (add-on engine)	no support (Monkey Brains was discontinued)
Unreal Engine (Epic Games)	behavior trees support
CryEngine (Crytek)	behavior trees support
RedEngine (CDProjektRED)	no information about AI support
Anvil, AnvilNext (Ubisoft)	no information about AI support

Fig. 4 AI support in chosen computer game engines

The list could be continued with quite some more engines with virtually no information about their support for AI algorithms. The reason is that many of these engines have proprietary licenses and are developed by a company specifically for the needs of the games created by that company.

The support of AI algorithms in game engines, if available, is concerned mostly with the decision-making process with very little or no mention of the movement layer. Currently the two most popular decision-making algorithms used for developing computer game AI are behavior trees and state machines - generic steering behaviors are very easy to integrate with both of these algorithms.

5.1 Our Solution and State Machines

Modelling an agent's decisions using finite state machines (FSMs) [6] is an obvious solution, as they are easy to implement and develop. Each state represents a certain decision of the agent that results in an action that will be performed until the FSM transitions into another state. Depending on the implementation, the states may encapsulate the movement code or they can only pass the decision back to the AI module for further processing. Our solution is designed to work in both situations: the FSM would simply create a steering behavior suitable for the decision based on the available knowledge and either call the `getSteering()` method directly, or pass the instance of the behavior to other classes. The movement that would be the result of each state could be therefore easily created using the provided algorithms.

The generic steering behavior algorithms are also suitable for more sophisticated FSMs, like fuzzy logic FSMs, as they are completely independent from the decision-making process.

5.2 Our Solution and Behavior Trees

Behavior trees [7] are a very powerful tool in modelling game AI. They can be seen as an evolution from decision trees, a simple algorithm that in many cases could be implemented as a series of if-else statements. With regards to implementation, a behavior tree consists of a huge number of small classes representing different stages of an agent's decision making process (a `TryToOpenDoor` class, which may need `OpenDoor`, `WalkThroughDoor`, `TryToBreakClosedDoor` classes etc.). These classes are used in a tree structure with additional control nodes (sequence nodes, selector nodes, inverter nodes and other). This approach allows engine designers to create behavior tree editors, where the AI designer simply builds such a tree using a graphical user interface. Each leaf of the tree contains a concrete action that can result in success, failure or no result at all, meaning that the action is still being carried out.

Generic steering behaviors that we propose are designed to blend perfectly with any implementation of the behavior tree algorithm, with the implementation of the leaves similar to the states of a FSM. Steering behaviors require only the knowledge about the agent's current goal and this is exactly what behavior trees compute. Again, it would be possible to create a steering editor tool similar to a tree editor - after creating all the branches, the AI designer would be able easily to create the movement layer out of provided basic, delegate and blended algorithms using a similar GUI tool.

5.3 Other Possible Applications of Our Solution

Considering a learning AI system, the agents would be able to fine-tune their movement capabilities based on their experience and knowledge, simply by creating instances of steering behaviors with updated parameters (or updating the parameters of the existing behaviors coded into any other AI logic, depending on the AI designer's convenience). Generic steering behaviors may also be used as the base of an experimental AI module, deplete of actual decision-making algorithms. Such a steering-oriented system could dynamically decompose an agent's goal into a series of steering behaviors [1].

6 Conclusion

In this paper we have proposed a generic set of low-level movement algorithms for computer game engines. Commercial game engines that include any AI support either contain mostly support for decision trees and pathfinding algorithms (often based on navmeshes and *Pathengine*) or have very engine-specific, non-reusable and poorly described movement algorithms. Therefore, as compared to existing approaches, our solution has advantages of being very easy to use and integrate with any game engine, while still being efficient. It enables the AI designer to quickly develop movement for an agent, regardless whether the AI system is based on decision trees, state machines or, in an extreme case, no decision making at all. The algorithms, seen as the base of the movement layer's implementation, effectively handle many problems that can be met when working on an agent-based AI system in different types of computer games. We are sure that these algorithms could be a great foundation for an effective, generic AI module that would produce very realistic movement animation and provide the flexibility needed by any computer game designer. We plan to further develop the algorithms into compound systems and experiment with ideas of cooperative steering and communication between higher levels of the AI module.

References

1. Millington, I., Funge, J.: Artificial Intelligence For Games, 2nd edn. Morgan Kaufmann Publishers, Massachusetts (2009)
2. Russell, S., Norvig, P.: Artificial Intelligence: A Modern Approach, 2nd edn. Prentice Hall, Upper Saddle River, NJ (2003)
3. Rabin, S.: AI Game Programing Wisdom. Charles River Media, Inc (2002)
4. Bourg, D.: AI for Game Developers. O'Reilly & Associates, Sebastopol (2004)
5. Patel, U.K., Patel, P., Hexmoor, H., Carver, N.: Improving behavior of computer game bots using fictitious play. Int. J. Autom. Comput. **9**(2) (2012)
6. Wagner, F.: Modeling Software with Finite State Machines: A Practical Approach. Auerbach Publications (2006)
7. Michele, C., Petter, g.: How behavior trees modularize hybrid control systems and generalize sequential behavior compositions, the subsumption architecture, and decision trees. In: IEEE Transactions on Robotics, vol. 33, pp. 99 (2016)
8. Schultz, C.P., Bryant, R., Langdell, T.: Game Testing All In One, Course Technology (2005)
9. Reynolds, C.W.: Flocks, herds, and schools: a distributed behavioral model. In: SIGGRAPH '87 Conference Proceedings (1987)
10. Reynolds, C.W. (1999) Steering Behaviors For Autonomous Characters (1999)
11. Gamma, E., Helm, R., Johnson, R., Vlissides, J.: Design Patterns: Elements of Reusable Object-Oriented Software. Addison-Wesley, Boston (1995). ISBN 0-201-63361-2
12. Cohen, J.D., Lin, M.C., Manocha, D., Ponamgi, M.K.: COLLIDE: an interactive and exact collision detection system for large scale environments. In: Proceedings of the 1995 Symposium on Interactive 3D Graphics (Monterey, CA) (1995)

The Dilemma of Innovation–Artificial Intelligence Trade-Off

Mieczyslaw Muraszkiewicz

Abstract Dialectic that confronts pros and cons is a long-time methodology pursued for a better understanding of old and new problems. It was already practiced in ancient Greece to help get insight into the current and expected issues. In this paper we make use of this methodology to discuss some relationships binding innovation, technology and artificial intelligence, and culture. The main message of this paper is that even sophisticated technologies and advanced innovations such as for example those that are equipped with artificial intelligence are not a panacea for the increasing contradictions, problems and challenges contemporary societies are facing. Often we have to deal with a trade-off dilemma that confronts the gains provided by innovations with downsides they may cause. We claim that in order to resolve such dilemmas and to work out plausible solutions one has to refer to culture *sensu largo*.

Keywords Technology · Innovation · Artificial intelligence · Intelligence explosion · Culture

1 Introduction

In 1937 Paul Valéry wrote in his essay "Notre Destin et Les Lettres": *L'avenir est comme le reste: il n'est plus ce qu'il était* (The future, like everything else, is no longer quite what it used to be). Indeed, people's outlook into the future constantly changes and is relative to the widely understood status quo, desires, aspirations, and dreams. Today, what is a specific feature of modernity, we often look at the future mainly through the lens of technology and its promises and assurances. We tend to believe that technology will free us from various nuisances, worries, material shortages and diseases, and that it will grant us access to the longevity, and provide us with an overwhelming and conspicuous betterment. Despite the warning by Neil Postman that "Technology is ideology and to maintain that technology is ideologically neutral

M. Muraszkiewicz (✉)
Institute of Computer Science, Warsaw University of Technology, Warsaw, Poland
e-mail: M.Muraszkiewicz@ii.pw.edu.pl

R. Bembenik et al. (eds.), *Intelligent Methods and Big Data in Industrial Applications*, Studies in Big Data 40, https://doi.org/10.1007/978-3-319-77604-0_3

is stupidity plain and simple" [1], the way many people perceive technology has morphed from technology-as-tool to technology-as-ideology with the conviction that technological determinism is the objective law that organises and drives the present and the future, and that it determines progress (whatever it might mean). In other words, technology is not merely a vehicle of change but it is also the driver. The wisdom of these people is like the wisdom of a man with a hammer for whom everything looks like a nail.

Within this mindset a special role to play has been given to innovativeness, which is considered a major mainspring of the engine of change, where consecutive innovations are the steps to the paradise of social peace, well-being and prosperity. Those who have faith in such utopian scenarios are above all afraid of two threats, which are bad politics and politicians, and environmental catastrophes; the rest, meaning other problems, they believe, will get resolved in the course of time by means of science and powerful technologies. We do share these two concerns; yet, we do not share the trust in technological determinism and do not follow the irritable reckless admiration of innovativeness, which is meant to cope with the swelling impediments, roadblocks and crises quickly and effortlessly, and will get us to the better times. This wide veneration of innovation by some politicians, economists, journalists and even scholars, often without an understanding of its nature and modus operandi might be dangerous and misleading. Interestingly enough, industrialists and technologists who are the main figures involved in the games of innovations are much more humble and moderate while talking about innovation and its impact on the society.

We argue that there is something more powerful and more determinant than technology and innovation, something that gives meaning to and purpose of our life and thereby influences our choices and decisions, and shapes the future. This something is culture; culture, which has so many times turned out particularly important and decisive in the inflection points of the human history. Now, we have again found ourselves in such a point. We are in crunch time for understanding that technology itself, even if its glittering innovations often indistinguishable from magic blunt our sense of criticism, is not a chief panacea for the difficulties and challenges we are faced with. The situation may even get worse when technology instead of being part of the solution becomes the problem. At this point we refer to the inevitably coming tension between artificial intelligence and humans. In the next section we shall briefly explain how we understand innovativeness and innovation, then, we present the pros (for) and cons (against) innovation endeavours. This will lead us to examining in the following section the relationship between technology, artificial intelligence, innovation, and culture. Also we present and discuss the trade-off dilemma that occurs when the increasing need for innovation that could be met by artificial intelligence is confronted with the risks related to the intelligence explosion. We conclude the paper with the opinion that culture offers a way to look for resolving this dilemma.

2 A Brief Note on Innovation

Although there is no doubt that innovation is a real game changer that impacts very many facets of life such as economy, education, security, environment, health, lifestyle, and entertainment, its meaning in a public discourse has become diluted because of inappropriate and promiscuous use. Therefore, for the sake of focusing our discussion, below we define how this notion is understood in what follows. Probably the simplest and concise yet still general enough and robust definition of innovation is this: It is a new product, process, or service that however, can comprise already existing and available components and addresses the actual demand or creates and satisfies an entirely new need.

Innovation, or better to say, innovativeness, should not be confused with discovering or inventing. It is neither about discovering new laws of nature or human behaviour (this is the role of science), nor about inventing new artefacts such as semiconductors or antibiotics. Obviously, there is no sharp borderline separating innovation from invention; it may also happen that in the course of an innovative endeavour something will be invented and become part of the innovative product. Innovation, as opposed to discovery and invention that are rather spontaneous, hardly controllable and manageable activities (sometimes being just a result of a serendipity), is in a sense a reverse engineering process that goes back from the innovation request (challenge) that actually is a specification of the final desired thing to its ultimate implementation relative to technological, ergonomic, economic, and social and psychological factors and means. In the corporate world, innovation is usually subject to management by scientific tools and methods, and best practices.

The fact that there exist a multitude of innovation frameworks and methodologies proves that none of them is general and suitable enough for any situation and organisation. The major lesson learned from practice and literature is that each and every innovation project is specific and requires a customised methodology relative to organisation's culture and resources as well as to general external trends in the marketplace and/or within the society. Yet, whatever the approach and methodology, the innovation project in order to succeed requires a comprehensively structured and rigorous execution plan, and a strong management alongside with a support of top executives of the organisation. It is worth noting that innovation projects carried out within an organisation often challenge engrained business processes, routine habits, narrow views and reluctance, and can trigger tensions and predicaments among the staff.

In [2] we presented a generic model of innovation that now is depicted in Fig. 1. This is a simple model of the innovation process (here, we emphasise the word "process", which is manageable and measurable) that starts with defining the innovation task and through consecutive steps such as scanning the environment and collecting information relevant to the task, proposing innovative solution(s) and prototyping, and eventually ends with commercialisation. Noteworthy, this model includes a feedback loop that allows innovators to come back to previous stages if such a need occurs as a result of an intermediary underachievement or negative scrutiny. Linear mod-

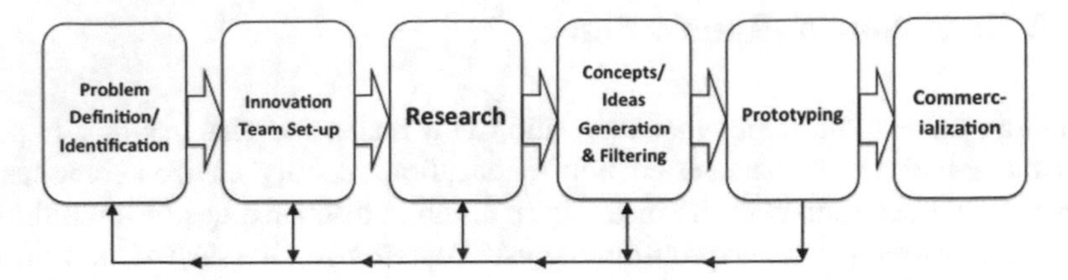

Fig. 1 A generic linear (with a feedback) model of innovation

els of innovation development have been widely applied for their conceptual and organisational simplicity as well as educational advantages. There might be many variations of the model displayed in Fig. 1. It can be implemented in various settings and situations, when innovation is managed and organised; below we shall mention three cases:

(a) Innovation is mainly a large scale R&D exercise in which focus is placed on developing a new concept requiring advanced laboratory research and a number of trails, often performed by researchers with solid scientific backgrounds. This approach has been practiced by large companies, which have at their disposal advanced R&D laboratories and often collaborating with academia. Since R&D projects are always subject to various risks, including financial ones, the innovation endeavours including strong R&D components are also exposed to risks and uncertainty. Incidentally, it should also be noted that thinking of innovation through the R&D lenses is emblematic for governmental agencies responsible for promoting and boosting innovation nationwide.
(b) Innovation is a process devised for incremental improvement of already manufactured products or provided services. In this case innovation is a kind of an on-going activity aimed at excelling the offer and maintaining the competitive/comparative edge in the marketplace. This is a typical conservative approach adopted by well-established companies that on the one hand have to compete in the marketplace, but on the other hand want to minimise risks related to investing in disruptive innovations laden with high uncertainty as far as commercial success is concerned.
(c) Innovation is a grass-root initiative, a proverbial "garage-born" innovation. Individuals, start-ups and SMEs are typical actors that follow the simplest version of the linear model of innovation. In this case it is not a huge financial and/or organisational investment that drives the innovation, it is rather eagerness, creativity and perseverance that drive the endeavour. However, due to limited resources and high costs of commercialisation these actors often look for venture capital and/or for selling their innovative solutions, sometimes a whole company, to stronger market players.

It has to be noted that the linear model of innovation has turned out not comprehensive and powerful enough to understand, capture and cover more sophisticated situations, also including chaotic moments, which take place in contemporary companies where new business models and processes exceeding traditional settings and boundaries are implanted and executed. Today, it is clear that pro-innovation attitudes and innovation propensity should become part of company's culture, permeating business processes and the company as a whole on a daily basis. In 1994 Peter Senge introduced the concept of a "learning organisation" that features five attributes "systems thinking, personal mastery, mental models, building shared vision and team learning" [3]. It has really been an influential and fruitful concept that reshaped many companies and businesses since its introduction. Now, we badly need a counterpart of this approach towards innovation, an "innovation organisation", in which the set of five features will be enhanced by the sixth one – innovation propensity. Such organisations already exist and owing to their tremendous commitment to innovation they have become icons of success-through-innovation approach. The idiosyncratic innovative styles of Amazon, Starbucks, PayPal, Google, Flipboard, Facebook, Instagram, WhatsApp, Twitter, Kenzo, Apple, Samsung, Salesforce, Uber and many others are subject to thorough analysis, dissections and scrutiny as well as careful observations by academic researchers and all those who want to replicate the success stories on their own sites.

3 Innovation Pros and Cons

We do not conceal our generally positive attitude towards innovation. But we worry about what we are excited about, therefore, identifying and investigating the pitfalls innovation may cause is also part of our agenda. On the cover page of Scott Berkun's book [8] one can find a striking sentence by Guy Kawasaki: "The naked truth about innovation is ugly, funny, and eye-opening, but it sure isn't what most of us have come to believe."

In ancient Greece Socrates in order to get a deeper insight into old and new problems, and a better understanding of the current and expected issues engaged his interlocutors in dialogues, asking questions and formulating pro and con arguments of discussed matters. We followed this approach for the purpose of this paper and as a result of authors' talks with a dozen academic researchers, business people, civil servants who promote pro-innovation awareness among small and medium entrepreneurs, NGO activists, and students of a technical university the following lists of pros and cons have been compiled.

Pros

In what follows a set of points characterising positive sides of innovation, presenting innovativeness as a mainspring of progress, demonstrating its pro-development potential, and supporting efforts to invest in innovation endeavours are summarised.

1. Technical and/or organisational innovation increases performance and productivity; often, while achieving these goals, innovation improves the use of raw materials and available resources.
2. Innovation adds value to existing settings and makings, and thereby it can create wealth and provide better off.
3. Social innovation can significantly contribute to human collaboration and thereby build and enhance social capital that is a condition sine qua non for communities and nations to develop and thrive.
4. Innovation can provide a comparative and competitive advantage in business and make fortunes, and improve non-business processes such as governance, citizens' security, and social inclusion.
5. Innovation opens new business opportunities; it moves one from the waters of Red Ocean to the waters of Blue Ocean, according to the metaphor proposed by W. Chan Kim and Renee Mauborgne in their bestseller book entitled "Blue Ocean Strategy" [4], i.e. to identify and exploit new business niches.
6. Innovation is a game-changer that can rejuvenate people and a company, trigger staff's creativity and increase motivation.
7. Innovation can lead to the creation of new professions and/or jobs.
8. Innovative solutions can break deadlocks when it comes to technical problems, but also when social issues occur.
9. Innovation and innovativeness are the ways and means for capitalism to redefine itself towards what is called cognitive capitalism based on knowledge, and to alleviate present growing economic, social and cultural problems. This is because innovation tends to displace presently used products or services and thereby propels consumption and market that are the foundations of capitalism.
10. Innovative concepts and scientific apparatuses can inspire and stimulate scientific research endeavours (innovation as food for thought). An interesting example of an innovative concept in science is the notion of big data that can lead to a new paradigm in science [5].
11. Innovativeness and innovation can be a shortcut for developing nations to catch up with some opportunities and offers available in the developed world (e.g. the use of mobile phones and tablets to make business, provide health advice, and organise communities in African countries).
12. The focus on innovation transforms business models and work organisation from what might be called a conventional industrial model based on hierarchy, detailed regulations, and strict control towards the model where work is more self-organised, more about self-development, knowledge and creative problem-solving, and more about focus on meaning than on repeatable jobs and routine duties – the work becomes employee-centric leading to employees self-actualisation. Referring to Douglas McGregor theories of human motivation presented in [6], one can say that the pursuit of innovation moves work organisation from theory X to theory Y.

Cons

Further on, we present a collection of points, caveats, worries, and anxieties that invoke a certain dose of scepticism with regard to innovation.

1. Innovation causes or even creates obsolescence of products and services (as noted already by Joseph Schumpeter) that are still useful but are moved out under the catchword of novelty. This leads to exhaustive exploitation of natural resources and labour.
2. It may happen that innovation solves one problem, but creates another problem(s), even more severe. In this regard, nuclear plants are a good example, since they produce cheap energy, but disposal of nuclear waste and security are real problems.
3. We cannot predict and control all impacts, side effects and unintended use of innovative products. Examples: (i) Alfred Nobel's dynamite was not meant as a weapon and an instrument of destruction – he thought of peaceful purposes only; (ii) cars were devised to revolutionize transportation for making it faster, more efficient and comfortable, which they actually did, but at the cost of the negative environmental impact and accidents that threaten humans' health and life.
4. Innovation, through significant increases in performance and production can hinder or reduce jobs, or even eliminate certain jobs.
5. To innovate is a risky business. Noteworthy, even experienced innovators and venture capitalists are exposed to fatal estimates and errors that can result in failures from which recovery might be difficult.
6. Along the line mentioned in the previous bullet let us add that too much emphasis on innovation whose upper extremity is a blind innovation proselytism can lead to developing destructive attitudes of the sort "Innovate or Perish" that can ruin businesses and people.
7. Undoubtedly, innovation has a cost. The question then arises: Who pays for innovation? There is no doubt that in case of larger companies and organisations these are customers and users, who ultimately incur the costs of innovation efforts. In the case of start-ups and small companies that have less or do not have at all "buffer capital" these are their sponsors and/or owners who take a risk to invest in innovative ventures.
8. Even a rough review of companies and organisations that in one way or another are engaged in innovative endeavours proves that striving for innovation may disturb routine work. The modus operandi of innovation projects is entirely different from usual procedures whose features are repetitiveness and control. One should then not be surprised that innovation initiatives are typically not welcome by managers who are responsible for steady achieving pre-defined outputs. Even a separation of what is being produced on production lines from the execution of innovation projects is considered by quite a number of managers and employees a disturbance of and/or a threat to a well-established order. One should not be astonished by a manger, who publicly speaks for the need and

empowerment of innovation, but when it comes to daily work, s/he joins the clandestine club of its enemies. Incidentally, this interesting topic undoubtedly deserves a separate study and examination.

9. Innovation efforts and results can be offered to or even imposed on people under the label of providing stewardship and betterment based on a simple "cost-benefit" analysis or simple statistics. In most cases such actions are not a result of a bad will; for example it was known at the outset that Jonas Salk's polio vaccine might be harmful for a tiny percentage of people and causes health complications, but the "cost-benefit" analysis justified the introduction of this vaccine on a large scale. Here, what causes questions and doubts is that such "political" judgments and decisions are rarely supported through checking out and/or taking into account the public opinion.
10. In many cases a race to innovate has resulted in the opinion and practice that the focus and emphasis on quality of offered products and services is of secondary importance. The product development process is finished at the point of a beta prototype that is introduced to customers for attracting their attention and to seize the market niche without any worry about excellence.
11. Innovation, be it social or technological, undermines tradition on which depends the continuity and coherence of human communities, and thereby it can entail various types of tensions and disruptions that can lead to dramatic cultural changes, some of which might be negative and destructive.
12. Innovation often puts law regulators behind the curve which opens the door to social conflicts or wrongdoings. An illustrative example of such a situation is the conflict of internauts with content distributors on the Internet regarding the intellectual property rights (it was interesting to follow the wide protest of internauts on the occasion of the Anti-Counterfeiting Trade Agreement that ended with recognising internauts' stance by governments).
13. Thorsten Veblen in his seminal book [7] introduced the notion of conspicuous consumption arguing that people after having satisfied their actual needs are still actively engaged in purchasing and consumption, whose major purpose is to compete with relatives, friends, neighbours, and co-workers by showing them how high level of the social ladder they occupy. Innovative products or services are the fuel that impels this effect. An immediate example of this case is a hype that accompanies consecutive editions of Apple's devices such as iPones and iPads, another example is the whole haute-couture business.

4 Technology, Artificial Intelligence, Innovation, and Culture

Without entering upon a long academic discussion on the meaning and scope of the term *technology* we can assume that technology refers to the realm of tools and machines that men use for transforming both material and immaterial world by constructing devices, mechanisms, procedures and services. The objective of technology developments is, on the one hand, to make human's life better (effortless, trouble-free, painless, healthier, more interesting, and the like), and on the other hand, alas, to control people more efficiently, strengthen military capacities, use coercion, etc. This term also comprises expertise embedded in procedures that instruct one how to use these tools. Technology is perceived as a powerful and sometimes even a mysterious force that can help resolve problems individuals and societies are faced with. This generally positive attitude is commonplace despite dramatic events and negative implications technology entailed (e.g. nuclear bombs dropped on Hiroshima and Nagasaki, Fukushima power plant disaster, thalidomide, acid rains, greenhouse effect). So, we can legitimately ask why people's attitude in respect of technology is in general terms positive? Perhaps we should seek the answer to this question in Marshal MacLuhan's theory that technology extends and strengthens human attributes and capacities, for instance a shovel adds power to our muscles, eyeglasses correct our vision, a car makes displacements easier and faster, and a computer boosts our intellectual capabilities. Indeed, technology is a great facilitator and amplifier, but it features something even more appealing: Technology boosts our individual and collective ego.

There are people who tend to perceive technology mainly as a source of power and coercion. This may be an incitement to domination, to conquering, and to a constant transformation of the environment according to one's will and ad hoc needs, not infrequently based on sheer hedonism and selfishness. Of course, it is not to say that technology automatically reveals and invokes a dark side of the human nature. There is no fatalism that this must be the case; yet, it may happen and actually happens at a smaller or larger scale. McLuhan argued that media change people's cognitive structures, and added that the effects do not occur only at the level of opinions or concepts, but alter our patterns of perception steadily and practically without resistance. Without risking a mistake we may say that a similar phenomenon may occur when it comes to sophisticated technologies, especially those that interact with our intellectual and affective structures. Let us wrap up this thread by quoting Melvin Kranzberg's first law of technology: "Technology is neither good nor bad, nor is it neutral."

Having said that we arrive at the vital point of our narrative in which shows up one of the greatest and strongest technology ever, i.e. artificial intelligence and the phenomenon dubbed *intelligence explosion*. The latter was explained and predicted by Irving J. Good, a distinguished researcher who worked at Bletchley Park on cryptography issues with Alan Turing; he wrote: "Let an ultra-intelligent machine be defined as a machine that can far surpass all the intellectual activities of any man however clever. Since the design of machines is one of these intellectual activities, an ultra-intelligent machine could design even better machines; there would then unquestionably be an 'intelligence explosion' and the intelligence of man would be left far behind. Thus the first ultra-intelligent machine is the last invention that man need ever make." [9] Let us complement this old warning by a recent caution from Eliezer S. Yudkovsky: "I would seriously argue that we are heading for the critical point of all human history. Modifying or improving the human brain, or building strong AI, is huge enough on its own. When you consider the intelligence explosion effect, the next few decades could determine the future of intelligent life. So this is probably the single most important issue in the world." [10] Therein lies the rub: social and moral values and objectives of those ultra-intelligent machines, which in one way or another will be autonomous entities, will not necessarily be the same as the ones of humans. Prof Stephen Hawking, one of the world's pre-eminent scientists, warned in the interview given for the BBC on the 2nd of November 2014: "Efforts to create thinking machines pose a threat to our very existence" and then he added "It would take off on its own, and re-design itself at an ever increasing rate. Humans, who are limited by slow biological evolution, couldn't compete, and would be superseded." Similar concerns were expressed by other pundits and technologists among whom are Elon Musk and Bill Gates.

Now, the moment is right to express the major dilemma related to technology and innovation. The overpopulated world suffering environmental constraints (e.g. global warming, natural resources depletion), and political and social problems (e.g. inequality, unemployment, ethnic conflicts, immigration, wars) badly needs both social and technological innovations. Undoubtedly, science and technology are the most instrumental devices to address this need, and admittedly artificial intelligence is the most effective tool to empower innovativeness and provide innovation. It is this vehicle of innovation that can deliver quickly, efficiently, cheaply and widely smart solutions of the existing and emerging problems and challenges. However, as mentioned above, artificial intelligence may put mankind at enormous risk. We therefore have to recognise a trade-off between the increasing need for innovation that could be met by artificial intelligence and the risks related to the intelligence explosion. That said one is right to ask whether and how this trade-off dilemma could be resolved while not to strike a Faustian bargain that would admittedly produce impressive innovations but at a very high price, which for a long run would not be acceptable. Because the stake is tremendously high there is a pressing need to find a solution. We are afraid that given the unstoppable March of artificial intelligence technologies and the inexorable proliferation of their results there is not straight and easy solution. There is however a way out of this trap. This way goes through culture.

Leaving apart a meticulous and rigorous discussion on what culture is, we can briefly say, as we did in [11], that culture is a universe of beliefs, symbols, sophisticated relations, values, laws, responsibilities, and duties that have been constructed, established and developed over the long history of mankind. This blend is incredibly stable and remains almost unchanged in the course of time. Culture tends to avoid changes. By its very nature it is conservative. Culture is an abstract entity that is deeply rooted in a human's psyche. We admit that its various materialisations, or the so called material culture (obviously, technology is part of the material culture), taking the forms of artistic activities, art products, religions, educational systems, laws, rhetoric, etc. are visible items; yet, the culture itself like a shadow is deeply hidden somewhere in Plato's cave or like the air is not visible. If a vivid dynamo were a relevant metaphor for technology, obviously, the right metaphor for culture is a homeostatic device whose principal objective is to maintain the status quo. Culture is a controlling mechanism that silently imposes values and behavioural patterns. Culture's charisma is soft, it does not provoke immediate results; however, its efficacy is admirable. Culture transforms individuals and non-coherent groups into communities and societies. Generally, culture rewards caution and positive conservatism. It may moderate and harness individual and collective egos and help take and maintain risks under control. This is not to say that culture cheers and promotes laziness and keeps people in lethargy, that there is no movement and progress within its kingdom; it is rather to say that culture strictly determines, or better to say, constrains the fields of activities, encourages mindfulness and reflection. Culture is thus the place where we should look for solutions of the innovation–intelligence explosion trade-off dilemma and find the arguments and remedies against the darker sides of innovation mentioned in Sect. 3. Culture is the territory where we can balance the desire to innovate and the risks it entails, where we can seek benefits while dodging downsides.

5 Final Remarks

Innovativeness has always been part of human fate and has always exhibited its brighter and darker sides. Humans have learned to balance the both sides by measuring and scrutinising pros and cons and to adopt or reject novelties. These experiences and skills are in high demand these days because the marriage of innovativeness and artificial intelligence is becoming dramatically critical for the mankind, which has found itself on a tipping point of its history.

In [11] we claimed that the level of incertitude and existential anxiety are proportional to the scale of technological achievements and innovations. This paper refers to this proposition and is a voice in the debate on identifying and addressing challenges and opportunities related to robust technologies comprising artificial intelligence solutions. Our blunt message is that the axiological and ethical perspective has to be included and seriously taken into account while boosting and pursuing innovativeness and innovation; otherwise, uncontrolled technological developments will bring us to the uncharted waters of uncertainty.

Acknowledgements The author thanks Professor Jaroslaw Arabas and Professor Henryk Rybinski of Warsaw University of Technology for stimulating and inspiring discussions on artificial intelligence, its prospects and impact on society.

References

1. Postman, N.: Technoploy. The Surrender of Culture to Technology. Vintage Books, Division of Random House, Inc, New York (1993)
2. Jacobfeuerborn, B., Muraszkiewicz, M.: ICT based information facilities to boost and sustain innovativeness. Studia Informatica, Silesian University of Technology Press **33**(2B(106)), 485–496 (2012)

3. Senge, P.: The Fifth Discipline Fieldbook: Strategies and Tools for Building a Learning Organization. Barnes & Noble (1994)
4. Kim, Ch.F., Marbougne, R.: Blue Ocean Strategy: How to Create Uncontested Market Space and Make Competition Irrelevant, 1st (edn.). Harvard Business Review Press, Boston (2005)
5. Jacobfeuerborn, B. Muraszkiewicz, M.: Big data and knowledge extracting to automate innovation. An outline of a formal model. Advances in knowledge organization. In: Proceedings of 13 International ISKO Conference, "Knowledge Organization in the 21st Century: Between Historical Patterns and Future Prospects", Krakow, vol. 14, pp. 33–40 (19–22 May, 2014)
6. McGregor, D.: The Human Side of the Enterprise, 1st edn. McGraw Hill (1960)
7. Veblen, T.: Theory of the Leisure Class. Oxford University Press, Oxford (2008)
8. Berkun, S.: The Myths of Innovation. O'Reilly Media, Beijing (2010)
9. Good, I.J.: Speculations concerning the first ultraintelligent machine. In: Alt, F.L., Rubinogg, M. (eds.) Advances in Computers, vol. 6, 31–88. Academic Press, New York (1965)
10. Yudkovsky, E.S.: 5-Minute Singularity Intro. http://yudkowsky.net/singularity/intro (2017). Accessed 26 Feb 2017
11. Muraszkiewicz, M.: Mobile network society. Dialog Universalism **14**(1–2), 113–124 (2004)

Can We Build Recommender System for Artwork Evaluation?

Cezary Pawlowski, Anna Gelich and Zbigniew W. Raś

Abstract The aim of this paper is to propose a strategy of building recommender system for assigning a price tag to an artwork. The other goal is to verify a hypothesis about existence of a co-relation between certain attributes used to describe a painting and its price. The paper examines the possibility of using methods of data mining in the field of art marketing. It also describes the main aspects of system architecture and performed data mining experiments as well as processes connected with data collection from the World Wide Web.

Keywords Data mining · Recommender systems · Art market · Paintings

1 Introduction

In the era of emerging Internet sale connected with an art market, one could ask a relevant question about a potential value of an artwork being bought. It is particularly interesting from the point of view of an art collector and possible profits to be gained by his investment in paintings. Yet, it is not the only option. It is also easy to imagine that such information could be used in the process of art valuation itself. In this paper we seek for a solution to this issue.

C. Pawlowski · Z. W. Raś (✉)
Institute of Computer Science, Warsaw University of Technology, 00-665 Warsaw, Poland
e-mail: ras@uncc.edu

C. Pawlowski
e-mail: czarekpawlowski@gmail.com

A. Gelich · Z. W. Raś
Department of Computer Science, University of North Carolina,
Charlotte, NC 28223, USA
e-mail: agelich@uncc.edu

Z. W. Raś
Polish-Japanese Academy of Information Technology, 02-008 Warsaw, Poland

R. Bembenik et al. (eds.), *Intelligent Methods and Big Data in Industrial Applications*, Studies in Big Data 40, https://doi.org/10.1007/978-3-319-77604-0_4

The price of an artwork distributed through an online art gallery consists of a price proposed by an artist and, usually, a fixed profit margin dictated by the gallery. Additionally, the value of a painting distributed that way is placed in the low-middle price range, which makes it little prone to speculation. Taking into consideration that artist follows some rules while valuating his painting, such as size, technique, time, his popularity, subject, current trends in the art market, one could seek to find some regularity in painting prices. It would be useful if those regularities could be expressed in the form of reasonable rules.

Recommender systems are software tools and techniques providing suggestions for items to be of use to a user [1, 2]. In this case we plan to built a system that recommends into which paintings buyer should put his interest, because their value might exceed the market one. The second chapter named Art market as a problem area describes a general view of the art market and highlights problems connected with price regulation. The third chapter named Recommender System gives an insight into the system components and their bindings. The last chapter titled Experiments presents the performed analysis.

At the end we include a summary and comment on some experiments.

2 Art Market as a Problem Area

In this paper we focus on online art galleries used by artists to display and sell their artwork. The reason for choosing such galleries is the possibility to extract information in a multimedia format (image and text) about artworks from the Web (WWW). Unfortunately, such approach is also a subject to certain shortcomings. Data is less reliable and a part of exhibited paintings might not be sold at a price asked by artists. On the other hand, the price range is in low-middle, which makes them more available to everyone and hence prices should follow some reasonable rules. Art market is not regulated in many aspects. Online art galleries usually do not influence artists on a price, so it is up to them to valuate their painting. We assume that artists follow some rules based on common sense. Finding those rules and solving this problem might even lead to a partial market regularization.

A decision table for the purpose of this work has been built using Internet resources. Initially, the following were taken into consideration:

1. Online art galleries such as SaatchiArt [3], Artfinder [4], and UGallery [5]
2. Social networking service [6].

For each given resource we created a data collection process, but eventually the application of such selection criteria as having:

1. price (seen as decision attribute in our table)
2. comments on paintings
3. rich structure of attributes describing a painting
4. simple interpretation of attributes describing an author

has led us, by elimination, to select SaatchiArt gallery as the only source of data on the top of which our database (decision table) was built. Such approach has additional advantage - we did not have to deal with semantic differences between data sources. Also SaatchiArt gallery is one of the world's largest international online art galleries.

As for the volume of the data used for research presented in this paper, we collected a small sample of 20,543 most popular oil, watercolor, and acrylic paintings.

3 Recommender System

In designing the infrastructure for building the recommender system we have focused on distributed computing and scalability. Components were chosen to handle large volume of data. Particularly, we allowed distributed storage and processing. The system was built using mostly open-source software. Multiple components such as a database, web crawler, distributed computing platform and others were integrated to build a recommender system and find the answer to the raised questions.

Figure 1 shows the general system scheme. The system consists of multiple modules responsible for such tasks as data collection, data processing and data mining. Each of these tasks is briefly described in the following sections.

3.1 System Architecture

The SaatchiArt website [7] was indexed with the use of Apache Nutch (a highly extensible and scalable open source web crawler) in two steps. Firstly, web pages with description of oil, watercolor and acrylic paintings were fetched by popularity. Then after analysis, the corresponding authors pages were fetched. In order to perform this task, proper URL (Uniform Resource Locator) patterns and seed pages were needed to be specified. Nutch automatically supports de-duplication, sets appropriate request

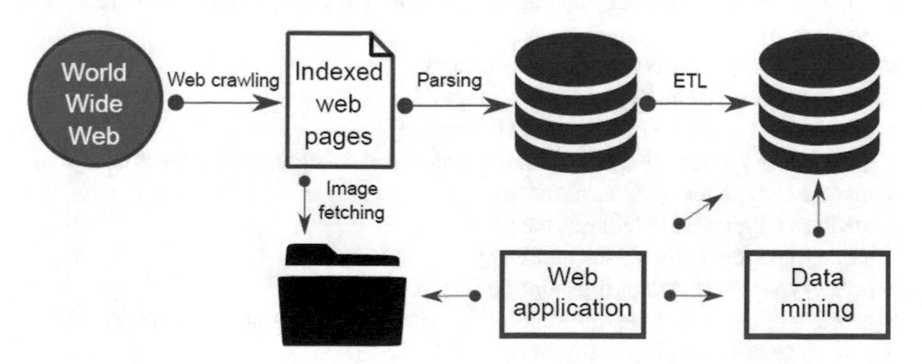

Fig. 1 System scheme

timeouts (so the process is not too intense for the target HTTP servers), offers rich configuration possibilities and can be run on a distributed Apache Hadoop cluster.

Indexed pages have been loaded into the full-text search engine called Apache Solr which allows filtering of documents and can also be run in the cluster mode.

Finally, web pages from Solr have been parsed by a software written in Scala. The access was gained to Solr documents from the application level with the use of the SolrJ library. JSoup, a HTML parser library, allowed extraction of data stored in HTML documents by DOM or CSS-selector syntax. The software was run in a distributed mode with the use of Master-Worker framework. Data extracted from HTML have been loaded into Apache Cassandra, a distributed database management system. Images have been fetched to a file system (distributed and POSIX-compliant in a production environment, e.g. GlusterFS).

The list of extracted features used to describe paintings in the resulting SaatchiArt table schema, their types, and descriptions are listed below.

- uuid, type uuid [Identifier generated from URL]
- author, type text [Author's name]
- author_about, type text [About section from the authors page]
- author_artworks_count, type int [Number of artworks painted by the author]
- author_country, type text [Country that author lives in]
- author_education, type text [Education section from the authors page]
- author_events, type text [Events section from the authors page]
- author_exhibitions, type text [Exhibitions section from the authors page]
- author_followers, type int [Number of author followers (users on SaatchiArt that follows given author]
- author_tstamp, type timestamp [Timestamp when authors page was collected]
- author_url, type text [Website address of the authors page]
- comments, type list/text [List of comments on the painting in JSON format (fields: author, time, text)]
- comments_count, type int [Number of comments on the painting]
- crawl_tstamp, type timestamp [Timestamp when painting page was collected]
- description, type text [Description of the painting]
- dimensions, type [Dimensions of the painting (format: "height H x width W x depth in")]
- favourites, type int [Number of favourites]
- img, type text [Remote path to the image]
- img_local, type text [Local path to the image]
- keywords, type text [Keywords describing the painting, filled by the author]
- materials, type text [Materials used]
- mediums, type text [Mediums used]
- name, type text [Title of the painting]
- price, type text [Price of the painting]
- processing_tstamp, type timestamp [Timestamp when painting was processed]
- sold, type boolean [If painting was sold?]
- styles, type text [Styles used]

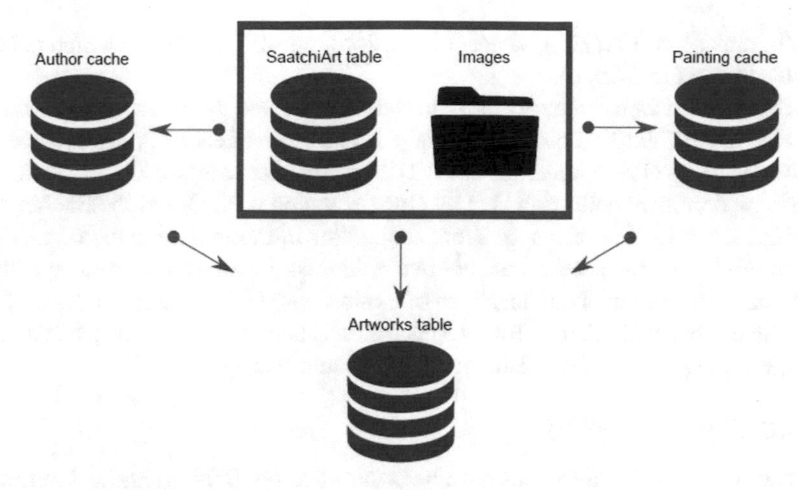

Fig. 2 Data processing

- subject, type text [Subject of the painting]
- thumb, type text [Remote path to the thumb]
- thumb_local, type text [Local path to the thumb]
- url, type text [Website address of the painting]
- views, type text [Number of views]
- year, type text [Year when artwork was painted].

3.2 Artworks Table

Data was processed by the software developed in Scala. As Fig. 2 shows, the processing splits into a few phases. Firstly, heavy calculations concerning paintings and authors have been cached into separate tables. Then, all was merged into resulting artworks table. Everything was performed by one configurable application, which runs in a distributed mode using Master-Worker framework. Data from Cassandra was mapped to objects using Phantom (Scala Object-Relational Mapping). All those tasks were performed by a simple ETL (Extract, Transform and Load) infrastructure written in Scala.

In the project we made an extensive use of APIs (Application Programming Interface) hosted on Google Cloud Platform. With Custom Search API we polled the Google search engine in order to build an external author's ranking. Additionally, with the use of Vision API we labeled images (paintings) and extracted dominant colors from them [8]. We used the extracted dominant colors to build a color scheme of a painting. Fourteen reference colors were chosen:

white (255, 255, 255), gray (128, 128, 128), black (0, 0, 0), red (255, 0, 0), green (0, 255, 0), blue (0, 0, 255), cyan (0, 255, 255), magenta (255, 0, 255), yellow (255,

255, 0), pink (255, 192, 203), violet (143, 0, 255), purple (128, 0, 128), orange (255, 127, 0), brown (150, 75, 0).

After that a distance between each dominant color and reference color was measured using CIEDE2000 color difference [9]. The score of each dominant color was transferred to the closest reference color. To be exact, there are two kinds of measures describing dominant color provided by Google Vision API. One is the fraction of a painting, second is the relevance score (e.g. color in the middle is more relevant), thus for each reference color there are two columns. We also introduced an authors index, based on geographical locations of exhibitions. Locations are extracted from text with use of NER (Named Entity Recognition) tagger. Each country has its own score, which is a scaled HDI (Human Development Index):

$$shdi(x) = 100 \cdot hdi^2(x)$$

Then scores of all exhibitions have been summed up. Exhibitions held in important cities such as capital cities are double scored. An assumption was made that because of low-medium painting prices, purchases will be made by middle class and, therefore, the use of economic indicator is justified. Apart from this we used OpenIMAJ library and MATLAB to extract global image features. MATLAB computations have been done offline and then loaded from CSV file to the painting cache using Cassandras COPY TO directive. All text processing tasks were performed with the use of Stanford NLP library. Eighty six new attributes (not listed in SaatchiArt table schema) have been extracted and added to the resulting Artworks table schema. The names of some of them with corresponding types and their descriptions are listed below.

- area, type double [Area of the painting ($\{height\} \cdot \{width\}$)]
- author_avg_exhibitions_count, type double [Average number of exhibitions per year]
- author_degree, type text [Authors degree (BA, MA) extracted by degree pattern search from {author_about} and {author_education}]
- author_exhibitions_count, type int [Number of author exhibitions extracted by year pattern search from {author_exhibitions}]
- author_popularity, type text [{author_followers} after equal-width discretization (low, medium, high)]
- author_hits, type bigint [Number of hits returned by the Google search engine with "$\{author\} \star art$" query]
- blurriness, type double [Fraction of non-edge area to overall painting area (edges are discovered with Canny edge detector [10])]
- brightness, type double [Average brightness of the painting, calculated using luminance]
- comments_popularity, type text [{comments_count} after equal-width discretization (low, medium, high)]
- comments_sentiment, type text [Average comment sentiment].

4 Experiments

After taking into account different discretization methods for the decision attribute price, we decided to manually discretize it by its density. A price distribution analysis was made in respect of groups of oil, acrylic and watercolor paintings, which resulted in the plots presented in Fig. 3. As the price is similarly distributed we decided to discretize this attribute regardless of a medium (treating all paintings as one group). Figure 2 (its right part) shows a price histogram along with the performed binning.

The values defining ranges of discretization were divided into two groups: basic and extended. In the price histogram (see Fig. 4), the orange lines (1020, 2335, 5100) indicate basic discretization and green lines jointly with orange lines (480, 1020, 2335, 5100, 10650, 19500) show the extended discretization cuts. Values were chosen in places where there is a gap in price attribute and the neighborhood is characterized by a relatively low density. Then, they were tuned a little bit in the neighborhood by observing the classifier performance.

Figure 5 presents the scores of chosen attributes in each step of the forward selection method for Decision Tree and Random Forest classifiers [8]. Both of them include the same set of attributes, though evaluated with a different measure. As we can see, the area of a painting is the most influential feature. Other attributes are mainly connected with the author, which seems to be justified as prices are very often influenced by the authors image.

We also used a random forest classifier (with 10 trees), although it seemed less reasonable because of a small number of attributes resulting from the use of forward selection. Nevertheless, the classifier used with all the features (without selection) did not yield any better results (precision: 0.557; weighted FS: 0.534). Considering the basic discretization with four values of the decision attribute [0; 1020), [1020; 2335), [2335; 5100), [5100; +1), the obtained results at the level of 0.560 are good for the Art Market domain. By comparison, a random classifier would yield 0.25.

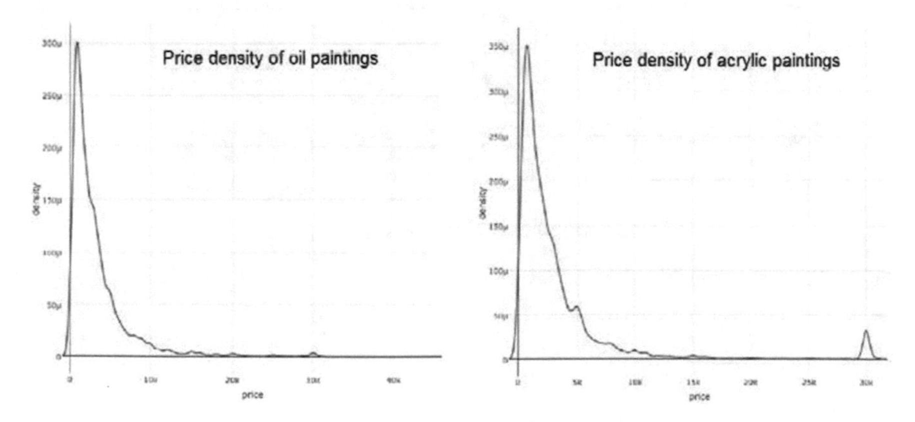

Fig. 3 Price density of oil and acrylic paintings

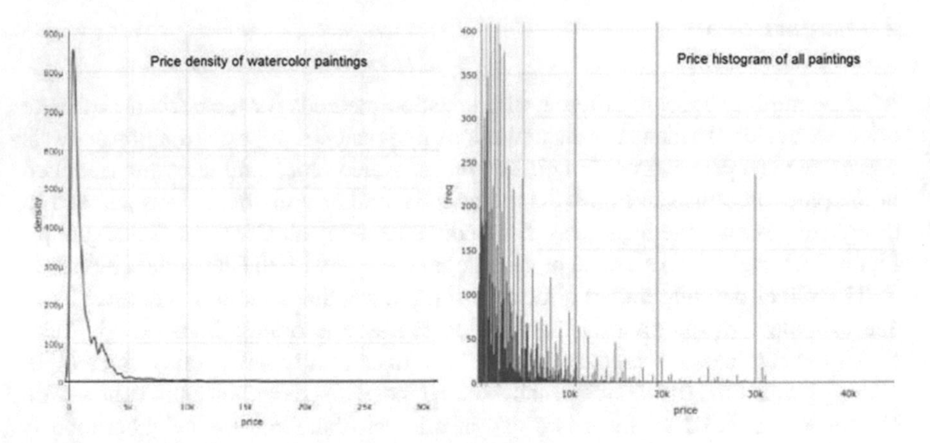

Fig. 4 Price density of watercolor paintings and price histogram of all paintings

Attribute	Step 1	2	3	4	5	6
area	0.540					
medium	0.377	0.550				
favourites	0.374	0.549	0.551			
author_artworks_count	0.362	0.544	0.549	0.560		
author_avg_...[1]	0.371	0.542	0.543	0.552	0.562	
author_degree	0.363	0.539	0.545	0.553	0.561	0.563

[1] author_avg_exhibitions_count_now

Decision tree (precision) [basic]

Attribute	Step 1	2	3	4	5
area	0.537				
author_country	0.374	0.553			
author_events_count	0.367	0.547	0.560		
author_followers	0.373	0.543	0.558	0.561	
views_popularity	0.363	0.537	0.551	0.558	0.565

Random forest (precision) [basic]

Fig. 5 Classifiers precision

Fig. 6 Multinomial Naive Bayes (left), Decision Tree (right)

	precision
keywords	0.420
labels	0.348

	precision
keywords	0.561
labels	0.556

Fig. 7 Decision Tree Classifier: 10-fold cross validation

a	b	d	c	← classified as	number
5977	1042	408	49	a = [0, 1020)	7476
1773	**1956**	1340	199	b = [1020, 2335)	5268
564	1128	**2713**	614	d = [2335, 5100)	5019
139	282	1107	**1252**	c = [5100, -)	2780

Fig. 8 Random Forest classifier: 10-fold cross validation

a	b	d	c	← classified as	number
5975	1007	418	76	a = [0, 1020)	7476
1304	**2585**	1139	240	b = [1020, 2335)	5268
461	947	**2972**	639	d = [2335, 5100)	5019
130	248	758	**1644**	c = [5100, -)	2780

Trying to boost these results, we thought of using text mining methods on keywords and image labels attributes, which we transformed into a term frequency vector. Afterwards, we ran a Multinomial Naive Bayes classifier (Fig. 6 (left part)).

We can observe that a keywords attribute might be slightly co-related with price. We evaluated influence of those vectors on the best Decision Tree model (Fig. 3), but it did not outperform the original model (Fig. 6 (right part)). We also performed 10-fold cross-validation using the best combinations of features for checking quality of classifiers. Validation was performed using Weka Explorer (see Figs. 7 and 8).

We observed that the results are slightly better - 0.58 is the weighted precision for Decision Tree Classifier and 0.64 for Random Forest.

Similar testing was also done for the extended discretization of the decision attribute price: [0; 480), [480; 1020), [1020; 2335), [2335; 5100), [5100; 10650), [10650; 19500), [19,500; +1). The results are shown on Figs. 9 and 10.

Comparing basic and extended discretizations, precision of the classifier is decreased when the number of decision values is increased (the intervals are smaller) −0.52 is the new weighted precision for Decision Tree Classifier and 0.55 for Random Forest. Also, it can be easily noticed that the confusion matrices for both classifiers show that paintings which cost is between $5,100.00 and $19,500.00 are overpriced. To have a better understanding about the level of paintings overprice, we introduce the notion of Left-Extended correctness and Right-Extended correctness of classifiers. It is directly computed from the confusion matrix of a classifier. To compute Left-Extended correctness of a classifier we add the val-

f	a	b	d	e	c	g	← classified as	Number
1863	731	264	71	2	0	2	f = [0,480)	2933
646	**2401**	1054	384	53	2	3	a = [480, 1020)	4543
250	1060	**2568**	1236	134	12	8	b = [1020, 2335)	5268
109	374	1135	**2915**	430	41	15	d = [2335, 5100)	5019
26	88	229	822	**676**	73	25	e = [5100, 10650)	1939
6	14	57	166	123	**78**	9	c = [10650, 19500)	453
4	12	18	85	44	11	**214**	g = [19500, -)	388

It looks like paintings within the range $5,100 – $19,500 are overpriced

Fig. 9 Confusion matrix for Decision Tree Classifier

f	a	b	d	e	c	g	← classified as	Number
1857	676	286	99	11	3	1	f = [0,480)	2933
576	**2467**	1026	400	61	6	7	a = [480, 1020)	4543
229	982	**2668**	1175	184	21	9	b = [1020, 2335)	5268
77	336	986	**3063**	493	43	21	d = [2335, 5100)	5019
19	76	227	709	**812**	70	26	e = [5100, 10650)	1939
6	12	49	151	118	**106**	11	c = [10650, 19500)	453
5	7	18	70	36	13	**239**	g = [19500, -)	388

Fig. 10 Confusion matrix for Random Forest classifier

Price Range	Correctness	Left Ext. Correctness	Right Ext. Correctness	Overpriced (?)
[0,1020)	78%	-	94%	-
[1020, 2335)	37%	71%	63%	YES
[2335, 5100)	54%	77%	66%	YES
[5100, -)	45%	85%	-	-

Fig. 11 Decision Tree Classifier: 10-fold cross validation

ues from the main diagonal of its confusion matrix to the values of its neighboring left-diagonal. To compute Right-Extended correctness of a classifier we add the values from the main diagonal of its confusion matrix to the values of its neighboring right-diagonal. So, for instance, from Fig. 8 we can build "Right Extended Correctness" column in Fig. 12 by calculating $((5975 + 1007)/7476) = 93$, $((2585 + 1139)/5268) = 71$, and$((2972 + 639)/5019) = 72$ (Fig. 11).

Similar testing was also done for the extended discretization of the decision attribute price: [0; 480), [480; 1020), [1020; 2335), [2335; 5100), [5100; 10650),

Price Range	Correctness	Left Ext. Correctness	Right Ext. Correctness	Overpriced (?)
[0,1020)	80%	-	93%	-
[1020, 2335)	49%	74%	71%	YES
[2335, 5100)	54%	78%	72%	YES
[5100, -)	59%	86%	-	-

Fig. 12 Random Forest classifier: 10-fold cross validation

Price Range	Correctness	Left Ext. Correctness	Right Ext. Correctness	Overpriced (?)
[0,480)	64%	-	88%	-
[480, 1020)	53%	67%	76%	NO
[1020, 2335)	49%	69%	72%	NO
[2335, 5100)	58%	**81%**	**67%**	YES
[5110, 10650)	35%	**77%**	**39%**	YES
[10650, 19500)	17%	**44%**	**19%**	YES
[19500, -)	55%	58%	-	-

Fig. 13 Confusion matrix for Decision Tree Classifier

Price Range	Correctness	Left Ext. Correctness	Right Ext. Correctness	Overpriced (?)
[0,480)	63%		86%	-
[480, 1020)	54%	67%	77%	NO
[1020, 2335)	51%	69%	73%	NO
[2335, 5100)	61%	**81%**	**71%**	YES
[5110, 10650)	42%	**78%**	**45%**	YES
[10650, 19500)	23%	**49%**	**26%**	YES
[19500, -)	62%	65%	-	-

Fig. 14 Confusion matrix for Random Forest classifier

[10650; 19500), [19,500; +1). The results are shown on Figs. 9 and 10 (Figs. 13 and 14).

5 Conclusions

The obtained results indicate that the three main factors triggering price are: time spent on creating the painting (area, medium, etc.), paintings popularity, and authors image. On the basis of performed experiments, it seems that when the price increases, rules are starting to get less determined. We believe that in these cases people follow more abstract and indeterminate rules. For example it is obvious that if price increases, the area of painting has lesser importance.

References

1. Ricci, F., Rokach, L., Shapira, B., Kantor, P.B.: Recommender Systems Handbook, 1st edn. Springer, New York (2010)
2. Kuang, J., Daniel, A., Johnston, J., Ras, Z.: Hierarchically structured recommender system for improving NPS of a company. LNCS, vol. 8536. Springer (2014)
3. SaatchiArt, Home page. http://www.saatchiart.com/
4. Artfinder, Home page. https://pl.artfinder.com/
5. UGallery, Home page. http://www.ugallery.com/
6. DeviantArt, Home page. http://www.deviantart.com/
7. Wikipedia, Saatchi gallery. https://en.wikipedia.org/wiki/Saatchi_Gallery
8. MathWorks, Image classification with bag of visual words. http://www.mathworks.com/help/vision/ug/image-classification-withbag-of-visual-words.html
9. Sharma, G., Wu, W., Dalal, E.N.: The CIEDE2000 color-difference formula: implementation notes, supplementary test data, and mathematical observations, Color Res. Appl. **30**(1):2130 (2005)
10. Canny, J.: A computational approach to edge detection. IEEE Trans. Pattern Anal. Mach. Intell. **8**(6), 679–698 (1986)

Modelling OpenStreetMap Data for Determination of the Fastest Route Under Varying Driving Conditions

Grzegorz Protaziuk, Robert Piątkowski and Robert Bembenik

Abstract We propose a network graph for determining the fastest route under varying driving conditions based on OpenStreetMap data. The introduced solution solves the fastest point-to-point path problem. We present a method of transformation the OpenStreetMap data into a network graph and a few transformation for improving the graph obtained by almost directly mapping the source data into a destination model. For determination of the fastest route we use the modified version of Dijkstra's algorithm and a time-dependent model of network graph where the flow speed of each edge depends on the time interval.

Keywords OpenStreetMap data · Time-dependent network graph · Fastest path

1 Introduction

In the literature the problem of determining the best path in a network graph has been extensively studied and at least several variants of that problem have been disused. The differences may concern: the detail objective (e.g. the way with the discussed minimal cost, the latest possible departure time), time invariant or time-dependent cost of edges, stochastic or static network etc. Many studies concern static network, with time-invariant data [12]. The time-dependent routing problem is mainly considered with respect to FIFO networks with discrete time.

In [6] a time-dependent model satisfying the FIFO property for the vehicle routing problem was introduced. In the model time-dependent travel speeds are calculated based on time-window. In the tests three time intervals and three types of roads were considered. In [1] several variants of the time-dependent shortest path problems are discussed. In the variants the ranges of the following variables are considered:

G. Protaziuk (✉) · R. Piątkowski · R. Bembenik (✉)
Institute of Computer Science, Warsaw University of Technology,
Nowowiejska 15/19, Warsaw, Poland
e-mail: gprotazi@ii.pw.edu.pl

R. Bembenik
e-mail: R.Bembenik@ii.pw.edu.pl

R. Bembenik et al. (eds.), *Intelligent Methods and Big Data in Industrial Applications*, Studies in Big Data 40, https://doi.org/10.1007/978-3-319-77604-0_5

starting node (one or many), destination node (one or many), departure time: the latest departure time from the source node in order to arrive at the destination node by a certain time t, arrival time: the earliest arrival time at the destination node if one leaves the source node at a certain time t. The basic problem (and the simplest) concerns situation in which we have one source node, one destination node, and one departure time. In [3] the approach proposed in [6] was applied to determine shortest paths between nodes in a network graph built based on historical real speed data in the north of England. The carried out tests showed that using real road traffic data and taking time-dependent journey times into account can significantly influence the distribution schedule and allow obtaining a better schedule. The problem of finding the minimum cost path in a time-varying road network was studied in [13]. Authors considered three parts of the cost: fuel cost influenced by the speed, driver cost associated with a travelling time, and congestion charge with respect to a scheme of surcharging users of a transport network in periods of peak. In order to solve the problem two heuristic methods incorporated into the Dijkstra's algorithm were proposed. A short review of time-dependent routing problems is provided in [4].

The stochastic network where the arc travel times are time-dependent random variables and the probability distributions vary with time is considered in [5, 8], and [9]. In the two first works the problem of finding a path with the shortest expected travel time is studied whereas the third work is focused on the problem of determination of the latest possible departure time with a route which allows one to arrive at the destination node not later than a specified time with a given probability. In [10] the shortest path problem is considered with respect to a non-deterministic network, in which some arcs may have stochastic or uncertain lengths.

There are many solutions on the market which may be directly used to determine the shortest path between two selected points (e.g. Google Maps,[1] AutoMapa[2]). However, there are significant difficulties with using them in further automatic processing; also the details of the applied algorithms are unknown. The second group of available solutions consists of programming components or libraries such as pgRouting[3] or libraries[4] in which the maps from the OpenStreeMap[5] (OSM) service are used. We decided to implement our own solution as we wanted to have the full control during experiments.

Nowadays, application of a time-dependent network graph, in which the travel speeds associated with edges depend on time, in solutions for determining the fastest path is an observed tendency as such model allows finding more reliable routes. However, the data needed to evaluate the travel speed is still not commonly available for most of the roads. In the paper we deal with such type of network. The main contributions of this paper is the introduced approach to build the network graph by using the OpenStreetMap data with methods for detail estimation of travel time for roads

[1]https://maps.google.com.

[2]http://www.automapa.pl/pl/produkty/automapa-traffic.

[3]http://pgrouting.org/.

[4]http://wiki.openstreetmap.org/wiki/Routing.

[5]http://www.openstreetmap.org.

of variant characteristics. We also provide some details concerning implementation of the network graph in the java language.

The remainder of the paper is organized as follows. In Sect. 2 we present our approach to finding the fastest route problem. In Sect. 3 we describe shortly the OSM data model and the way of transforming this data into a network graph. In Sect. 4 we introduce methods aiming at improvement of the network graph obtained based on the OSM data. In Sect. 5 we discuss ways of estimating the travel time for various segments of the roads. Section 6 is dedicated to implementation issues, whereas in Sect. 7 we discuss performed experiments. Section 8 concludes this paper.

2 Finding the Fastest Route

The proposed method of determining the fastest route was developed under the following assumptions:

- A route will be travelled by car.
- Travel times may vary depending on the date and the start time.
- The speed that a vehicle can achieve is limited by the law or a traffic congestion. In our analysis we assume driving with maximal speed possible under given circumstances.
- Average speeds in cities and outside of them usually vary widely.
- There is a strong dependency between average speeds and traffic congestion.

These assumptions concerning travel speed are not true e.g. for pedestrians when the speed is much more dependent on the human's abilities than traffic congestion.

2.1 The Data Model

In the system the routes are represented in the form of a direct, weighted graph, without loops and multiple edges. We defined a graph in a typical manner as a triple: $G = (V, E, w(e, t))$, where:

- V is a set of vertices,
- E is a set of edges - ordered pairs of vertices $(v1, v2)$, $v1$ is the head of an edge and $v2$ is the tail of an edge.
- $w(e, t)$ is a weighting function: $E \times R \rightarrow R$, where e is an edge and t is the time of reaching the head of the edge e.

As in this paper we focus on FIFO networks, where First-In-First-Out principle is effective for the entire traffic, we require that a weighting function has the following properties:

- **Property 1. Non-negative weights** - a weight of each edge in a graph has to be greater or equal to 0. Formally:

$$\forall e \in E\ \forall t\ w(e,t) \geq 0 \quad (1)$$

This constraint does not hinder the modelling of routes; there is no case where a travel time through a given road is negative.

- **Property 2. Unprofitability of waiting**. This requirement is a bit more complex. To illustrate it let us consider a journey where the starting and destination points are located on the opposite sides of the city, and one may start not earlier than the afternoon communication peak. Probably, the travel time understood as the difference between the arrival time and the departure time will be shorter off peak. However, with a given departure time, if one wants to reach the destination as soon as possible she or he should leave as early as possible (in such cases the travel time is not very important). As the rule, the earlier one leaves, the earlier one arrives. If this rule is not fulfilled it could happen that a driver A departing later overtakes a driver B who left earlier or the driver A had a better route. Nevertheless, if overtaking was possible, it means that the driver B could travel faster, assuming that vehicles were moving below their maximum speeds (what is usually met in the city), what leads to contradictions. If we assume the full knowledge about the road network (also about travelling time) the driver B could also choose the route selected by the driver A and arrive earlier than A or at least at the same time. Equation 2 defines this requirement formally.

$$\forall e \in E\ \forall t, t_0\ w(e, t_0 + t) + t \geq w(e, t_0) \quad (2)$$

2.2 The SPGVW Algorithm

For the determination of the fastest path we use the **S**hortest **P**ath in a **G**raph with **V**ariable **W**eights (SPGVW) algorithm which is an adaptation of the Dijkstra's algorithm [2]. The pseudo-code of the algorithm is presented in Procedure 1. The changes to the original algorithm concern calculation of weights of paths and they are underlined.

The proof. The SPGVW algorithm returns one from the correct paths with the minimal departure time for a destination vertex if the following statement is true: *if the vertex v is removed from Q, then $t_{min}[v]$ is the earliest possible at this moment* (finally we removed the destination vertex). Below, we prove this statement.

Let's assume that the statement is not true and consider the vertex v, which has been removed from the Q queue. In order to reduce $t_{min}[v]$, there must be a neighbour u for which the algorithm finds the new $t_{min}[v]$. The reduction occurs if $t_{min}[v] > t_{min}[u] + w(e, t_{min}[u])$. However, since v has been removed from the queue earlier, $t_{min}[v] < t_{min}[u]$. So, in order to reduce $t_{min}[v]$ the weight $w(e, t_{min}[u])$ has to be negative (what is forbidden) and we obtain contradiction.

Now, let's assume that in some iteration the calculated $t_{min}[u]$ is less than $t_{min}[v]$. However, to lower $t_{min}[u]$ below $t_{min}[v]$, we have to find some neighbour of u, let's

Procedure 1 Shortest Path in a Graph with Variable Weights

Input: graph G, the source vertex v_p, the destination vertex v_k, starting time of a journey t_start
Output: The path for which the arrival time to the destination vertex is the shortest.

```
Q: Priority queue of "open" (not yet visited ) vertices
t_min[v]: An array containing the earliest times of reaching each of the vertices
path[e]: An array containing the path ensuring that the arrival time is equal to t_min[e]

Q ← all vertices of the graph G
for all vertices v of G do
   t_min[v] ← ∞
end for
t_min[v_p] ← t_start
path[v_p] ← empty path

while Q is not empty do
   v ← vertex from Q with the minimal t_min
   remove v from Q
   if t_min[v] = ∞ then
      return there is no path from v_p to v_k
   end if
   if v = v_k then
      return t_min[v_k] and s[v_k]
   end if
   for all neighbours v, denoted as v_1, reachable by the edge e do
      if t_min[v_1] > t_min[v] + w(e, t_min[v]) then
         t_min[v_1] ← t_min[v] + w(e, t_min[v])
         path[v_1] ← path[v] + v
      end if
   end for
end while
```

assume u_1, at $t_{min}[u_1]$ less then $t_{min}[v]$. However, such vertex u_1 would be removed from the queue before v and we obtain contradiction.

The algorithm visits the starting vertex s at the earliest as possible - at the start of the journey. Then, vertices are removed when their t_{min} is minimal. Thanks to the Property 1 and Property 2, it is known that t_{min} obtained for these vertices will not be earlier for any later start time.

In summary: when the destination vertex v_k is removed from the queue, its t_{min} is the earliest possible for a given start time and $path[v_k]$ is the fastest path.

3 Data Modelling

The basic format of OpenStreetMap data is XML. In the described research we used a file concerning Poland provided by Geofabrik.[6]

[6]http://download.geofabrik.de/.

3.1 OSM Nodes

A node, which describes a single point, is the basic object in the data model. From the perspective of the graph of routes the most important properties of nodes are:

- id - an identifier - attribute `id`
- latitude and longitude - attributes `lat` and `lon` respectively,
- information that a given node represents a crossroad with a traffic light (the `tag` tag and its attributes).

A sample of a OSM node is provided in Example 1:

Example 1 OSM node representing the crossroad of Świętokrzyska and Marszałkowska streets in Warsaw:

```
<node id="3378411269" visible="true" version="2"
changeset="30070926" timestamp="2015-04-08T18:24:39Z"
user="kocio" uid="52087"
lat="52.2352389" lon="21.0086316">
<tag k="highway" v="traffic_signals"/>
</node>
```

3.2 OSM Ways

A way is composed of nodes. Ways may represent roads/streets or big buildings. A road is represented as an ordered set of points (nodes); in some cases the first node and the last node can be identical. Additional information may be associated with a road such as a name, a category, maximum allowed speed. A sample OSM wary representing a road is given in Example 2.

Example 2 Sample description of a road in OSM data (fragment):

```
<way id="203484442" (...)>
(<nd ref="2134999713"/>
(...)
<nd ref="321395177"/>
<tag k="highway" v="tertiary"/>
(...)
<tag k="maxspeed" v="50"/>
<tag k="name" v="Handlowa"/>
<tag k="ref" v="5615W"/>
(...)
</way>
```

3.3 Network Graph

Conceptually, the OpenStreepMap model is similar to a graph structure: vertices can be created based on nodes and edges can be created based on ways. The simple method for performing such one-to-one conversion is presented in Procedure 2.

Procedure 2 OSM data into graph: one-to-one transformation algorithm

1. Create a vertex for each OpenStreetMap node.
2. For each way representing a road iterate through its nodes starting from the first one and for consecutive pairs of nodes create:
   - one edge starting from a vertex corresponding to the first node of the considered pair of nodes - for one ways rouads and roundabouts,
   - two edges between vertices corresponding to the nodes of the considered pair of nodes (in both direction).

After execution of the one-to-one conversion one obtains the graph which represents a traffic network, without weights of edges. For determining the fastest route weights should reflect travel time not distance. The travel time can be calculated based on the maximal allowed speed for a given segment and its length. We computed that data by using the procedure presented in Procedure 3.

Procedure 3 Procedure for computing the maximal allowed speed

1. The length of a given segment is calculated as the shortest distance between its start and end nodes.
2. The maximal speed is determined in the following manner:
   - use the maximal speed directly given in the description of the road including the considered segment, else
   - if a category of the road is given assign the maximal speed allowed by the law for a given category, else
   - use the maximal speed which is valid for most of the roads.

---

4 Improvement of the Traffic Network Graph

In this section several ways for improvement of the traffic network graph obtained by the direct mapping of the OpenStreepMap data are provided. They concern removing unnecessary vertices and more accurate estimation of travel times.

4.1 Removing Vertices

The graph obtained by means of the procedure presented in Sect. 3.3 reflects the spatial structure of the original data. It means that in the graph there are vertices which do not represent cross-roads, but only reflect nodes included in poly-lines used for modelling the roads. Such vertices should be removed from the graph because they do not increase the number of available routes and inversely affect the time needed to finding the fastest path. The procedure for removing such vertices is presented in Procedure 4.

Procedure 4 Removing surplus vertices

***One-way roads*:**
A vertex is removed if it has only one incoming edge and one outgoing edge. A new edge is created which connects the starting vertex of the incoming edge with the ending vertex of the outgoing edge. The weight of this edge (the travel time) is the sum of weights of the incoming and the outgoing edges. This situation is shown in Fig. 1a.

***Two-way roads*:** A vertex is removed if:

1. it has exactly two incoming edges and two outgoing edges;
2. the ending vertex of an outgoing edge is the starting vertex of an incoming edge;
3. it is possible to create two pairs of edges with the following properties:
   - one edge is incoming, one edge is outgoing;
   - the starting vertex of the incoming edge and the ending vertex of the outgoing edge are different.

   For each pair a new edge (also its weight) is created in the same manner as in the case of the one-way road. This situation is presented in Fig. 1b.

4.2 Turns

In order to estimate the travel time accurately the turns should be taken into consideration, especially in the roads of a low category. The turns with small radius can significantly affect the travel time. OpenStreetMap data in many situations lacks of information concerning maximal safe speed for such turns, however, such maximal speed can be estimated based on geometry of the turn. For that purpose we used the Eq. 3.

$$v_{smax} < \sqrt{Cr} \tag{3}$$

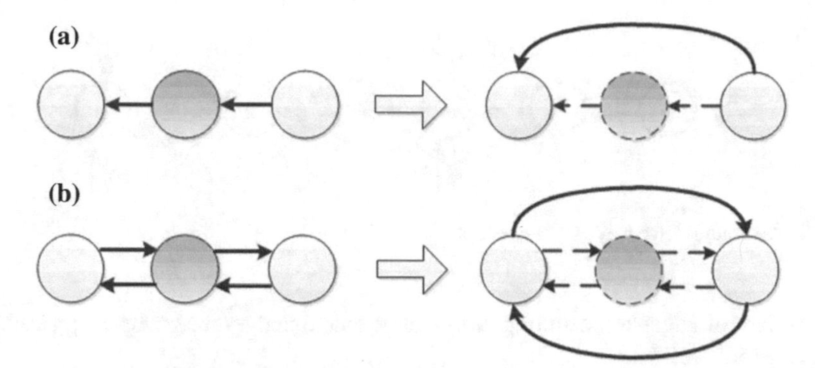

Fig. 1 Two possible schemas of removing a single node as described in Procedure 4: **a** - for one way roads, **b** - for two-way roads

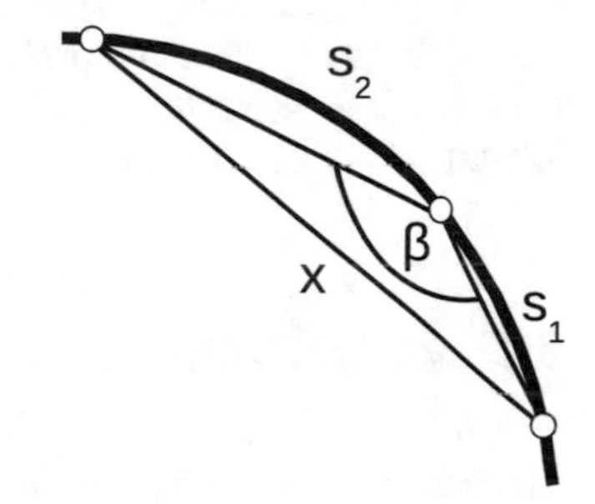

Fig. 2 Estimation of the radius of a turn. White points stand for the points obtained from a GPS receiver

where v_{smax} is the maximal safe speed, r is the radius of the turn, and C is the factor which value was determined by using of Polish road construction standards: ($C \approx 4225\,\frac{\mathrm{km}}{\mathrm{h}^2}$ for speeds measured in $\frac{km}{h}$ and radius in kilometres).

The radii of turns are not directly given in the source data and they were computed based on courses of roads (approximated by poly-lines). In order to determine lengths of radii the following assumptions were made:

- Turns are the perfect arcs. This does not have to be fulfilled in reality - but this assumption seems to be good enough for the available data.
- Points included in a given poly-line are resultant from GPS track recorded in a vehicle travelling through the considered road.
- A vehicle travels on arcs of radius r.
- Turns are determined by three points of poly-lines (see Fig. 2).

We calculated the radii by using the standard equation 4 for calculation of a radius of a circumscribed circle of the triangle.

$$r = \frac{x}{2sin\beta} \tag{4}$$

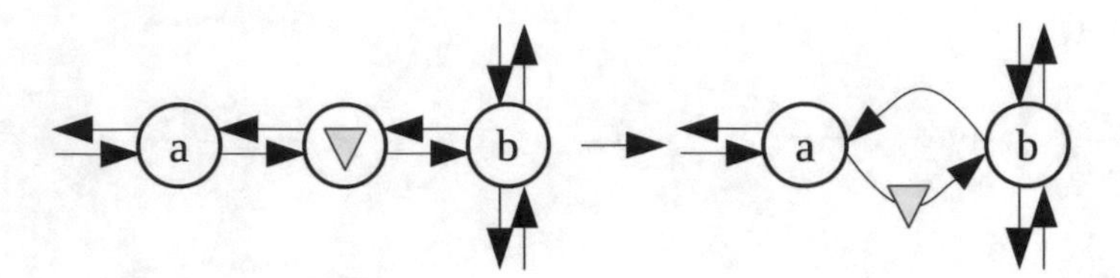

Fig. 3 Assigning "give way" label to edges

The weight of edges representing turns can be calculated by means of the procedure presented in Sect. 5.1.

Example 3 Let us consider a road with the maximal allowed speed $v_{max} = 90$ km/h. On the road there is a turn with the maximal safe speed $v_{safe} = 30$ km/h. The weight of an edge representing this turn is equal to: $2 \cdot sH(30, 90)$, where $sH(v_{safe}, v_{max})$ stands for the procedure presented in Sect. 5.1. If such a turn joins two roads with $v1_{max} = 90$ km/h and $v2_{max} = 60$ km/h, the weight of the edge is equal to: $sH(30, 90) + sH(30, 60)$.

4.3 *Right-of-Way*

The information about priorities is included in OpenStreetMap data in the following two ways:

- by road category - tag *highway* for ways,
- by pointing out localizations of "Give way" signs by means of tags *highway* with values *give_way* or *stop* for nodes.

In the former case the interpretation is straightforward. In the latter that information should be associated with edges. While executing the procedure removing surplus vertices (see Sect. 4.1) a newly created edge is labelled "give way" if its end vertex represents a cross-road, see Fig. 3.

4.4 *Cross-Roads*

The next type of obstacles on roads are cross-roads. In this case loss of travel time depends on types of roads on which one travels. The both roads: entering at and exiting from a given cross-road should be taken into consideration. A network graph in which cross-roads are represented by one node does not allow for modelling delays in travel caused by cross-roads. The typical approach to solve this problem is to represent the cross-road by many nodes and edges between these nodes. In order to obtain such representation for each vertex representing cross-road the following procedure was applied:

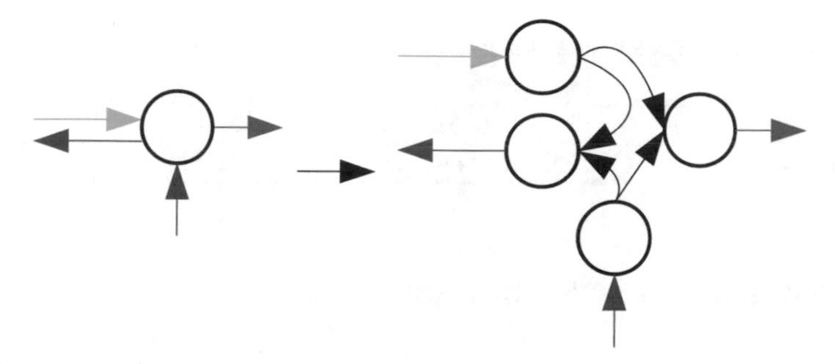

Fig. 4 Creation of a network graph that allows modelling time needed for passing cross-roads

1. Create a new vertex for each edge (incoming or outgoing) associated with the considered vertex.
2. For each incoming edge set its end vertex to the newly created vertex for that edge.
3. For each outgoing edge set its starting vertex to the newly created vertex for that edge.
4. For each possible crossing through the cross-road (represented by the considered vertex) create an edge connecting appropriate newly created vertices. Values returned by the weighting function for a newly created edge should reflect the time required for passing the crossing represented by that edge.

The procedure is illustrated in Fig. 4. In the context of cross-roads, there is a problem of detection of a direction in which the vehicle turns at a cross-road (left, right, straight, U-turn). This information is needed to determine which vehicles should be given the right-of-way while passing a cross-road. To solve that problem a method based on the calculation of the difference between the azimuth of the last segment of a road which a vehicle enters the cross-road and the azimuth of the first segment of a road which the vehicle leaves a cross-road. The details of the applied method are provided in Procedure 5.

Procedure 5 Determination of travel direction at cross-roads

1. Compute how many degrees left ($left\_angle$) a vehicle has to turn in order to change the travelling azimuth from that for starting road to the one for the final road.
2. Compute such an angle ($right\_angle$) for the right turn.
3. Determine the smaller angle: $small\_angle = min(left\_angle, right\_angle)$
4. If the smaller angle is lesser than a certain threshold corresponding to an almost imperceptible turning, return: straight.
5. If the smaller angle is greater than a certain threshold indicating U-turn, return: U-turn.
6. If $left\_angle < right\_angle$ return: left, otherwise return right.

Table 1 Example average speed for 1.5 distance in time intervals

Time	Avg speed	Travel time
18:45–18:59	30 km/h (0.5 km/min)	3 min
19:00–19:14	60 km/h (1 km/min)	1.5 min

5 Estimation of Travel Time for Edges

The values of the weight function $w(e, t)$ can be determined by a method for estimating the travel time for a given part of the route. The simple approach is based on average speed: if for a given time t for a given segment v of the length s the average speed v_{avg} is assigned then $w(e, t) = v_{avg} \cdot t$. However, with the discrete time and taking into consideration Property 2 the weighting function should be carefully selected. The main problem with selection of such a function is shown in Example 4.

Example 4 Let us consider a 1.5 km distance and a 15 min interval for which the average speed is calculated. The example data for that case is provided in Table 1. The direct usage of travel time presented in Table 1 does not fulfill Property 2 of a weighting function because for the discussed case when leaving at 18.59 it is better to wait one minute (the time travel will be shorter). Without waiting the travel time (tt) is equal to the time of travelling with the 0.5 km/min speed plus the time of travelling with the 1 km/min speed[7]: $tt = 1\,\text{min} + \frac{1.5\,\text{km} - (0.5\,\text{km/min}) * 1\,\text{min}}{1\,\text{km/min}} = 2\,\text{min}$, whereas with the waiting the travel time is equal to the time of travelling with speed of 1 km/min: $tt = \frac{1.5\,\text{km}}{1\,\text{km/min}} = 1.5\,\text{min}$.

In our system we adapted the method introduced in [11]. We used the simpler version of that method, as we allowed only one change of time range while travelling through a given segment represented by a single edge. This approach can be justified by the following observation: typically, time ranges last at least several minutes, so in practice one time range is almost always enough to pass the entire segment, so hardly ever more than one change of ranges occurs. Another reason is that such an approach is much easier to implement.

5.1 *Braking Before an Obstacle*

Calculation of the travel delay associated with braking before an obstacle is done by comparing travelling with the maximal allowed speed (v_{max}) with travelling with a safe speed (v_{safe}). For that purpose the Procedure 6 was applied. The loss of time due to acceleration from an obstacle can be calculated in the analogous way.

[7]In the calculation we applied the standard equation for speed computation: $V = \frac{S}{T}$, where V - speed, S - distance, T - time.

Procedure 6 Calculation of loss of time caused by braking before an obstacle

Input: Initial speed v_{max}, final speed v_{safe}, deceleration a
Output: Loss of time associated with braking

$t_h = \frac{v_{max} - v_{safe}}{a}$ - time needed for braking

$s_h = v_{max} t_h - \frac{a t_h^2}{2}$ - the distance between the starting point (the point in which the braking starts) and the obstacle

$t_n = \frac{s_h}{v_{max}}$ - travelling time with the speed v_{max} from the starting point to the obstacle

return Loss of time: $t_h - t_n$.

5.2 *Passing Cross-Roads*

The following issues should be taken into consideration during estimation of the time needed for passing cross-roads:

- necessity of slowing down in case of turning;
- given a right-of-way;
- traffic lights.

Give Way The first step in estimating the delay in travel related to giving way is determination to whom the way should be given. It is done by comparing an edge by which a vehicle arrives with each other edge representing entering roads for an analysed cross-road. On the basis of the recorded priority information, or if these are not decisive, the legal basis, it is determined to whom a considered vehicle should give way. In order to estimate the considered delay we applied the following procedure:

- Each "priority" edge causes a delay equal to the time needed to pass that edge with the maximum speed.
- In the case in which more than one edge has priority, the maximum of delays caused by each single edge is taken as the resultant delay.

One of the natural alternative method for the solution presented in the second point consists of calculation of the least common multiple, e.g. if one has to give way to two direction where in the first there is a gap (opportunity to pass) every 4 min, and in the second every 6 min, then after maximum 12 min ($4 \cdot 3$ and $6 \cdot 2$) the gap occurs for both directions at the same time. The main disadvantage of this method is that the resulting values seem to be too big. Also, some additional solution should be proposed in case in which waiting times are given as real numbers.
Traffic Lights Delays in travel caused by traffic lights vary quite widely: with the green light there is no delay, when the red one appears one has to stop, wait and accelerate again. In order to calculate such delay we introduced a probabilistic model of passing through traffic lights. The following assumptions were made:

- only discrete time is considered;

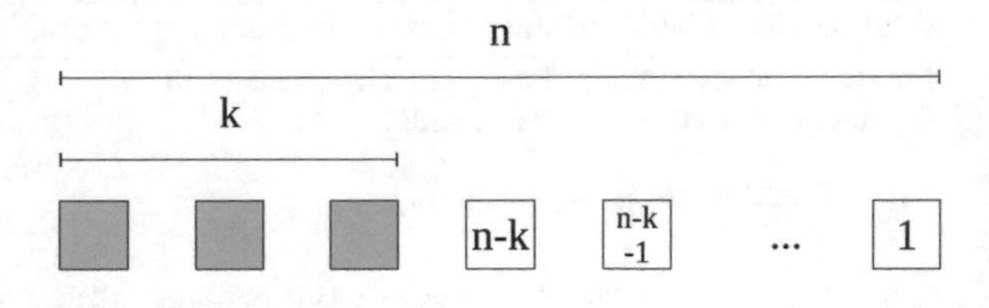

Fig. 5 Model of traffic light work

- a cycle of light changing takes n units of time;
- a probability of hitting each cycle unit is equal to $\frac{1}{n}$;
- only red and green lights are considered;
- a green light continues uninterrupted by k units, a red by the remaining;
- at the green light a delay is equal to 0;
- at the red light, a delay is equal to a certain value of C summed up with a red light waiting time.

The Fig. 5 presents the idea of the model of a traffic lights. Based on the introduced assumptions the delay in travel (the expected value) can be calculated by using one of the following equations:

$$EX = \frac{1}{n}\left(\sum_{i=1}^{n-k} i + (n-k)\cdot C\right) \; or$$
$$EX = \frac{1}{n}\left((n-k)\cdot\frac{1+(n-k)}{2} + (n-k)\cdot C\right) \tag{5}$$

where n - number of units in one cycle, k - number of units with the green light, C - a delay associated with the red light.

Example 5 Given $C = 1$ s, cycle time $n = 60$ s, time of the green light in one cycle $k = 20$ s. The result is equal to $\frac{43}{3} = 14.(3)$.

5.3 Closing/Opening of Roads

The phenomenon that occurs in reality, and there is no simple counterpart in the presented model, is the accessibility of certain roads only for some period of time. An example of such situation may be the tenable closing of the Pancera viaduct in Warsaw or restricted traffic (especially in the centres or historical areas) in many European cities. A simple solution would be to set the value of the weight to ∞. However, if for some t_0 the value is ∞, and for t_1, H which may be high but finite, it is worth waiting from t_0 to t_1 what in turn is contradictory with (2). This problem may be solved by using the following way of calculation of the travel time:

- $w(e, t)$ if t means time when the road e is open.

- $w(e, t_p) + t_p - t$ if t means time when the road e is closed, t_p - the time of opening the road ($t_p >$ time of calculation)

This method of travel time calculation meets the Property 2 of a weighting function.

Proof In the second case after substitution to (2) we obtain:

$$\begin{aligned} w(e, t_p) + t_p - (t_0 + t) + t \geq w(e, t_p) + t_p - t_0 \\ -t_0 - t + t \geq -t_0 \\ 0 \geq 0 \end{aligned} \tag{6}$$

The required inequality is met.

6 Implementation

The introduced approach was implemented in the experimental system. The system consists of two applications:

1. Application transforming OpenStreetMap data into a traffic network graph.
2. Application, which uses the already built network graph for determination of the fastest route and travel time.

The implementation was made in Java, version 8, which allows one to use the lambda expressions. The main elements of programming data model are nodes (representing vertices in the graph) and edges (representing edges in the graph). The node has the following properties:

- the unique identifier consisting of the corresponding node identifier in OpenStreetMap (*osmId*) and auxiliary *internalId*. It allows creating additional nodes when processing cross-roads;
- the geographic location (longitude and latitude);
- information whether a node represents a traffic signal or the "give way" sign. The 1-byte attribute *flags* is used to store that information. According to [7] that solution is more cost-effective than using multiple *boolean* variables;
- outgoing and incoming edges. This information is stored by means of *ArrayList* structure with a default size of 2 that corresponds to a node associated with a one-way road.

We distinguish the following type of edges: *simple*, *complex*, and *latency* which differ in the amount and detail of stored information. Each type is represented by the dedicated programming class:

- *RpLatencyEdge* - the simplest class storing information about a delay, useful for edges representing obstacles, which passing requires travelling relatively long time through a (very)short distance.

- *RpSimpleEdge* - a class representing edges created in the process of the direct mapping of the OpenStreetMap data to the graph, before the elimination of vertices. An instance of this class represents one segment of a road.
- *RpComplexEdge* - the most complex class for representing edges resultant from the elimination of vertices process.

The edge has the following properties:

- the starting and the end nodes;
- weights (modelled by means of an auxiliary programming class)
- information about the road:
  - id (number);
  - category;
  - points list containing points corresponding to nodes that are eliminated in the process of graph improvement(see Sect. 4.1);
  - flags.

Edges corresponding to real roads have additional properties dedicated to storing spatial information such as:

- *length*: the length of road/segment in meters;
- *timeToDrive*: minimum time needed to cross the segment in hours;
- properties of a road (a way in OpenStreetMap data) and its segments represented by the edge:
  - *startAngle*/*endAngle*: azimuths in degrees, respectively the beginning of the first segment and the end of the last segment;
  - *startSpeed*/*endSpeed*: the maximum allowed speed, respectively for the first segment and the last segment;
  - *startLength*/*endLength*: the length in meters, respectively of the first segment and the last segment;

7 Tests

The system was launched on a Lenovo X201t laptop with an Intel Core i7 L640 and 8GB of RAM. A 64-bit Java virtual machine running under Windows 7 was used. The tests concerned the efficiency of OpenStreetMap data transformation and the quality of the computed shortest paths.

7.1 Processing OpenStreetMap Data

In Table 2 the results of processing OpenStreetMap data XML files for two selected regions are provided. The results show that the processing took reasonable amount of time and the proposed procedures for improvement of the initial network graph allow reducing significantly the size of that graph.

7.2 Determination of the Shortest Path

The aim of tests concerning determination of the fastest path. For the test purpose we applied the following simple model of ways:

1. Roads are divided into the three following categories:
   - *fast*: main roads, motorways, and highways;
   - *slow*: local roads (*tertiary* and lower);
   - *other*: others.
2. For each category of roads, non-overlapping time ranges are defined. The ranges covers one day.
3. Each range has assigned a multiplier. In order to calculate travelling time in a given range time the minimum value for a given edge should be multiply by the assigned multiplier.

In the tests the following ranges and multipliers were defined:

1. fast roads: 0–7: 1.0/7–10: 2.0/10–24: 1.0
2. slow roads: 0–24: 1.0
3. other roads: 0–14: 1.0/14–17: 1.5/17–24: 1.0

The data indicates that:

1. for the fast roads: at the morning rush hour travelling takes two times more than in other time ranges.
2. for slow roads: the travelling time is the same for all ranges.
3. for main roads: in the afternoon travelling takes one and half times more than in other time ranges.

In order to evaluate the practical usefulness of the proposed approach we determined several ways with two starting time: at night (0.00) and in the morning (6.30, 7.00, and 8.00 depending on the test). We tested short (a few kilometres in the city) and average (about 50 km) and long (about 200 km) routes. The computed routes did not raise any important reservations - they could be considered as good routes to travel. We observed changes in paths according to start times. In the cases of the morning start of travelling the computed routes included limited overall distance of “city” roads (or such roads were omitted at all) due to the morning rush hour.

Table 2 Processing OpenStreetMap

Area	Processing time (s)	Nodes			Edges		
		Beginning number	End number	Beginning/final in %	Beginning number	End number	Beginning/final in %
Warsaw	24	214193	189770	88.6	389015	336441	86.49
Mazovia	41	706970	492233	69.63	1377253	897524	65.17

8 Conclusions

In the paper we presented the complete approach to build the system for finding the fastest path in time-dependent network graph based on data available in OpenStreetMap service. The introduced solution includes: the method for modelling a network graph based on the mentioned above source data, methods of estimation travel time associated with edges with taking into consideration different types of obstacles such as turns or traffic-lights, and the simple method for computing the fastest time in time-dependent network practically satisfying FIFO principle. The performed tests showed that proposed approach may be used in practice. However, as it is still not mature solution, there is space for further research and significant improvement and development.

References

1. Dean, B. C.: Shortest paths in fifo time-dependent networks: theory and algorithms. Rapport technique, Massachusetts Institute of Technology (2004)
2. Dijkstra, E.W.: A note on two problems in connexion with graphs. Numerische mathematik **1**(1), 269–271 (1959)
3. Eglese, Richard, Maden, Will, Slater, Alan: A road timetabletm to aid vehicle routing and scheduling. Comput. Oper. Res. **33**(12), 3508–3519 (2006)
4. Gendreau, Michel, Ghiani, Gianpaolo, Guerriero, Emanuela: Time-dependent routing problems: a review. Comput. Oper. Res. **64**, 189–197 (2015)
5. Huang, He, Gao, Song: Optimal paths in dynamic networks with dependent random link travel times. Transp. Res. Part B Methodol. **46**(5), 579–598 (2012)
6. Ichoua, Soumia, Gendreau, Michel, Potvin, Jean-Yves: Vehicle dispatching with time-dependent travel times. Eur. J. Oper. Res. **144**(2), 379–396 (2003)
7. Lindholm, T., Yellin, F., Bracha, G., Buckley, A.: The Java Virtual Machine Specification. Pearson Education, New Jersey (2014)
8. Miller-Hooks, E.D., Mahmassani, H.S.: Least expected time paths in stochastic, time-varying transportation networks. Transp. Sci. **34**(2), 198–215 (2000)
9. Nie, Y.M., Wu, X.: Shortest path problem considering on-time arrival probability. Transp. Res. Part B Methodol. **43**(6), 597–613 (2009)
10. Sheng, Yuhong, Gao, Yuan: Shortest path problem of uncertain random network. Comput. Ind. Eng. **99**, 97–105 (2016)
11. Sung, K., Bell, M.G.H., Seong, M., Park, S.: Shortest paths in a network with time-dependent flow speeds. Eur. J. Oper. Res. **121**(1), 32–39 (2000)
12. Toth, P., Vigo, D.: Vehicle Routing: Problems, Methods, and Applications. SIAM (2014)
13. Wen, Liang, Çatay, Bülent, Eglese, Richard: Finding a minimum cost path between a pair of nodes in a time-varying road network with a congestion charge. Eur. J. Oper. Res. **236**(3), 915–923 (2014)

Evolution Algorithm for Community Detection in Social Networks Using Node Centrality

Krista Rizman Žalik

Abstract Community structure identification has received a great effort among computer scientists who are focusing on the properties of complex networks like the internet, social networks, food networks, e-mail networks and biochemical networks. Automatic network clustering can uncover natural groups of nodes called communities in real networks that reveals its underlying structure and functions. In this paper, we use a multiobjective evolution community detection algorithm, which forms center-based communities in a network exploiting node centrality. Node centrality is easy to use for better partitions and for increasing the convergence of evolution algorithm. The proposed algorithm reveals the center-based natural communities with high quality. Experiments on real-world networks demonstrate the efficiency of the proposed approach.

Keywords Social networks · Complex networks · Multiobjective community detection · Centrality

1 Introduction

Community structure identification is an important task of research among computer scientists focusing on the structural and functional properties of complex networks like the Internet, e-mail networks, social networks, biochemical networks and food networks. A complex network systems can be represented by graphs consisting of nodes and edges. Nodes are members of the network and edges describe relationships between members. As an example, in a friendship network, a node is a person and an edge represents relationship between two people.

Community structure, which is an important property of complex networks, can be described as the gathering of nodes into groups such that there is a higher density of edges within groups than between them. Community detection is discovering natural

K. Rizman Žalik (✉)
Faculty of Electrical Engineering and Computer Science, University of Maribor, Maribor, Slovenia
e-mail: krista.zalik@um.si

R. Bembenik et al. (eds.), *Intelligent Methods and Big Data in Industrial Applications*, Studies in Big Data 40, https://doi.org/10.1007/978-3-319-77604-0_6

sub-groups in a graph [1], which are called clusters, modules or communities. Graph clustering has become very popular with many applications in informatics, bioinformatics and technology. Communities have different number of nodes, different density of connections and functionality. While the objects of these communities have the most relations with the objects of their own communities, they have few relations with the objects of other communities.

Community detection in networks, also called graph or network clustering, is not well-defined problem. There is no common definition of community. This gives a lot of freedom to different approaches for solving the problem. There are a lot of concepts and different definitions, on which a large number of methods are based. There are no common guidelines how to compare and evaluate different algorithms with each other.

Most of the community detection algorithms require knowledge of the structure of the entire network and are classified as global algorithms in contrast to local algorithms which need only local information. Gargi et al. [2] emphasized that global algorithms do not scale well for large and dynamic networks. Aggarwal [3] noted that different parts of social networks have different local density and a global methodology for community detection may discover unreal communities in sparse sub-parts of the network. Some hybrid approaches [5] efficiently joins a local search with global search on a reduced network structure. Using local node centers enables local community detection.

Community detection algorithms can uncover different partitions while they implement different definitions of communities. The most often used is the definition of community in a strong sense, which requires that each node should have more connections within the community than with the rest of the graph [6]. One widely used quality metric is network modularity proposed by Girvan and Newman [28]:

$$Q = \sum_{i \in C} \left[\frac{E_i^{in}}{E} - \left(\frac{2E_i^{in} + E_i^{out}}{2E} \right)^2 \right] \tag{1}$$

where C is the set of all communities, i is one community in C, E_i^{in} is number of internal edges in community i, E_i^{out} is number of external edges to nodes outside of community i and E is number of edges. Modularity has been defined also as:

$$Q = \frac{1}{2E} \sum_{ij} \left[A_{ij} - \left(\frac{k_i k_j}{2E} \right) \right] \delta_{c_i, c_j} \tag{2}$$

where k_i is the degree of node i, A_{ij} is an element of the adjacency matrix (with elements $A_{ij} = 1$, if there is an edge between nodes i and j, and $A_{ij} = 0$ otherwise). δ_{c_i, c_j} is the Kronecker delta symbol, and c_i is label of the community to which node i is assigned.

Modularity suffers a resolution limit, while it is unable to detect small communities [8]. Moreover, methods have to deal efficiently with the most real complex networks which have largely heterogeneous degree distributions beside heterogeneous community sizes. Because the natural partition as a ground truth does not exists for the most real complex systems, the evaluation of community detection algorithms is even more difficult. The problem of determining communities by using modularity optimization is known as an NP-hard combinatorial optimization problem [9]. Therefore different approximation methods have been used such as spectral methods [10] or genetic algorithms [11, 12], which belong to evolution algorithms.

In the last decade, many new methods based on evolutionary computation for the community detection problem have been proposed. The reason is the efficiency of evolutionary computation in solving any tasks formulated as optimization problems, while requiring the definition of a suitable representation for the problem, population evolution strategy and the function to optimize.

Some knowledge which can be used in community detection process derives from observing each nodes' neighborhood [13] or links among nodes exploited in different label propagation methods [14] and can also be used in evolution algorithms to spead up the convergence process and to reduce the search space of solutions. Genetic Algorithms (GAs), and all evolutionary strategies, using also other nature inspired approaches, such as particle swarm [15] and ant colony optimization [16] and bat methods [17] for finding communities have given a significant contribution to community detection.

Evolution computation techniques for community detection present many advantages. The number of communities is automatically determined during the process of community detection. Knowledge about the problem can be incorporated in initialization or a variant of operators instead of random selection (see examples of problem-specific operators in [18]), which spreads up the community detection process also by reducing the possible solution space. Evolution methods are population based methods, that are naturally parallel and offer efficient implementations suitable also for large networks.

Genetic algorithms [13] are optimization methods in artificial intelligence. Genetic algorithms are one of the best ways to solve a problem for which little is known and a way of quickly finding a reasonable solution to a complex problem. Genetic algorithms are inspired by nature and evolution. Genetic algorithms use the principles of selection and evolution to produce several solutions to a given problem. In genetic algorithm, more individuals, which are called chromosomes, form a solution. Each chromosome represents a possible solution and genetic algorithm finds the best chromosome for one or more chosen objective functions. All chromosomes are randomly initialized at the beginning and then in more iterations good chromosomes survive the evolution process. Fitness function assigns a fitness value to each chromosome and estimates how good a chromosome is to solve the problem. In each iteration, chromosomes for new population are generated by performing the following genetic operations: cross-over between chromosomes, mutation and selection.

Most of the proposed community detection algorithms optimize one objective function, while recently multiobjective optimization is often used for community detection. Different optimization functions are proposed for multiobjective community detection such as Community Score and Community Fitness in the MOGA-Net [19], which uses NGSA II [20] optimization framework, or modularity Q and Community Score used by BOCD multiobjective optimization method [4].

Both components of modularity have also been used as optimization functions in multiobjective genetic algorithm named MOCD [12], which uses the optimization framework PESA-II [21].

Beside these approaches we propose a multiobjective solution named Net-degree using criteria function based on the node centrality. We try to uncover partitions in social networks which do not optimize any artificial property like modularity or community score, but identify real groups in social networks. Therefore, we use centrality, while the idea of the centrality of individuals and organizations in social networks was proposed by social network analysts long ago [24] with the sociometric concept of the central person who is the most popular in group and who has the greatest influence on others.

Center-based functions and algorithms are not uncommon in clustering of data objects. Center-based functions are very efficient for clustering large databases and high-dimensional databases. Clusters found by center-based algorithms have convex shapes and each cluster is represented by a center. Therefore, center-based algorithms are not good choices for finding clusters of arbitrary shapes. In this paper, we study the efficiency of center-based community detection in networks.

2 The Net-Degree Algorithm

Net-Degree is multiobjective optimization that uses the node centrality for community detection and NGSA-II [20] algorithm as multiobjective optimization framework. NGSA-II is a nondominated sorting genetic algorithm. It is a multi-objective optimization algorithm which finds more solutions, while a single-objective optimization algorithm finds one optimum solution that minimizes one objective function. NSGA-II calculates fitness measure using all optimized objectives. First it assigns a fitness value to each chromosome in the solution space and then creates a number of fronts with a density estimation. NGSA-II sorts the population into a hierarchy of sub-populations based on the ordering of Pareto dominance. Evaluation of similarity between members of each sub-group on the Pareto front gives inverse front of solutions. While in each generation, the best chromosomes survive, a good chromosome is never lost. In a multi-objective optimization problem, many Pareto optimal solutions form a Pareto optimal front [22]. Each Pareto optimal solution is good according to at least one optimization criteria. Solutions can be worse than another solutions according to one objective, but they should be better than another solution in at least one other objective. Each Pareto optimal solution is good according to one objective function, but there is only one solution that maximizes the whole fitness function.

We used the fitness function based on the node centrality and we modified the steps of the genetic algorithm to satisfy the needs of genetic algorithm for community detection. Initialization, mutation and cross-over operators are modified.

The locus-based adjacent representation [23] is used for representation of chromosomes and an array of integers is used for each chromosome. For a network of size n, each chromosome consists of n genes $< g_1, g_2, \dots g_n >$. The value j of i-th gene means, that nodes i and j are connected with a link in the graph. This also means that, node i is in the same community with node j. In a locus-based adjacent representation, all the connected components from graph G belong to the same community.

The algorithm starts with creation of initial population. Each node belongs to the same community as one random neighbor. This is achieved with assigning the neighbor label as a node community number. If two nodes are in the same community, they should be neighbors connected by an edge. This limits the possible solution space and reduces the invalid search.

The evolution starts with a set of solutions represented by chromosomes of initial population. Evolution is performed in an iterative process. The solution from previous iteration is taken and used to form a new population. The population in each iteration is called a generation. During each iteration, the algorithm performs crossover between members, and mutations. First initial population of N chromosomes is generated. Then in more iterations new and better populations are generated. For each chromosome the optimization function values are calculated and chromosomes are ranked. Mutation and crossover is performed to get M child chromosomes. Proportional cloning to current population is performed and then the usual binary tournament selection, mutation and modularity crossover operator on the clone population creates an offspring population. A combined population is formed from current and offspring population. Each solution is assigned a fitness rank equal to its nondomination level by NGSAII algorithm (1 is the best level, 2 is the next-best level, and so on). Minimization of both optimization functions is performed. Then, the combined population is sorted according to nondomination by NGSAII. Crowding distance values of all nondominated individuals is calculated and first N individuals are chosen as a new population. Dominant population is updated. All steps except initialization are repeated until the iteration number reaches some predefined threshold.

2.1 Objective Functions

In the proposed multiobjective community detection we use two functions, the first is based on the node centrality and the second on the ratio of external edges. Both functions have to be minimized to get the best partition.

The first function is based on the view of a community as a central node which is a node with high degree surrounded by neighbors with lower densities. Centrality concepts were first developed in social network analysis. Central person is close to everyone else in the group or network. Central person communicates with many others directly and can easy gets information and can also spread information to many other people in the network.

Facts about centrality pointed out by Freeman in 1979 are still true: "There is certainly no unanimity on exactly what centrality is or on its conceptual foundations, and there is little agreement on the proper procedure for its measurement." In graph theory, central nodes are the most important nodes within a graph with a lot of direct neighbors. Centrality indices allow ranking nodes and identifying the most important nodes. We use the node degree centrality, which is defined by the degree - number of neighbors of node.

The first used function counts unconnected nodes with central node of community to which the node belongs and calculates the ratio of sum of unconnected nodes and number of nodes in network. It should be low for a good community structure, i.e. highly separable communities that are densely connected internally but only sparsely connected to each other.

$$F1 = \frac{\sum_{i=1}^{k} n_i - sameCom_i}{n}; \quad sameCom_i = \sum_{i,j \in C_i; j \in N_i} sameCom(i,j); \quad sameCom(i,j) = \begin{Bmatrix} 0 \; comm(i) \neq comm(j) \\ 1 \; comm(i) = comm(j) \end{Bmatrix} \tag{3}$$

where n is the number of all nodes of a community, n_i is the number of nodes of the community C_i and k is number of communities and N_i is a set of neighbors of the node i.

While the first function evaluates the quality of each community, we propose and use the second function that measures the quality criteria of each node in a community. We use the strong criteria of node quality calculated from the ratio of node neighbors which do not belong to the same community as the node and the number of all neighbors of the node. The neighbors of each node should mostly be inside the same community with it to achieve that a community should contain more internal links among nodes inside the community than external links with other communities. Average ratio of external edges of each node is used for the second criteria function (F2) which should be low for a good community structure, i.e. highly separable communities that are densely connected internally but only sparsely connected to each other.

$$F2 = \frac{\sum_{i=1}^{n} F2(i)}{n}; \quad F2(i) = \frac{\sum_{i,j \in E; j \in N_i} diffCom(i,j)}{deg(i)}; \quad diffCom(i,j) = \begin{Bmatrix} 1 \; comm(i) \neq comm(j) \\ 0 \; comm(i) = comm(j) \end{Bmatrix} \tag{4}$$

where $deg(i)$ is degree (number of neighbors) of node i and E is set of edges.

NSGA-II algorithm searches the minimum of both optimization functions. F_o should be high for a good community structure consisting of communities that are densely connected internally but only sparsely connected to each other.

$$Fo = 1 - F1 - F2 \tag{5}$$

2.2 Objective Functions Using Number of Discovered Communities

We used another pair of functions which includes also the number of identified communities in calculation of fitness value. Both functions have to be minimized to get the best partition.

The first function of the pair (F3) uses the ratio of unconnected nodes with central node of community to which the node belongs and degree of node. It should be low for a good communities which are densely connected internally but only sparsely connected to each other.

$$F3 = \frac{\sum_{i=1}^{k}(n_i - sameCom_i)/n_i}{k};$$
$$sameCom_i = \sum_{i,j \in C_i} sameCom(i, j);$$
$$sameCom(i, j) = \begin{Bmatrix} 0 \; comm(i) \neq comm(j) \\ 1 \; comm(i) = comm(j) \end{Bmatrix} \tag{6}$$

where n is the number of all nodes of a community, n_i is the number of nodes of the community C_i and k is the number of communities.

While the first function of the function pair evaluates the quality of each community, we propose the second function of the pair that measures the quality criteria of each node in a community. We use the same strong criteria of quality as in Function F_2 with using the ratio of node neighbors which do not belong to the same community as the node and number of all neighbors. The neighbors of each node should mostly be inside the same community as the node to achieve that a community should contain more internal links among nodes inside the community than external links with other communities. Average ratio of external edges of each node is used for the second criteria function (F4). Function F4 should be low for a good community structure with densely connected communities internally and sparsely connected with other communities.

$$F4 = \frac{\sum_{i=1}^{n} F2(i)}{k};$$
$$F2(i) = \frac{\sum_{i,j \in E} diffCom(i, j)}{deg(i)};$$
$$diffCom(i, j) = \begin{Bmatrix} 1 \; comm(i) \neq comm(j) \\ 0 \; comm(i) = comm(j) \end{Bmatrix} \tag{7}$$

where n is the number of all nodes of a community, k is the number of communities and E is set of edges and N_i is a set of neighbors of the node i.

NSGA-II algorithm searches the minimum of both optimization functions and the resulted function Fk should be high for a good community structure with sparsely connected communities with other communities and densely connected nodes internally.

$$Fk = 1 - F3 - F4 \tag{8}$$

2.3 Mutation

Mutation generates random changes in chromosomes and puts a new genetic material to the population. The offspring is created by randomly changing part of existing individuals. Mutation should not be performed often and not on a lot of genes of chromosomes as it would lead to a random search and chromosomes with many mutated genes usually do not survive the selection. Mutation is used to push a chromosome of a population out of a local minimum and so provides to discover a better minimum.

A neighbor-based mutation operation is used which mutates genes considering only the effective connections [31]. The community label of this node is replaced with one of its neighbors' label. The mutation on each node of individual is performed if a generated random value is smaller than the predefined mutation probability. We use mutation probability 0.95.

2.4 Two-Way Crossover

Crossover operates on two chromosomes. It is executed if a generated random value is smaller than the predefined crossover probability. We use crossover probability 0.8. It takes two parent chromosomes and creates two new child chromosomes by combining parts from parents. So the offspring generated by the crossover contains the features of both parents. We use crossover operator performing two-way crossing introduced in Ref. [30]. First, two chromosomes are selected randomly and one node is chosen randomly in the first chromosome. The community of chosen node is determined and all nodes of the first chromosome, which belong to the community of chosen node are also assigned to the same community label in the second chromosome. Then the first and the second chromosome are swapped, and the same procedure is repeated.

2.5 Crossover Based on the Objective Function

We use crossover to inherit the best communities with the highest optimization function fitness values from one generation to the next to faster reach a local minimum. Therefore we perform this crossover only on the nodes with the higher fitness value as predefined threshold 0.95. We use tournament selection of an individual from a population of individuals for selection of parents for crossover operation. All communities from both parents are sorted by the optimization fitness values. First we create the community that contains the same nodes as community with the highest fitness value in the offspring. Only nodes that do not belong to this community can be assigned to the other community. Then we create the community with the second highest value in the offspring, but only from nodes not assigned before. The process of forming communities is stopped when all nodes in the offspring are assigned. After that, some mutations are performed on the offspring, so that each node belongs to the community to which belong the most neighbors.

3 Experiment Results

Zachary karate club network is a well-known social network [25]. Because of conflict between the administrator and instructor, the club splits into two parts. Two communities are identified in the resulted partition shown in Fig. 1 with modularity 0.358 and value of the optimization function Fo 0.741. The communities discovered by our algorithm agree exactly with the result given by Zachary. Partition with 4 communities have the best modularity value 0.415, but this partition has smaller value of the proposed optimization function Fo (0.508) than partition with 2 communities.

Maximal value of function Fk (0.735) gives three communities shown in Fig. 2 with modularity 0.375. For the highest modularity value 0.417, 4 communities are identified shown in Fig. 3, but the obtained value of Fk 0.65 is smaller than for partition with 3 communities shown in Fig. 2. The correctness of such results can be proved by visually exploring the networks. If we compare partitions with 3 and 4

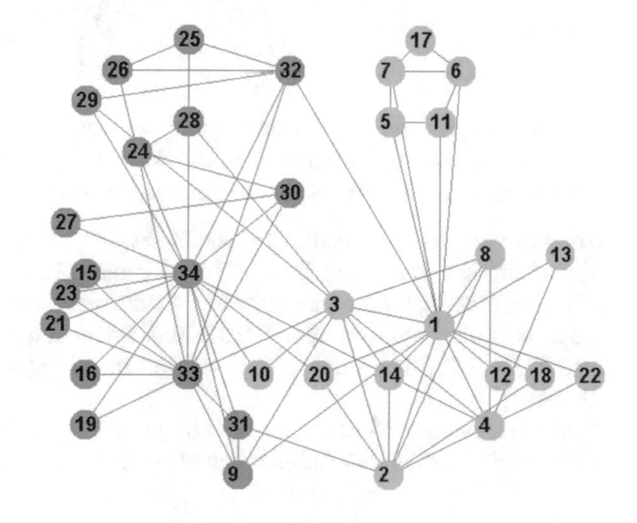

Fig. 1 Zachary karate club network partitioned into 2 communities using optimization function Fo (0.741)

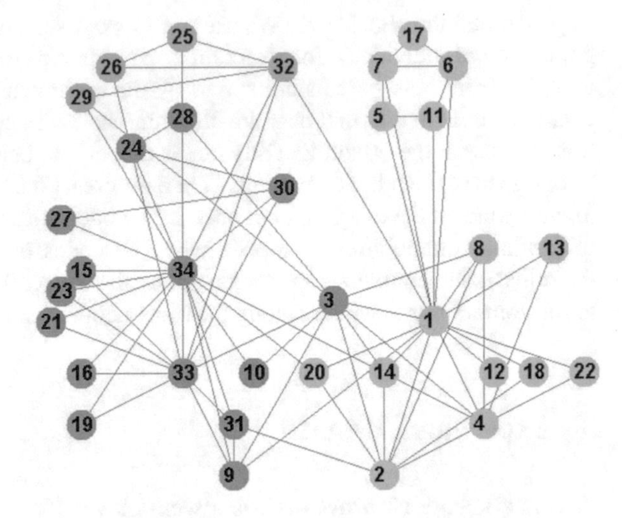

Fig. 2 Zachary karate club network partitioned into 3 communities using optimization function Fk (0.735)

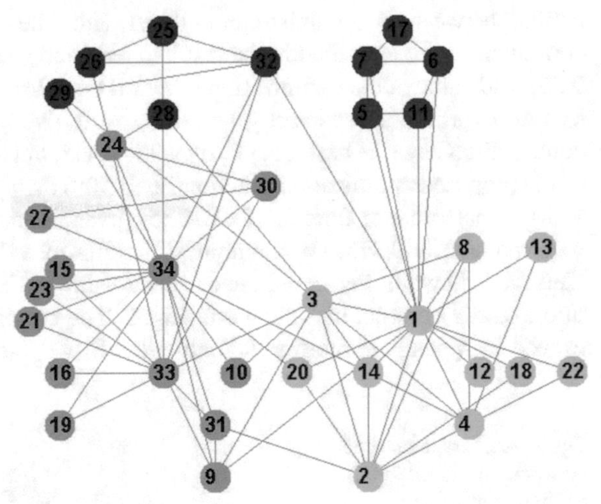

Fig. 3 Zachary karate club network: 4 communities

communities, we can see, that community with center 1 in partition with 3 communities is split into two communities. But 4 nodes (5, 6, 7, 11, 17) of a new formed community are connected with the center (node 1) of the other community.

Dolphin network is an undirected social network of frequent associations between 62 dolphins in a community living in New Zealand [26]. 4 communities are identified in the resulted partition shown in Fig. 4 with modularity 0.522 and value of the proposed optimization function Fo (0.3995). Partition with 5 communities has higher modularity 0.526, but it has smaller value of the proposed optimization function Fo (0.346).

Maximal value of function Fk (0.466) gives the partition with 5 communities and a little higher modularity (0.524) than obtained by function Fo (Table 1).

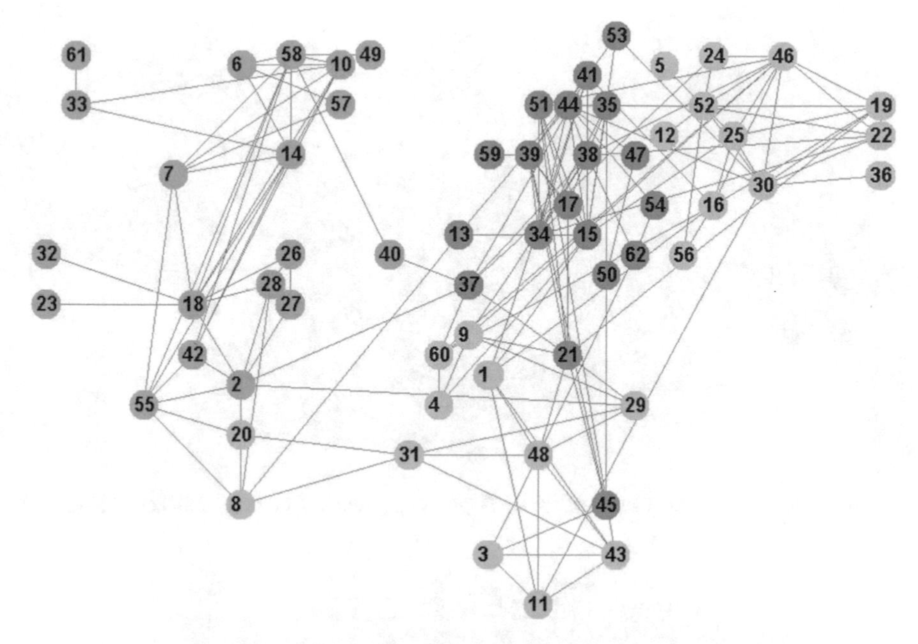

Fig. 4 Dolhin network partition obtained by Degree-Net using optimization function Fo

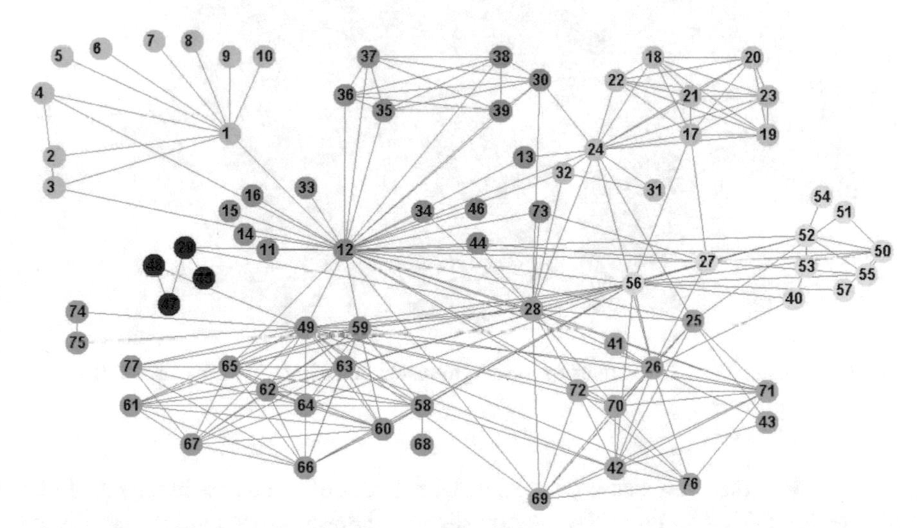

Fig. 5 Les miserables network partition obtained by Degree-Net using optimization function Fo

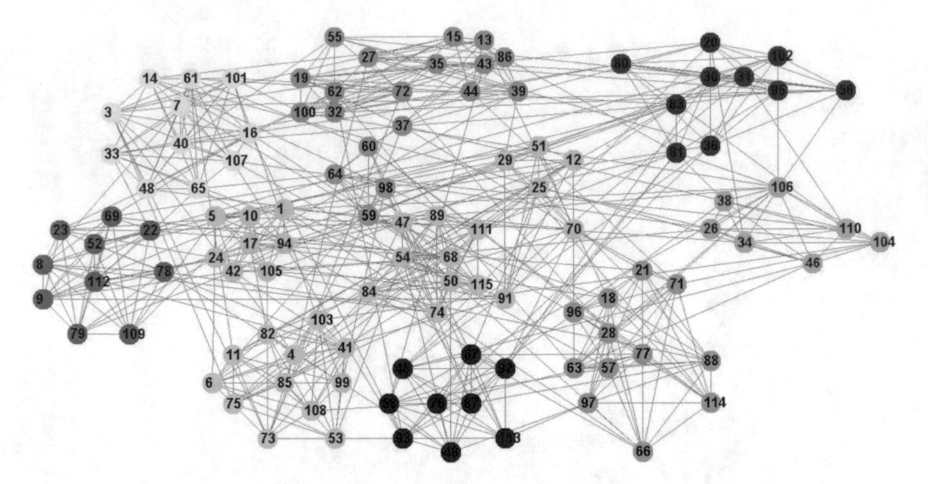

Fig. 6 10 communities are identified by Net-Degree using function Fo in football network data set

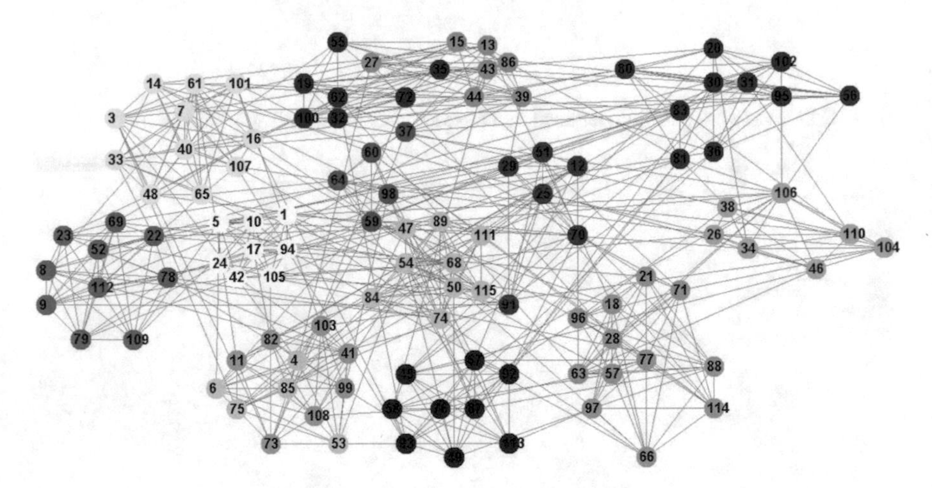

Fig. 7 13 communities are identified by Net-Degree using function Fk in football network data set

Les Miserabes network is coappearance network of characters in the novel Les Miserables [27]. The proposed method identified seven communities (see Fig. 5) with value of modularity function 0.557 and value of optimization function Fo 0.68. Maximal value of function Fk gives partition with seven communities with the similar modularity 0.5579 as obtained using function Fo (Table 1).

Football network is a network of American football games between Division IA colleges during regular season Fall 2000 [28], where nodes represent football teams and edges between two nodes represent regular season games. The proposed method identified ten communities with value of modularity function 0.6 and value of opti-

Table 1 Modularity obtained by Louvian method and Net-degree method using functions: F_o and F_k on some often used real-world networks for evaluation of community detection methods

–	Louvian	Net-degree with F_o	Net-degree with F_k
Karate	0.42	0.358 ($F_o = 0.741$)	0.387 ($F_k = 0.693$)
Dolphin	0.52	0.52 ($F_o = 0.3995$)	0.52 ($F_k = 0.465$)
Polbooks	0.52	0.52 ($F_o = 0.4839$)	0.512 ($F_k = 0.541$)
Jazz	0.44(4)	0.442 ($F_o = 0.538$)	0.44 ($F_k = 0.638$)
Polblogs	0.43	0.38 ($F_o = 0.3995$)	0.397 ($F_k = 0.754$)
Football	0.6	0.6 ($F_o = 0.456$)	0.581 ($F_k = 0.641$)

mization function Fo 0.456. Football teams are organized into 12 conferences and the Net-Degree using function Fo identified 10 communities shown in Fig. 6. The reason is that this is not center-based network, while each football team play all regular season games with different teams. Net-Degree using function Fk identified 13 communities shown in Fig. 7 with smaller modularity value 0.558 than using function Fo (Table 1).

Discussion The modularity valucs and values of optimization functions Fo and Fk of Net-Degree for six data sets with known structure are shown in Table 1. The polbooks data set is a network of frequent co-purchasing of political books by the same buyers from Amazon.com [32]. Books can be divided regarding political option into liberal and conservative books.

Jazz network is a network of 198 Jazz musicians with 2,742 edges. They can be separated into three communities that are densely connected [34]. Each node is a Jazz musician and an edge denotes that two musicians have played together in a band.

Polblocks network models hyperlinks between web blogs about US politics in 2005 [33]. Value attributes indicate political option left or liberal and right or conservative.

For most datasets more central-based communities are identified using function F_k than using function F_o while modularity values are usually higher than using function F_o.

The modularity is used to measure the quality of community detection results. It can be seen that the modularity values of our partitions are lower than those published for the Louvain [35] algorithms. This can be expected because our method is not specifically designed to optimize the modularity as Louvain algorithms does.

4 Conclusion

In this paper, the automatic network community detection is formulated as an multiobjective optimization problem facilitated by node centrality and variance of communities of node's neighbors. We modified initialization, mutation and crossover for efficient community detection and to reduce the solution only on possible solution space. We used crossover based on objective function to improve the convergence and efficiency of the algorithm. Experiments on real-world networks demonstrate the efficiency of the proposed approach in identifying center-based communities.

Acknowledgements This work was supported by the Slovenian Research Agency (grant no.: J2-8176, P2-0041).

References

1. Schaefer, S.E.: Graph clustering. Comput. Sci. Rev. **1**(1), 27–64 (2007)
2. Gargi, U., Lu, W., Mirrokni, V.S., Yoon, S.: Large-scale community detection on youtube for topic discovery and exploration. ICWSM (2011)
3. Aggarwal, C.C., Xie, Y., Philip, S.Y.: Towards community detection in locally heterogeneous networks. In: SDM, pp. 391-402 (2011)
4. Aggrawal, R.: Bi-objective community detection (bocd) in networks using genetic algorithm. Contemp. Comput. **168**(1), 5–15 (2011)
5. Shang, R., Bai, J., Jiao, L., Jin, C.: Community detection based on modularity and improved genetic algorithm. Stat. Mech. Appl. Phys. A **392**(5), 1215–1231 (2012)
6. Radicchi, F., Castellano, C., Cecconi, F., Loreto, V., Parisi, D.: Defining and identifying clusters in networks. Proc. Natl Acad. Sci. U.S.A **101**(9), 2658–2663 (2004)
7. Newman, M.E.J., Girvan, M.: Finding and evaluating community structure in networks. Phys. Rev. E **69**(026113) (2004)
8. Fortunato, S., Barthélemy, M.: Resolution limit in community detection. Proc. Natl. Acad. Sci. U.S.A **104**(1), 36–41 (2007)
9. Brandes, U., Delling, D., Gaertler, M., Gorke, R., Hoefer, M., Nikoloski, Z., Wagner, D.: On modularity clustering. IEEE Trans. Knowl. Data Eng. **20**(2), 172–188 (2008)
10. Nadakuditi, X.R., Newman, M.: Spectra of random graphs with community structure and arbitrary degrees. Phys. Rev. E **89**(4), 042816 (2014)
11. Pizzuti, C.: Ga-net: a genetic algorithm for community detection in social networks. PPSN, 1081–1090 (2008)
12. Shi, C., Yan, Z., Cai, Y., Wu, B.: Multi-objective community detection in complex networks. Appl. Soft Comput. **12**, 850–859 (2012)
13. Rizman, Ž.K.: Maximal neighbor similarity reveals real communities in networks. Sci. Rep. **5** 18374, 1–10 (2015)
14. Rizman, Ž.K.: Community detection in networks using new update rules for label propagation. Computing. **7**(99), 679–700 (2017)
15. Kennedy, J., Eberhart, R.: Particle swarm optimization. In: IEEE International Conference on Neural Networks, pp. 1942–1948, (1995)
16. Dorigo, M., Caro, G.D.: Ant colony optimization: a new meta-heuristic. In: Proceedings of the Congress on Evolutionary Computation. IEEE Press. pp. 1470–1477 (1999)
17. Yang, X.S.: A new metaheuristic bat-inspired algorithm. In: Nature Inspired Cooperative Strategies for Optimization-NICSO 2010, pp. 65–74 (2010)
18. Rizman, Ž.K., Žalik, B.: Multi-objective evolutionary algorithm using problem specific genetic operators for community detection in networks. Neural Comput. Appl. 1–14 (2017). https://doi.org/10.1007/s00521-017-2884-0
19. Pizzuti, C.: A multiobjective genetic algorithm to find communities in complex networks. IEEE Trans. Evol. Comput. **16**(3), 418–430 (2012)
20. Deb, K., Pratap, A., Agarwal, S.A., Meyarivan, T.: A fast and elitist multiobjective genetic algorithm: NSGA-II. IEEE Trans. Evol. Comput. **6**(2), 182–197 (2002)

21. Corne, D., Jerram, N., Knowles, J., Oates, M.: PESA-II: Region-based selection in evolutionary multiobjective optimization. GECCO, 283–290 (2001)
22. Soland, R.: Multicriteria optimization: a general characterization of efficient solutions. Decis. Sci. **10**(1), 26–38 (1979)
23. Park, Y.J., Song, M.S.: A genetic algorithm for clustering problems, In: Proceedings 3rd Annual Conference on Genetic Programming (GP'98), Madison, USA, pp. 568–575 (1998)
24. Freeman, L.C.: Centrality in social networks: conceptual clarification. Soc. Netw. **1**(3), 215–239 (1979)
25. Zachary, W.W.: An information flow model for conflict and fission in small groups. J. Anthropol. Res. **33**, 452–473 (1977)
26. Lusseau, D., Schneider, K., Boisseau, O.J., Haase, P., Slooten, E., Dawson, S.M.: Behav. Ecol. Sociobiol. **54**, 396–405 (2003)
27. Knuth, D.E.: The Stanford GraphBase: A Platform for Combinatorial Computing. Addison-Wesley, Reading (1993)
28. Girvan, M., Newman, M.E.J.: Community structure in social and biological networks. Proc. Natl. Acad. Sci. USA **99**, 7821–7826 (2002)
29. Deb, K.: Multi-Objective Optimization using Evolutionary Algorithms. Wiley, Chichester (2001)
30. Gong, M.G, Fu, B., Jiao, L.C., Du, H.F.: Memetic algorithm for community detection in networks. Phys. Rev. E 006100 (2011)
31. Pizzuti, C.: A multiobjective genetic algorithm to find communities in complex networks. IEEE Trans. Evol. Comput. **16**, 418–430 (2012)
32. Krebs, V.: The network was compiled by V. Krebs and is unpublished, but can found on Krebs' web site. http://www.orgnet.com (Accessed: 10 December 2016)
33. Adamic, L.A., Glance, N.: The political blogosphere and the 2004 US Election, In: Proceedings of the WWW-2005 Workshop on the Weblogging Ecosystem, 36–43 (2005)
34. Gleiser, P., Danon, L.: Adv. Complex Syst. **6**(4), 565–573 (2003)
35. Blondel, V.D., Guillaume, J.L., Lambiotte, R. Lefebvre, E. Fast unfolding of communities in large networks. J. Stat. Mech. Theor. Exp. (2008)

Part II
Complex Systems

High Performance Computing by the Crowd

Nunziato Cassavia, Sergio Flesca, Michele Ianni, Elio Masciari, Giuseppe Papuzzo and Chiara Pulice

Abstract Computational techniques both from a software and hardware viewpoint are nowadays growing at impressive rates leading to the development of projects whose complexity could be quite challenging, e.g., bio-medical simulations. Tackling such high demand could be quite hard in many context due to technical and economic motivation. A good trade-off can be the use of collaborative approaches. In this paper, we address this problem in a peer to peer way. More in detail, we leverage the idling computational resources of users connected to a network. We designed a framework that allows users to share their CPU and memory in a secure and efficient way. Indeed, users help each others by asking the network computational resources when they face high computing demanding tasks. As we do not require to power additional resources for solving tasks (we better exploit unused resources *already* powered instead), we hypothesize a remarkable side effect at steady state: energy consumption reduction compared with traditional server farm or cloud based executions.

N. Cassavia · S. Flesca · M. Ianni
DIMES, University of Calabria, Rende, Italy
e-mail: nunziato.cassavia@icar.cnr.it

S. Flesca
e-mail: flesca@dimes.unical.it

M. Ianni
e-mail: mianni@dimes.unical.it

N. Cassavia · E. Masciari (✉) · G. Papuzzo
ICAR-CNR, Rende, Italy
e-mail: elio.masciari@icar.cnr.it

G. Papuzzo
e-mail: giuseppe.papuzzo@icar.cnr.it

C. Pulice
UMIACS, University of Maryland, College Park, MD, USA
e-mail: cpulice@umiacs.umd.edu

R. Bembenik et al. (eds.), *Intelligent Methods and Big Data in Industrial Applications*, Studies in Big Data 40, https://doi.org/10.1007/978-3-319-77604-0_7

1 Introduction

Computer Science is a really young discipline whose several branches grew up at impressive rates. Hardware cost is continuously decreasing while its performances are reaching unprecedented levels. Based on the availability of a plethora of powerful (portable) computing devices, new computing tasks have been designed and new data management paradigm emerged (e.g., the well known Big Data metaphor [1, 3, 6, 14]). Even though the advances in computer sciences lead to quite effective solution implementations, several problems still require too much computational resources for a single device execution. Indeed, since the introduction of Linux operating system, the idea of gathering from the crowd the resources for completing a task, have been widely leveraged.

The word Crowdsourcing was first introduced by Howe [10] in order to define the process of outsourcing some jobs to the crowd. It is used for a wide group of activities, as it allows companies to get substantial benefits by solving their problems in effective and efficient way. Indeed, crowd based solutions have been proposed for a wide range of applications as in image tagging [12], or sentiment analysis [4] to cite a few.

Many attempts have been made to properly define the very nature of collaborative distributed systems, however the solution to this apparently trivial task is far from being easy. As a matter of fact, all the successful systems (e.g., Wikipedia,[1] Yahoo! Answers,[2] Amazon Mechanical Turk[3]) rely on some assumptions: They should be able to involve project contributors, each contributor should solve a specific task, it is mandatory to effectively evaluate single contributions, and they should properly react to possible misconduct. The above mentioned points calls for a good trade-off between openness of the system and quality of service when designing a collaborative environment.

A well known open source framework that is widely used (mainly) for scientific purposes is BOINC (Berkeley Open Infrastructure for Network Computing).[4] It allows volunteers to contribute to a wide variety of projects. The computing contribution is rewarded by credits for getting to a higher rank in a leaderboard. On the opposite side in recent years a new category of collaborative approaches is born for cryptovalue mining such as Bitcoin [13]. Users aiming at mining new Bitcoins contribute to a decoding task and are rewarded with a portion of the gathered money that is proportional to the effort put in the mining task. Our approach can be seen as a trade-off between BOINC framework implementation and Bitcoin mining.

More in detail, the novelty of our project is the exploitation of a peer to peer approach for reusing resources that users do not fully exploit to provide services that are currently offered by centralized server farm at a high cost with no customization. In particular, current limitation to high performance computing access (high price

---

[1] https://www.wikipedia.org.

[2] https://answers.yahoo.com/.

[3] https://www.mturk.com.

[4] https://boinc.berkeley.edu/.

and difficult customization of resources), could be overcome by using our framework. Users who may need computing power, can simply ask other users that are not fully exploiting the potential of their computing devices (PC, smartphone, smart TV, etc.).

Our solution does not require the continuous purchase of new servers, as users continuously provide (up to date) computational power. Finally, the better re-use of already powered resources could induce a beneficial systemic effect by reducing the overall energy consumption for complex tasks execution. We plan to further investigate our conjecture in a future work as we are not able at this stage to generalize our early results.

Our Contribution. To summarize, we make the following major contributions: (1) We design and implement a hybrid P2P infrastructure that allows collaboration among users by sharing unexploited computational resources; (2) We discuss our communication protocols that allow higher performance for our network.

2 System Architecture

Our goal is to build a reliable infrastructure for computational resource sharing easy to use and secure for the end user.[5] To this end, we implemented an hybrid peer to peer framework [8] where some functionality is still centralized as we aim at guaranteeing continuous service availability.

Figure 1 shows the interactions among system components. Herein: *Client Node* refers to those users that want to execute a high computing demanding *task* (also referred to as *project* in the following). *Server Node* refers to those users who are willing to share computational resources that are not fully employed on their devices. Finally, *Central Server* is the core of our system that is in charge for the allocation of the nodes, the displacement of the messages and the scheduling of tasks issued from client nodes that have to be distributed among server nodes.

Our system is general purpose and it is implemented by software agents in order to abstract from the low level processes and/or threads that are specific of the task being executed on the network. More in detail, for each component we deploy several agents as shown in Fig. 1 that depicts the overall system architecture.

Aside the Client and Server Nodes two new components complete the architecture: the DB Server and the Web Server. The first consist of a classical relational DB which store information about the network status, the users and devices. The web server, instead, allows users to interact with the system via web interface. As subtasks are completed, we check their correctness and reward the participating peers. In the following, we describe our model for subtask assignment that guarantees efficient

[5]Our framework has to be robust against attacks from malicious users, analogously to every distributed computing systems [2] or distributed storage systems [7]. More in detail, we need to guarantee secure communication between clients and server, trusted software for remote execution and privacy for the intermediate computation. As regards the web communication between client and server are guaranteed by *Secure Sockets Layer* (SSL - https://tools.ietf.org/html/rfc6101) [11] in order to prevent Man-in-the-Middle Attacks [5].

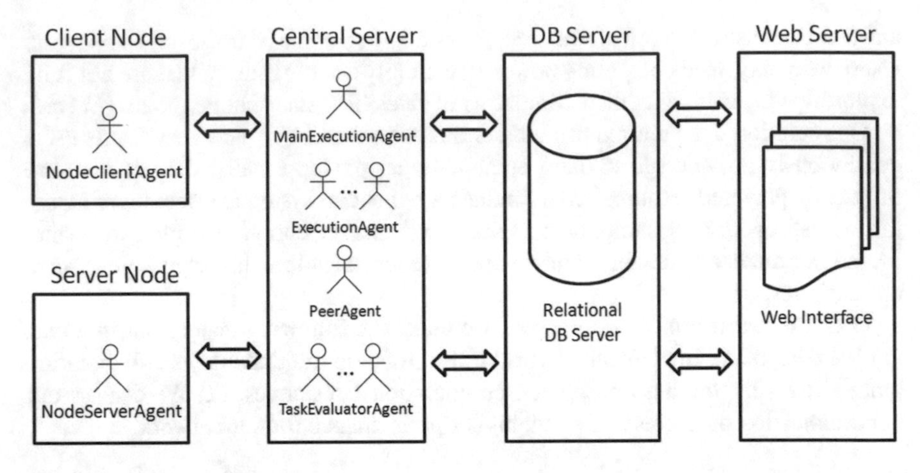

Fig. 1 Our system architecture

execution for clients and gave to all peers (even if they have limited computing power) the possibility to be rewarded. Interesting enough, users that are not going to ask for task execution have the chance to accumulate credits that can be redeemed by coins or gadgets. As it will be shown in the experimental section, the latter features makes our framework a more convenient choice w.r.t to other collaborative systems such as cryptovalue miners. Indeed, in the early stage of development of our system we performed some preliminary user analysis by asking 500 students at University of Calabria their interest in joining the network. After specifying our rewarding model all of them agreed to join the network and they are currently beta testing the software.

2.1 Client Nodes

The Client Node component is the component that resides on the machines of the users that want to submit new projects on our network. It allows the execution of the agent *NodeClientAgent* whose role is that of creating the project to be submitted to the network. Once created a project is submitted to an agent residing on Central Server component. A *NodeClientAgent* agent allows users to perform the following actions after authentication: (1) *Create and send new project to Central Server*. Depending on the type of task to be performed, the agent can create a new project by specifying the execution data and parameters; (2) *List all active project of an given user.* This component allows the user to have an overview of all projects that are currently running; (3) *Stop and Resume the remote Execution of a project*. A key feature of our system is the ability to stop and subsequently resume the project from where you left off without work or credit loss; (4) *Delete one project.* The cancellation of a project can take place only after that it has been stopped; (5) *Partial Updates.* Another key feature of the system is the Partial Update Management on the client node. This way, if the output of the project is huge, it can be downloaded while it is still running.

2.2 Server Nodes

This component lies on the machines of the users who want to share their computing resources making them available to other users (*resource providers*). It allows the execution of the *NodeServerAgent* agent, whose job is that of receiving and executing subtasks sent by the Central Server. The agent is able to retrieve the characteristics of the machine on which it runs and allows the user to choose how many computing resources share with the network. The resources that you can currently share is: CPU or GPU power, RAM and disk space. Depending on the amounts of shared resources, the central server may decide to assign or not to assign a subtask to a server node.

2.3 Central Server

The Central Server component adopt a clustered architecture and manage the execution of the task over the network. This is the component that centralize some core functionality on the peer to peer network and allows the (concurrent) execution of different types of agents.

MainExecutionAgent This agent is responsible for receiving the requests for remote execution of a project performed by NodeClientAgent agents. For each request it spawn the execution of a new agent of type *ExecutionAgent* that handles it. The primary role of MainExecutionAgent is to manage the load balancing of internal servers by assigning the ExecutionAgents to different machines.

ExecutionAgent It takes care of the management of remote execution of a single project. More in detail this agent is responsible for: (i) *Subtask execution coordination for a single project.* This activity includes the initial assignment of subtasks to NodeServerAgents, receiving execution status information and the final subtask result from each NodeServerAgent, periodically allocating new NodeServerAgents to subtasks in the case that the assigned NodeServerAgents are running slowly; (ii) *Result Quality Control.* Specifically, since a subtask is assigned to more than one resource provider, it is crucial for the ExecutionAgent to validate the obtained results. As our protocol is general purpose, the validation step is tied to the type of process being executed and is done by comparing the different results exploiting a task evaluation metric by using a $TaskValidation$ API that is fully customizable; (iii) Credits Management. For each subtask st, when at least three NodeServerAgents ended the execution of the assigned subtask it assigns credits to all the NodeServerAgents which have worked on st (as mentioned above this feature is not of interest for this paper so we will not describe it in detail); (iv) *Data Transfer Management (Upload/Download project/subtask data)*; (v) *Data Project Encryption.*

PeerAgent The PeerAgent is responsible for monitoring resource providers that became available for execution subtasks of a project. Moreover, it gather execution statistics from resource providers and update them in the database. This agent interact on one side with the Resource Providers and on the other side with the Execution-Agent providing it a pool of available resource providers when the ExecutionAgent request it.

TaskEvaluatorAgent The main role assigned to this agent is to estimate the total duration of a project in order to partition it into multiple tasks of similar duration. Each TaskEvaluatorAgent run on a single machine (denoted as reference machine) and provides and split a task in several subtask, each of them requires a fixed amount of computation on the reference machine. The reference machine adopted in the current implementation has the following characteristics:

- CPU: Single core - Single thread 4.0 GHz Intel Processor
- RAM: 32 Gb DDR3 1600 MHz
- HDD: SATA SSD 1 Tb 550 Mb/s peak

The TaskEvaluatorAgent estimates the project total time considering this reference architecture and partitioning it into sub-tasks by performing evaluations of each task computation time. Observe that, even if the reference architecture need a large amount of RAM as have to deal with possibly big projects, the typical amount of RAM needed is quite adequate for typical personal computer equipment.

2.4 Communication Protocols

This section describe the main communication protocols between agents to carry out the main functions on the network. Each project (task) Pj is a triple $\langle type, D, pms\rangle$, where $type$ is the type of the task to be performed, D are the data on which the execution should be carried out and pms are the execution parameters. Each message M exchanged on the network is of the form $\langle crd, mtype, content\rangle$ where crd is the credentials of user which is using the service (typically is a pair of string: username and password), $mtype$ is a string that defines the type of message and $content$ is a generic object that specifies the content of the message.[6]

In the following, we report the details of two communication protocols we leverage in order to enable the communication between the Central Server and Client/Server nodes: Adding New Project Protocol and Task Assignment and Execution Protocol. These protocols are open specification in order to allow the developer community to develop new Client/Server node implementations.

[6]Note that the credentials are included in each message, but as mentioned above we use the SSL protocol to avoid eavesdropping, tampering, or message forgery.

Adding New Project Protocol The main agents involved in the submission of a new task to the system are: (a) ***NodeClientAgent***: Create the project choosing $type$, sending D (typically via FTP upload) and choosing the values of one or more pms for the selected project $type$; (b) ***MainExecutionAgent***: Listen for remote execution requests, creates a new ExecutionAgent and deploy it to a selected machine to balance the system load; (c) ***ExecutionAgent***: manage the project evaluation interacting with the TaskEvaluatorAgent, the PeerAgent and the NodeServerAgents; (d) ***TaskEvaluatorAgent***: Receive projects from ExecutionAgent and estimate the Pj total cost splitting it in (uniform cost) subtasks.

To add new project, a NodeClientAgent sends to a MainExecutionAgent a message of type $RemoteExecutionRequest$. The credentials provided in the message are checked and, if are valid, the MainExecutionAgent create and deploy a new instance of ExecutionAgent trying to balance the system load. After the new ExecutionAgent is created, a new message of type $ExecutionAgentAddress$ are sent to NodeClientAgent with a $content$ containing the following information: address and port of the ExecutionAgent, address and port of a FTP server and credentials for the FTP server (Figs. 2 and 3).

At this point the NodeClientAgent talks directly to the ExecutionAgent that will handle both the creation and the remote execution of the project. Each Execution-

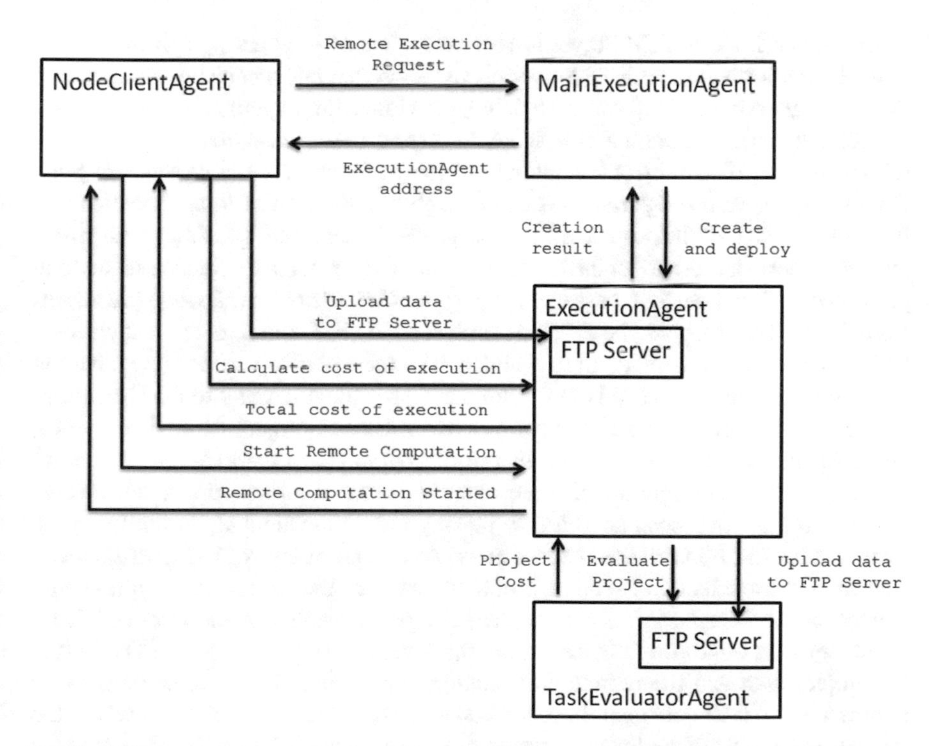

Fig. 2 Adding new project protocol

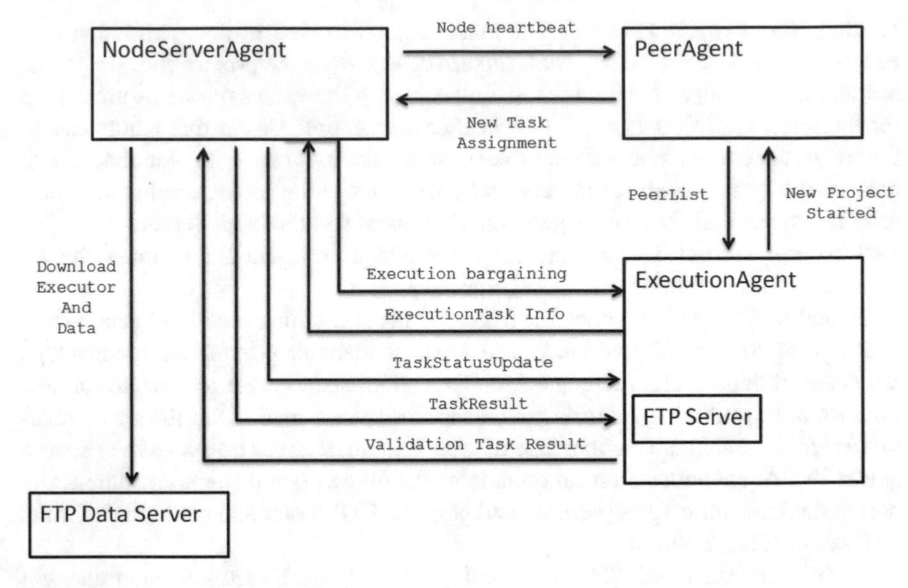

Fig. 3 Task assignment and execution protocol

Agent has an internal FTP Server to receive D. The user agent upload the data D to the ExecutionAgent via FTP Server using the credentials received with the previous message. After upload was completed, the NodeClientAgent can send a request of cost calculation through a message with $mtype$ $CalculateCostExecution$ and with a $content$ of an object containing the values for the project parameters pms. Obviously, the value of parameters can change the duration of total execution and therefore the cost of the project. This message can be sent multiple times with different parameters in order to let the user choose the preferred parameters based on total project cost. When the first message with $mtype$: $CalculateCostExecution$ arrives from the ExecutionAgent, the D value is uploaded via FTP and pms are sent via message to the TaskEvaluatorAgent to evaluate the total cost of Pj. After the evaluation is completed a new message is sent from the TaskEvaluatorAgent to the ExecutionAgent with the total cost and the number of subtask of project. Note that an ExecutionAgent executes only one Pj, therefore a triple $\langle ExecutionAgent, Pj, pms\rangle$ uniquely identifies a request of Pj evaluation. Once the ExecutionAgent receive the message of total cost, send a new message to NodeClientAgent notifying the total cost for the Pj with the selected pms. At this point the NodeClientAgent can start the remote execution sending a new message to ExecutionAgent with $mtype$ $StartRemoteComputation$ and $content$ the pms selected. If the User on NodeClientAgent is eligle for task execution, the ExecutionAgent writes on DB Server the project data and the remote computation start; otherwise an error message is returned to NodeClientAgent. In order to save resources on the machines where the ExecutionAgents are deployed a session timeout is used. When an ExecutionAgent is created a timeout (10 min in our implementation) is started and if the data D is not sent in this time range the agent is closed and the resource released. Furthermore after D is uploaded another timeout (30 min in our implementation) is started to allow users to select preferred *pms* (via $CalculateCostExecution$ messages) and start remote computation. Again if this timeout is not respected the ExecutionAgent is closed.

Task Assignment and Execution Protocol Each Coremuniti project Pj is divided into several subtask $(st_1, \ldots, st_n)$; this protocol describes the interaction between a Server Node (*resource providers*) and the Central Server to retrieve and execute a subtask of a project Pj. The main agent that interact with the system in this case are:

- ***NodeServerAgent***: Receive the subtask to execute, download the data and upload the results to the ExecutionAgent,
- ***PeerAgent***: Provide a pool of NodeServerAgent to the ExecutionAgent,
- ***ExecutionAgent***: Manages the remote execution of Pj.

The communications between agents are always made via messages of the form $\langle crd, mtype, content\rangle$ but in this case the crd are different from the NodeClient-Agent credentials to increase the security level. Each registered user has a personal credential for access to site and the use of a NodeClientAgent and multiple devices each with own credentials to use with NodeServerAgents.

When a NodeServerAgent start, it send a message to PeerAgent with $mtype$ $NodeHeartBeat$ to notify his presence and ask for tasks to perform. The $content$ of this message is a list of information about the architecture and the status of the device on which the agent is running:

- ***Device Name***: The name of computational device;
- ***Device Type***: The type of device, currently type are *CPU* or *GPU*;
- ***Core Number***: The number of cores of the device (optional);
- ***Available RAM (Mb)***: The total amount of RAM of the device in Megabytes;
- ***Available HD (Mb)***: The total amount of disk space of the device in Megabytes;
- ***Currently Used RAM (Mb)***: Currently used RAM of the device in Megabytes;
- ***Currently Used Device Power (%)***: Currently used Device Power the device in Percentage;
- ***Currently Used HD (Mb)***: Currently used disk space of the device in Megabytes;
- ***Max Shared RAM (Mb)***: The maximum amount of RAM the user wants to share in Megabytes;
- ***Max Shared Device Power (%)***: The maximum amount of device power the user wants to share in Percentage (i.e. percentage of CPU);
- ***Max Shared HD (Mb)***: The maximum amount of disk space the user wants to share;
- ***Public signature key***: A public key to uniquely sign the results sent to the central server.

This message is cyclically sent to PeerAgent to notify the peer availability. A node is considered alive if

$$LTU < CST - \Delta t$$

where LTU is the timestamp of the $NodeHeartBeat$ message, CST is the timestamp of the current server time and Δt is an interval of 60 s. Currently the supported GPU are only the Nvidia card CUDA-Enabled [9, 15].

When an ExectionAgent start a remote project send a message with $mtype$: $NewProjectStarted$ to PeerAgent. This agent selects a pool of available NodeServerAgents that can be assigned to the project. Afterwards it send a message to ExecutionAgent with $mtype$: $PeerList$ and $content$ a list of nodes id that is authorized to execute on Pj. Subsequently, a message with $mtype$: $NewTaskAssignment$ is sent to all the NodeServerAgent that have been chosen by the ExecutionAgent. This message contain the ExcutionAgent address and port. A this point, the NodeServerAgent contact the ExecutionAgent via message with $mtype$: $ExecutionBargaining$ and if it is recognized as authorized node will receive a message of type $ExecutionTaskInfo$. This message contain all the information needed by the agent to execute the subtask and send the result to the ExecutionAgent:

- ***Executable Links***: Links to download the executable file for the assigned task. This is a list of link, each for a different system architecture and Operating System. Currently OS supported are Windows (x86 and x64), Mac Os (x64) and Linux (x86 and x64);
- ***Params***: Parameter to pass to the executable;
- ***Upload Server Data***: These are the data to use for connection on internal ExecutionAgent FTP Server. Specifically it is: address, port, username and password.

After received the $ExecutionTaskInfo$ message, the NodeServerAgent download the data D and start the execution of the assigned subtask. Furthermore, cyclically it sends a message with $mtype$: $TaskStatusUpdate$ to notify the local subtask execution progress. The $content$ field of this message contain a couple $\langle ES, Perc\rangle$ where ES is the Execution Status and $Perc$ is the percentage of execution. The ES variable can take only the values: *DOWNLOAD,EXECUTION* or *UPLOAD*. When the execution of the subtask end, the agent upload the results (typically a file) to Execution Agent FTP Server. Moreover, the NodeServerAgent sends a message with $mtype$: $TaskResult$ to notify the ExecutionAgent the completion of the task.

3 Conclusion and Future Work

In this paper, we proposed a hybrid peer to peer architecture for computational resource sharing. By our network users can provide their unexploited computational resource to those users who run computational demanding tasks and they are rewarded for their collaboration. In order to guarantee the efficiency and effectiveness of the computation process, we design a task partitioning and assignment algorithm that reduce the execution times while allowing satisfactory revenues for resource providers.

References

1. Agrawal, D., et al.: Challenges and opportunities with big data. A community white paper developed by leading researchers across the United States (2012)
2. Bhatia, R.: Grid computing and security issues. Int. J. Sci. Res. Publ. (IJSRP) **3**(8) (2013)
3. Borkar, V.R., Carey, M.J., Li, C.: Inside "Big Data Management": ogres, onions, or parfaits? In: International Conference on Extending Database Technology, pp. 3–14 (2012)
4. Brew, A., Greene, D., Cunningham, P.: Using crowdsourcing and active learning to track sentiment in online media. In: Proceedings of the 2010 Conference on ECAI 2010: 19th European Conference on Artificial Intelligence, pp. 145–150 (2010)
5. Chen, Z., Guo, S., Duan, R., Wang, S.: Security analysis on mutual authentication against man-in-the-middle attack. In: 2009 First International Conference on Information Science and Engineering, pp. 1855–1858 (2009)
6. T. Economist: Data, data everywhere. The Economist (2010)
7. Firdhous, M.: Implementation of security in distributed systems - a comparative study. CoRR (2012). arXiv:abs/1211.2032
8. Franklin, M.J., Kossmann, D., Kraska, T., Ramesh, S., Xin, R.: CrowdDB: answering queries with crowdsourcing. In: Proceedings of the 2011 ACM SIGMOD International Conference on Management of Data, pp. 61–72
9. Harris, M.: Many-core GPU computing with NVIDIA CUDA. In: Proceedings of the 22nd Annual International Conference on Supercomputing, ICS '08 (2008)
10. Howe, J.: Crowdsourcing: Why the Power of the Crowd is Driving the Future of Business, 1st edn. Crown Publishing Group, New York (2008)
11. Liu, N., Yang, G., Wang, Y., Guo, D.: Security analysis and configuration of SSL protocol. In: 2008 2nd International Conference on Anti-Counterfeiting, Security and Identification, pp. 216–219 (2008)
12. Liu, X., Lu, M., Ooi, B.C., Shen, Y., Wu, S., Zhang, M.: CDAS: a crowdsourcing data analytics system. Proc. VLDB Endow. **5**(10), 1040–1051 (2012)
13. Nakamoto, S.: Bitcoin: a peer-to-peer electronic cash system. Freely available on the web (2008)
14. Nature: Big data. Nature (2008)
15. Nickolls, J., Buck, I., Garland, M., Skadron, K.: Scalable parallel programming with CUDA. Queue **6**(2) (2008)

Zero-Overhead Monitoring of Remote Terminal Devices

Jerzy Chrząszcz

Abstract This paper presents a method of delivering diagnostic information from data acquisition terminals via legacy low-throughput transmission system with no overhead. The solution was successfully implemented in an intrinsically safe RFID system for contactless identification of people and objects developed for coal mines in the end of 1990s. First, the goals and main characteristics of the application system are described, with references to underlying technologies. Then transmission system and the idea of diagnostic solution is presented. Due to confidentiality reasons some technical and business details have been omitted.

Keywords Automatic identification · Safety monitoring · RFID · TRIZ

1 Introduction

Many digital systems have distributed structure, with devices being installed in several distant locations and exchanging information through communication interfaces. This organization is typical in data acquisition applications, where multiple terminal devices located in proximity of data sources collect and transmit input data to remote central unit that supervises operation of the system. In order to support reliability and maintainability of such distributed system, the central unit should be aware of the operational status of all the terminals. In spite of its conceptual simplicity, the task of continuous monitoring of remote devices might be challenging because of technical constraints resulting from low bandwidth of communication links as well as limitations of transmission protocols, which likely may be the case in industrial installations. Innovative solution described in the paper overcomes these limitations.

Majority of coal mines in Poland are exposed to risk resulting from emission of methane, which mixed with air may create conditions for originating and propagat-

J. Chrząszcz (✉)
Institute of Computer Science, Warsaw University of Technology, Nowowiejska 15/19, 00-665 Warsaw, Poland
e-mail: jch@ii.pw.edu.pl

R. Bembenik et al. (eds.), *Intelligent Methods and Big Data in Industrial Applications*, Studies in Big Data 40, https://doi.org/10.1007/978-3-319-77604-0_8

ing massive explosions in underground excavation spaces. Therefore any electronic device intended to be used in such conditions must be designed as intrinsically safe and pass formal certification process to assure, that hazard of initiating an explosion of gas mixture has been sufficiently reduced. This implies several strict requirements regarding the design and the construction of underground equipment and also influences other phases of product lifecycle, because testing, maintaining and servicing of intrinsically safe devices is more complicated and more expensive. Hence, commodity solutions are either unfeasible for underground usage or they need fundamental redesign to adapt them for this specific environment.

One of the biggest concerns in heavy industry is safety of the people. This aspect is especially important in coal mines, where the list of risk factors is longer than in other cases. Access to reliable information about the location of all workers is priceless for supporting decisions on evacuation or planning rescue missions. Delivering such complete and timely information was the main goal of bringing the GNOM system to life in 1996.

2 System Description

The system has been developed by Citycom Polska using patented Radio Frequency Identification (RFID) technology licensed by Identec Ltd [1, 2] for providing identification of people and objects in mines and other industrial installations where explosion hazard existed. The Identec technology was selected due to its unique – at that time – feature of proper identification of multiple tags detected in the sensing area and because it used active tags, which did not need to harvest energy from the electromagnetic field. Primary application of the system was to monitor hazardous zones and keep track of people in real time, with additional ability to track and record circulation of machines, materials and other mobile objects.

In 1997 components of GNOM system had been examined in "Barbara" Experimental Mine and received certificate of being intrinsically safe. Chief Inspectorate of Mines decided on admission of system components for use in underground sites of coal mines with hazard of methane and coal dust explosion.

2.1 Functionality

The system consisted of stationary readers and portable identifiers (tags) carried by people or fastened to devices. The structure of the system is schematically depicted in Fig. 1. Basic configuration comprised readers installed at the entries to hazardous zones and it could be extended with readers installed inside the zone to allow for finer tag positioning. Additional readers could be also located at the main entrance and other important places in order to perform complete site and crew monitoring.

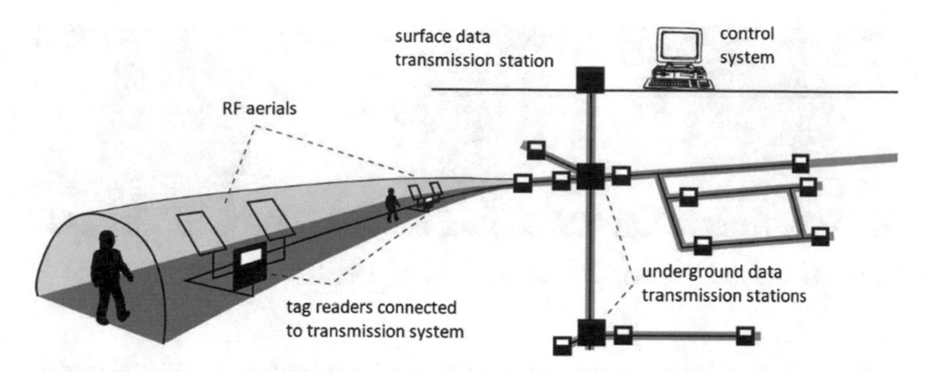

Fig. 1 Schematic structure of GNOM system with tag readers and data transmission stations

Readers were connected to data transmission system that passed information about tag movement to the host computer. Due to radio transmission the identification process was contactless, so that people did not have to touch the readers with tags. Consequently, the tags could be worn in arbitrary way, for instance attached to personal equipment and the identification was performed automatically and unobtrusively when the tags passed through reading area. Employed technology guaranteed reliable identification of tags within a few meters and therefore there was no need to install any barriers or gates that could be seen in other identification systems. Large sensing distance combined with capability of reading many tags simultaneously supported desired system operation during evacuation and other emergency conditions.

2.2 Technology

Identification was performed using low frequency electromagnetic waves and involved reading unique codes stored in tags placed in RF field. Each reader was equipped with two aerials with partially overlapping sensing areas (as shown in Fig. 2), so that tag movement direction could be detected by reader and reported in addition to tag code as “entry” or “exit” to/from the monitored zone. In case of prolonged tag presence in the reading area additional “inside” event was reported.

Identification protocol employed challenge-password scheme and required two-way data transfer between the tag and the reader. The aerials were bidirectional, with separate transmitter and receiver RF channels. Actual reading range depended on size and orientation of the aerials as well as emission power and was about four meters in typical conditions. Tag data was transmitted in scrambled format and communication sequence was changed every time to secure the identification process. Estimated error rate resulting from used cryptographic techniques did not exceed one for one hundred millions.

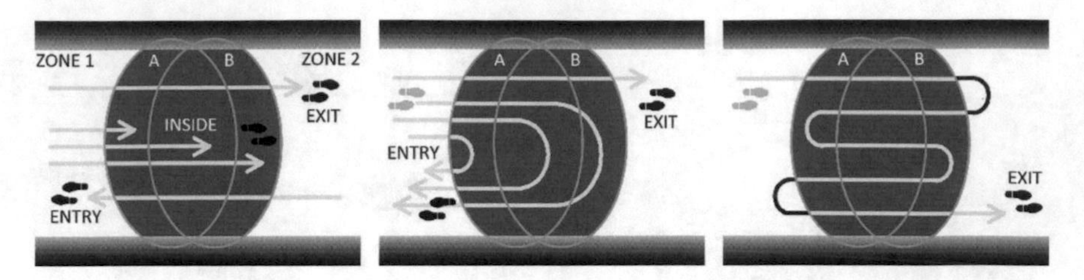

Fig. 2 The rules of distinguishing logical events from tag behavior

Readers were carefully protected against loss of data. In case of malfunction of the data transmission system, identification records were buffered in local memory until the communication was restored. The power supply was backed with accumulators and it allowed for uninterrupted recording of tag movement for several hours during power outage periods. And after total depletion of the accumulators, memory contents were sustained by backup batteries. Therefore complete information regarding tag circulation might be retrieved even after prolonged power failure, what could be of great importance for rescue operation planning.

There was a real-time clock installed in each reader, periodically resynchronized with the master clock, and all recorded events were labelled with current date and time. The records were stored in memory, reformatted accordingly to transmission system requirements and sent to the host computer. Each of hermetically sealed tags contained active transponder with fixed, read-only 64-bit identification code, aerial and lithium battery. The battery ensured correct tag operation for several years, depending of usage statistics. If the battery voltage dropped below predefined threshold, then "low battery" condition was reported to the reader along with the identification code. Such early warning procedure allowed for successive and scheduled exchange of used up tags.

System firmware offered possibility of extending reader capabilities with additional peripheral devices, for example displays showing number of persons in the monitored zone. Provisions have also been made for remote monitoring and configuring of the readers from the host computer (clock synchronization, parameter setting, firmware updates, peripheral control etc.).

2.3 *Data Transmission*

The reader unit could be connected to data transmission system using interface chosen among RS-232, RS-485 and TTY current loop with optical isolation. In order to minimize transmission overhead firmware implemented simple data compression. Communication was supervised using standard DUST 3964R industrial protocol [3], supporting transparent binary transfers with checksums, retransmission on errors and automatic communication buildup after link failure. Data integrity was additionally

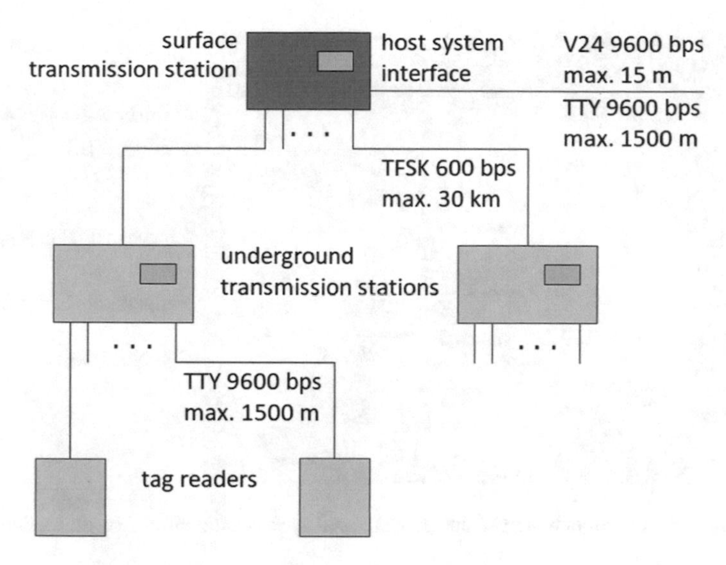

Fig. 3 Schematic structure of the transmission system

guarded with timestamps and sequential record numbers generated by readers. Structure of the transmission system is depicted in Fig. 3.

As it can be seen in the figure, data links formed simple tree-like structure without any redundancy and the transmission stations featured separate interfaces for local communication (up to 1.5 km) and for distant communication (up to 30 km).

2.4 *Application Software*

Recorded events were reported to the host computer, which buffered, processed and managed the data. Each record included date, time, reader number, event number, event type code and tag id code. Information about tag circulation was stored and archived.

Application program maintained data tables relating readers to zones and tags to crew members, machines and/or other objects. Operator might update these definitions anytime without interfering with data acquisition process. The program continuously updated information about all tags and all zones in real time. Detailed information concerning current status or history of events regarding particular zone or particular tag could be retrieved, viewed and printed under operator control. Sample snapshot of the main application window is presented in Fig. 4.

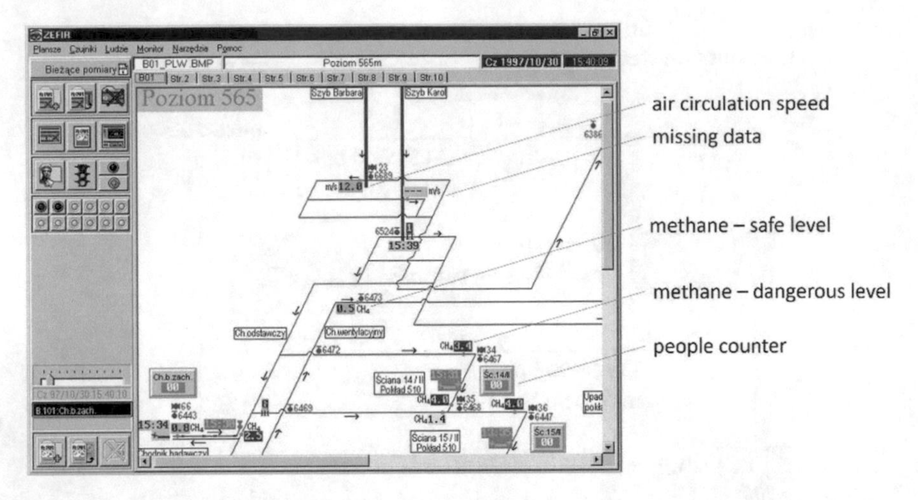

Fig. 4 Application program merged data from several systems (identification, ventilation, etc.)

3 Engineering Challenges

High-level objective of the project was to adapt basic RFID solution available from Identec for industrial use, with dedication to Polish coal mines as the primary market. And because that particular market appeared tightly populated with big players, we decided to minimize the resistance against new solution by positioning it as a sales enabler for existing vendors. Therefore we started cooperation with local (i.e. Silesian) system integrator and focused efforts on building intrinsically safe tag reader to be used with a third-party transmission system and application software familiar to decision makers. With such assumptions, the engineering problem was approached by controlling reader operation with a dedicated single-board microcomputer capable of:

- acquiring data from original RFID controller,
- providing required data handling and reader diagnostics,
- protecting recorded data during prolonged power outage periods,
- interfacing with selected data transmission systems.

3.1 Reader Construction

At the hardware side, the control microcomputer should meet functional requirements (e.g. contain battery-powered backup memory) as well as environmental requirements (e.g. provide galvanic separation of transmission interfaces). Software side covered data manipulation (buffering, formatting, timestamping, etc.) and the supervision of operational status of the building blocks contained in the reader unit or

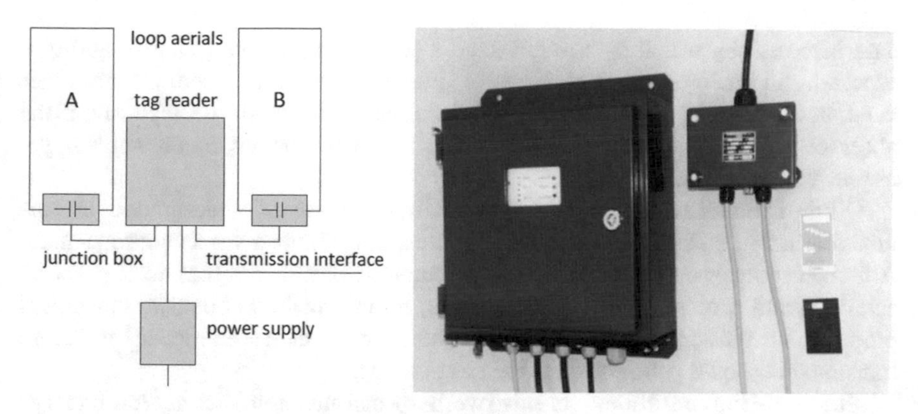

Fig. 5 Interconnection scheme and view of the reader unit

connected to it (external power supply, internal accumulator, backup battery and transmission cable). Heavy-duty housing provided mechanical and environmental protection for the electronic circuits and it was connected via feeder cables to junction boxes with connectors of loop aerials and capacitors required for their tuning. Connection scheme of a reader is shown in Fig. 5, together with a picture of the device presented during SIMEX trade show in Katowice in 1997.

3.2 Reader Diagnostics

Because of the critical mission of the system as well as difficult access and limited possibility of on-site maintenance of underground devices resulting from strict safety regulations, the problem of remote monitoring of reader operability was particularly challenging. Typical fault conditions are briefly described below, with some remarks on possible causes and feasible remedies.

External power interruption might result from several causes, including failure of power supply or cable, and detection of such event should switch the reader into accumulator mode, designed to guarantee a few hours of operation. External supply should be restored within this period to sustain reliable system performance.

Depletion of internal accumulator below preconfigured threshold voltage caused decreasing of tag reading range, which deteriorated reliability of identification process (as the movement direction might not be recognized properly, or a tag might be not detected at all). Further discharge of the accumulator disabled communication with the reader, while previously acquired data was protected in backup memory and became available upon restoration of power.

Reader transmission interruption might be caused by faulty interface in the reader or in the transmission station or faulty connection between the two. Such condition did not affect reliability or completeness of the identification, but it made data inaccessible, which also might have a serious impact on system usability – especially in emergency, when rescue operations must be planned. On the other hand, however, delayed delivery of data from one reader might be neglected if the observed tags were detected by other readers at a later time, giving incomplete, yet usable track of records.

Other transmission interruptions, such as failure of underground or surface transmission stations or respective cables, caused much worse consequences. Although some or even all of the readers might be operable at that time, properly acquiring and processing identification data, no information about tag movement propagated to master control system. This was the worst case scenario, and therefore transmission should be restored as soon as possible.

Abnormal tag conditions were also worth special attention, such as "low battery" flag transmitted with identification code and tag repeatedly detected in sensing field. The former was important for proper tag servicing and it required replacing the old tag with a new one and updating application database to sustain correct mapping between tags and persons. The latter situation might indicate a lost tag (hence, also a person circulating in excavation spaces without identification), or a person remaining in the vicinity of the reader over prolonged time period (e.g. unconscious or disabled).

Incorrect sequence of events might be caused by broken aerials or feeders as well as insufficient or imbalanced sensing ranges of the reader aerials, leading to improper recognition of "entry" as "exit" or vice versa. Moreover, different errors of local clocks might result in labelling one event with a timestamp showing later time than timestamp given to the following event by another reader. Finally, if some events were missing (or temporarily unavailable because of power or transmission problems), then observed sequence of events described by received records was incomplete. Such errors might be detected, and in many cases also corrected, at the application level using sequential record numbers together with topological and temporal constraints implied by structure of zones comprising the monitored space.

4 Innovative Solutions

4.1 Hardware and Software Design

Several minor innovative concepts born during the project have been published in a world-wide technical magazine as "design ideas", including software [4, 5], hardware [6, 7] and hybrid hardware-software solutions [8]. Original algorithmic methods were also developed and used internally (without publication) for verification and correction of tag trajectory within a hierarchical structure of real and virtual zones as well as for avoiding conditions resulting in repeated DUST 3964R communication sequences, which increased transmission channel utilization and power consumption.

4.2 Zero-Overhead Monitoring

The most promising idea developed during that time addressed delivery of diagnostic information from terminal devices to central unit without communication overhead [9]. Demand of avoiding unnecessary communication in such systems is doubly important:

- low bandwidth of data transmission links must be used with particular care,
- transmission decreases period of proper operation of battery-powered devices.

Scanning (polling) of the terminal devices by central unit is impractical in that situation and known solution employs terminal-initiated periodical transmission of so-called “heartbeat” records sent during prolonged breaks in regular communication regarding sensor data. Monitoring of such records allows the central unit to distinguish between two states of a terminal: “operable and accessible” versus “inoperable or inaccessible”. This binary categorization appeared insufficient for the purposes of monitoring GNOM readers. We needed information regarding errors of several categories together with timestamps, while limitations of the transmission system precluded extending communication scheme beyond heartbeat diagnostics. From current author’s perspective, as a TRIZ practitioner, this situation could be recognized as a “physical contradiction” [10]:

- new record type MUST be used to send diagnostic information, BUT
- new record type MUST NOT be used to comply with transmission standard.

TRIZ (Teoriya Resheniya Izobretatelskikh Zadach), sometimes abbreviated as TIPS (Theory of Inventive Problem Solving), is a methodical approach aimed at identifying and solving problems, which cannot be solved with standard engineering techniques. It was originally developed in Soviet Union in 1950s and its newer versions have been adopted by many key players of global economy, such as Samsung, Intel and Siemens. Contradiction is one of the central concepts of the classical TRIZ, being perceived as a symptom of inventive situation. Several types of contradictions have been categorized by TRIZ researchers as patterns of problems and patterns of powerful solutions for those generic problems were identified as well. Unfortunately, we did not know TRIZ during project time, and developed solution quite unconsciously followed the pattern known as “separation in system level”.

Indicated conflicting requirements were separated by embedding small fragments of diagnostic data into unused fields of heartbeat records generated by the microcomputer. Such scattered and hidden data was extracted, gathered and interpreted by API function providing record stream to application at the other end of the transmission channel. Diagnostic information was organized as collection of error counters with a timestamp of the latest entry in each category. Little pieces of this collection were “smuggled” in heartbeat records one by one and the transmission system was not aware of sending additional information. By using regular heartbeat records as envelops, detailed and periodically updated diagnostic data was delivered to the central unit with literally zero transmission overhead. For critical errors, however, extra heartbeats were generated asynchronously to signal possible failures as soon as possible.

5 Conclusions

This retrospective paper presents background and some details of an inventive solution developed for transmitting important diagnostic data in mission-critical system without communication overhead. Described solution was built upon three basic concepts:

- partitioning diagnostic data to decrease required bandwidth and power consumption,
- encapsulating diagnostic data into standard records to preserve format compatibility,
- generating asynchronous heartbeats to enhance system usability.

Such approach allowed for solving the contradiction of sending non-standard records using standard communication format. Regular transmission stations were not aware of hidden contents, while it could be easily decoded by intended recipient of the message. Unfortunately, in spite of its industrial genesis, respective patent application [9] was rejected because of allegedly "non-technical nature" of the proposed solution.

Acknowledgements Author gratefully acknowledges Rafał Gutkowski, ex-President of Citycom Polska for his kind permission for adapting some materials created during GNOM project for the purpose of this paper.

References

1. Dodd, H., Stanier, B.J.: Programmable transponder, WO Patent Application, PCT/GB1989/001,391, 1990/05/31
2. Dodd, H., Stanier, B.J.: Access control equipment and method for using the same, US Patent 5,339,073, 1994/08/16
3. DUST 3964R protocol description. https://de.wikipedia.org/wiki/3964R (in German). Accessed 26 April 2017
4. Chrząszcz, J.: Software speeds 8051 RS-232C receiver. (Design Idea 2125), EDN, 1997/12/04, 130
5. Chrząszcz, J.: Scheme speeds access to uP's real-time clock. (Design Idea 2187), EDN, 1998/06/04, 118
6. Chrząszcz, J.: Circuit translates TTY current loop to RS-232C. (Design Idea 2230), EDN, 1998/08/03, 120
7. Chrząszcz, J.: Use 8051's power-down mode to the fullest. (Design Idea 2557), EDN, 2000/07/06, 138

8. Chrząszcz, J.: PC monitors two-way RS-232 transmission. (Design Idea 2661), EDN, 2001/02/01, 126
9. Chrząszcz, J.: Method for monitoring terminal devices in distributed measurement and control systems. PL Patent Application P-384040, 2007/12/14 (in Polish)
10. Gadd, K.: TRIZ for Engineers: Enabling Inventive Problem Solving, Wiley (2011). ISBN: 9780470741887

Asynchronous Specification of Production Cell Benchmark in Integrated Model of Distributed Systems

Wiktor B. Daszczuk

Abstract There are many papers concerning well-known Karlsruhe Production Cell benchmark. They focus on specification of the controller—which leads to a synthesis of working controller—or verification of its operation. The controller is modeled using various methods: programming languages, algebras or automata. Verification is based on testing, bisimulation or temporal model checking. Most models are synchronous. Asynchronous specifications use one- or multi-element buffers to relax the dependence of component subcontrollers. We propose the application of fully asynchronous IMDS (Integrated Model of Distributed Systems) formalism. In our model the subcontrollers do not use any common variables or intermediate states. We apply distributed negotiations between subcontrollers using a simple protocol. The verification is based on CTL (Computation Tree Logic) model checking integrated with IMDS.

Keywords Production Cell · Distributed system specification · Distributed system verification · Asynchronous systems · Communication duality

1 Introduction

The Production Cell is a real-world system developed in Karlsruhe. It was a benchmark for several research projects on specification, design, synthesis, testing and verification of a controller for the Production Cell [1]. After the original book was released, many additional research papers on the benchmark were published. In several papers the Production Cell was used as the example among other benchmarks [2–5]. Similar real-world systems were used in a number of papers [6, 7].

The original benchmark is a set of devices for delivering of blanks (metal plates) with a feed belt, putting them into a press by a two-armed rotary robot, withdrawing

W. B. Daszczuk (✉)
Warsaw University of Technology, Institute of Computer Science,
Nowowiejska 15/19, 00-665 Warsaw, Poland
e-mail: wbd@ii.pw.edu.pl

R. Bembenik et al. (eds.), *Intelligent Methods and Big Data in Industrial Applications*, Studies in Big Data 40, https://doi.org/10.1007/978-3-319-77604-0_9

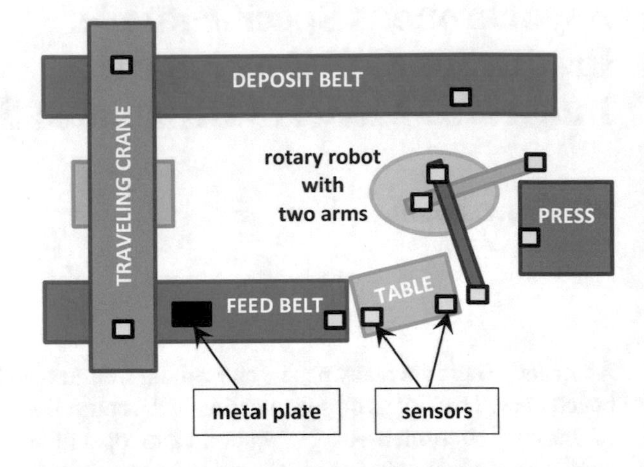

Fig. 1 The physical structure of the Production Cell

forged plates back and moving them to a deposit belt (Fig. 1). Equipping the robot with two arms allows to transport two plates simultaneously: a blank one to the press and a previously forged one from the press. A difference in the levels of the feeder belt and the robot's arm is bridged by a elevating rotary table between them.

In the physical model of the Production Cell, the press only pretends to forge the plates, they remain blank and may be turned back in a loop. It is just the purpose of a traveling crane which moves the plates between the belts.

The benchmark was specified, analyzed and verified in numerous papers, which is discussed in Sect. 2.

We propose a modeling formalism IMDS (Integrated Model of Distributed Systems—the formal definition is given in [8, 9]). An IMDS model of a system consists of four sets: *severs*, *values* of their states, offered *services* and *agents* (distributed computations). The server's *states* are pairs *(server, value)*, while *messages* sent between servers are triples *(agent, server, service)*. The behavior of a system is defined in terms of *actions*, which form a relation Λ = *{(message, state)* λ *(new_message, new_state)}*. This means that an action is caused by a *matching* pair *(message, state)* and produces a new pair *(new_message, new_state)*. A special kind of action is an *agent-terminating* action, in which a new message is absent on the output. A *configuration*[1] of a system consists of states of all servers and messages of all agents (except terminated ones). The system starts with *initial configuration* containing *initial states* of all servers and *initial messages* of all agents.

In any action a *server* component of a matching pair: state *(server, value)* and message *(agent, server, service)* must be equal, which means that the *action* is the execution of the *service* invoked by the *agent* in the context of the *server's* state. The action transforms its *input configuration*, containing a matching pair *(message, state)*, to its *output configuration* containing *(new_message, new_state)*. In the output configuration the *new_message* replaces the *message*, while the *new_state* replaces the *state*. All states of other servers and all messages of other agents remain unchanged.

[1] We do not use the term "system state" to avoid ambiguity.

The *new_state* concerns the same server as the *state*, therefore the *server* component of both states is equal. The *new_message* concerns the same agent as the *message*, therefore the *agent* component of both messages is equal. The *server* component is typically different in *message* and *new_message*, which models sending the *new_message* to another server.

Processes in the system are defined as *sets of actions*. The system may be decomposed to *server processes* if we group actions basing on a *server* component. This is the *server view* of the system, in which servers' states are the carriers of the processes while messages are communication means. Alternatively, the system may be decomposed to *agent processes* if we group actions basing on an agent component. This is the *agent view* of the system, in which agents' messages are the carriers of the processes while states are communication means.

In such a system, several temporal formulas for automatic deadlock detection are defined [9]. The formulas are given for individual processes, therefore partial deadlocks—not concerning all processes in a system—may be identified (which is rare in automatic deadlock detection mechanisms). A deadlock in the server view is a *communication deadlock*, while a deadlock in the agent view is a *resource deadlock*. Not every resource deadlock is a communication deadlock (for example if a deadlock results from a lack of resources, while all servers run).

IMDS highlights some real and natural features of distributed systems:

- Communication duality: a physical structure is reflected in server processes communicating by messages, while distributed computations are modeled as traveling agents. Both views are included in a single, uniform specification. The views are obtained by the two projections of system's actions: one projection onto the servers and the other one onto the agents.
- Locality of actions: an action depends only on a state of a server and messages pending at this server.
- Autonomy of servers: a decision, which action will be executed and when, is made autonomously by every server, independently of a situation in other servers.
- Asynchrony of actions: a message waits for matching server state, or a server state waits for matching message which invokes a service.
- Asynchrony of communication: messages are passed over unidirectional channels, without acknowledgments. A response may be sent over a separate channel in opposite direction.

Such an approach is rare among modeling approaches (mentioned in Sect. 2). Most of specifications use synchronous communication, which is unrealistic in really distributed systems, where no common variables like buffers or shared sensors exist. Even those that introduce asynchrony, apply intermediate variables like buffers or queues.

Modern solutions based on IoT (Internet of Things) paradigm should apply simple asynchronous communicating protocols between servers. The protocols are used to agree on coordinated behavior between servers. IMDS fits to this kind of distributed cooperation.

A contribution of this paper is the application of asynchronous specification method with communication duality (message passing/resource sharing) to model the operation of Production Cell benchmark. This allows to observe the system from the perspective of cooperating devices in the server view or from the perspective of processed blanks in the agent view (both in a single model).

Many papers deal with verification of the Production Cell, using temporal model checking or bisimulation. Desired features are modeled as temporal formulas or graphically specified structures, both in terms of elements of a verified system. In our approach only a limited set of features may be verified, namely a deadlock and a distributed termination, but it is achieved automatically, in a "push the button" style. It is achieved by using universal temporal formulas, not related to specific features of a model. The formulas concern the behavior of individual servers/agents, therefore both partial and total deadlocks/termination may be checked. The verification helps to identify unneeded situations and to follow paths leading to them.

A verification environment Dedan was developed in ICS, WUT [10]. It is constructed on the basis of IMDS, using an own temporal verifier TempoRG [11]. External verifiers like Uppaal [12] may be also used. Dedan offers additional engineering tools like graphical representation, simulation and structural analysis.

Our goal in this paper is to show improper operation in a specific case of one blank passing through the cell, and to prove the correctness in a case of higher number of blanks. We also wanted to learn how many blanks may be modeled in internal and in external verifier used in our modeling tool.

The paper is organized as follows: Sect. 2 covers previous modeling and verification tasks applied to the Production Cell. In Sect. 3 IMDS model of the benchmark is described. Section 4 presents the verification using the Dedan tool. In Sect. 5 a version of the model with increased parallelism of devices behavior is presented. An attempt to verify the Production Cell with time constraints is given in Sect. 6. Section 7 concludes the research.

2 Background

2.1 Centralized Versus Distributed Modeling

Some approaches assume centralized controller of the Production Cell: Tatzelwurm [1]-ch.14 and TLT [1]-ch.9. Yet, most of research concern distributed control of the devices inside the Production Cell (i.e., belts, table, robot arms etc.). The distribution is modeled as common variables between the subcontrollers of the devices [13] or synchronous message passing [14–16].

2.2 Synchrony Versus Asynchrony of Specification

Most models are built on synchronous paradigm, i.e., the devices agree on their states in common actions [14–20]. Many approaches are based on well-known formalisms: CSP (Communicating Sequential Processes [21]) or CCS (Calculus of Communicating Systems [22]). Some papers concern real time modeling [23–27]. Synchrony requires a kind of nonlocality, for example sensors common to the controllers of both neighboring devices.

Only a few papers concern asynchronous modeling: SDL (Specification and Description Language [1]-ch.10), provides asynchronous, simple protocols between devices. Focus [1]-ch.11, [28] covers asynchronous network of components (agents) that work concurrently and communicate over unbounded FIFO channels, translated into either a functional or an imperative programming language. It is admitted that each arrow corresponds to an asynchronous event, i.e. the event will be sent even though the receiving agent is not ready to accept. A Promela model in [29] uses asynchronous channels. The Model Checking Kit [30] allows to prepare models in various languages, including IF (Interchange Format): a language proposed in order to model asynchronous communicating real-time systems. Architecture Analysis and Design Language [31] provides asynchronous specification, like pairs of events `feedBelt.InFeedBeltReady -> loader.InFeedBeltReady`.

Nonblocking Supervisory Control of State Tree Structures [17] provides Hierarchical Statecharts description. Synchronous system is modeled directly in BDD (with controlled order of BDD variables), features are written as formulas over states (similar to invariants). A combination of AND-states (Cartesian products of component states) and OR-states (exclusive-or between component states) model asynchrony of structures similar to messages pending at servers. We may say that AND states model the structure (hierarchically) and parallelism, while OR states model the behavior (the dynamics).

In some papers asynchronous communication is modeled by data structures: Multiparty Session Processes with Exceptions [32] use queues of messages. In [13] asynchrony is introduced by a place "in between" in a Petri net (1-element buffer).

2.3 Real-Time Modeling

Most papers on the Production Cell modeling deal with timeless control sequences. Yet, several of them deal with real-time constraints, generally in synchronous models. In HTTDs (Hierarchical Timed Transition Diagrams [1]-ch.15) sequences of timed transitions are used. In [33] Networks of Communicating Real-Time Processes are described: various levels of abstraction, I/O-interval structures and input sensitive transitions. In [27], systematic transformation from a UML-RT [34] to CSP + T (time) [35] is applied. Timed Game Automata [36] and Timed Games [23] are used for on-the-fly verification. A timed specification language SL^{time} is described in [24], while graphical specification with time is possible in [37]. Timed Automata are used in [26] and Cottbus Timed Automata in [38].

2.4 Message Passing/Resource Sharing

Synchronous modeling usually is based on variables, as in Lustre [1]-ch.6, or synchronous channels [14–16]. Communication based on resource sharing in distributed environment is presented in Graphical Communicating Shared Resources [37].

2.5 Synthesis and Verification

Some papers concern synthesis of Production Cell controller (or a set of distributed controllers of individual devices). Such are Esterel [1]-ch.5, SDL [1]-ch.10, Deductive Synthesis using first order predicate logic [1]-ch.17, Fusion method [28] using regular expressions (# denotes receiving):

```
lifecycle Table: initialize . #feed_table .
        (go_unload_position . #pick_from_table . go_load_position . #feed_table)*
```

Modal Sequence Diagrams and Feature Diagrams [14] are used for incremental synthesis exploiting the similarities between individual device specifications.

Testing and simulation as validation methods are addressed in several papers: Symbolic Timing Diagrams [1]-ch.18 are used for testing in waveform monitoring. Formal testing method is applied in [39] and [4]. Statistical testing of Reusable Concurrent Objects is described in [40]. A simulation is used as informal validation in SDL [1]-ch.10 and Graphical Communicating Shared Resources [37].

In general, safety conditions are checked, like the avoidance of machine collisions, and liveness properties: if two blanks are inserted into the system, one of them will eventually arrive at the travelling crane.

In our approach, only deadlock detection and distributed termination checking is possible. The strong point is identification of partial deadlocks (in which not all processes participate) and differentiation between communication deadlocks in the server view and resource deadlocks in the agent view.

The modeling in IMDS and verification using Dedan differs from previous attempts mainly in using really asynchronous approach, which exploits locality and autonomy of individual subcontrollers. Following the IoT paradigm, simple protocols are used for devices cooperation. Verification is performed automatically, without a necessity to express partial/total deadlocks in terms of the features of a verified model. Automatic verification supports the fast prototyping principle.

3 Modeling Production Cell in IMDS

The IMDS specification of Production Cell benchmark consists of definition of servers (devices) and agents (blanks traveling through the cell). Every pair of servers negotiate their behavior using simple protocol, being a sequence of three messages: a supplier sends a message `try`, which tests the readiness of a receiver to accept a blank. When the receiver responds with a message `ok`, the supplier may deliver the blank. Then a message `deliver` is issued by the supplier, which has two meanings: completion of the protocol and physical passing a blank from the supplier to the receiver. For example, a protocol between the rotary table (`T`) and the robot (`R`), seen from the perspective of the robot is as follows (in IMDS notation, an action *(message, state)* λ *(new_message, new_state)* is denoted as *{agent.server.service, server.value}* → *{agent.ohter_server.some_service, server.new_value}*):

```
<i=1..N>{A[i].R.tryIn, R.arm1_at_T} -> {A[i].T.ok, R.arm1_at_T},
<i=1..N>{A[i].R.deliverIn, R.arm1_at_T} -> {A[i].R. rotate_ccw_P, R. R.rot_ccw_a2_P},
```

The meaning of this specification is: for every agent (`<i=1..N>`), if a message `tryIn` is received by the robot and its state is staying at receiving position (`arm1_at_T`), the a message `ok` is issued and the robot remains in `arm1_at_T` state, waiting for `deliverIn` message. This simple protocol (*try*→*ok*→*deliver*) is used by every pair of devices, it is illustrated in a trail in Fig. 2 (box). As the message `deliverIn` arrives, the robot starts to rotate counterclockwise, issuing a message `rotate_ccw_P` to itself and entering the `rot_ccw_a2_P` state, which means that the robot rotates until its `arm2` points at the press (`P`). All of the above is performed in the context of an agent `A[i]`, modeling the `i`th blank.

It is assumed that the belts do not stop, therefore they deliver blanks to receiving devices without using any protocol, simply a `deliver` message is issued.

Agents `A[1]..A[N]` model `N` blanks travelling through the cell. However, sometimes operations are desirable like returning of the table to `at_FB` position after delivering a blank to the robot. It is modeled by additional agents internal to some devices, for example an agent `LT` in the table (`T`):

```
{LT.T.return, T.at_R_free} -> {LT.T.go_to_FB, T.mov_to_FB},
{LT.T.go_to_FB, T.mov_to_FB} -> {LT.FBC.ready, T.waits},
{LT.T.ok_ready, T.waits} -> {LT.T.return, T.at_FB},
```

The robot is equipped with two arms which extend for picking/leaving the blanks, and which retract before the rotation of the robot. For example, if the robot reaches the utmost position at the table and it reaches the `tryIn` message, it extends to a proper length and then issues the `ok` message. After picking a blank, the arm retracts and the robot starts to rotate. This changes the previous specification (a symbol ■ will be described in the next sections):

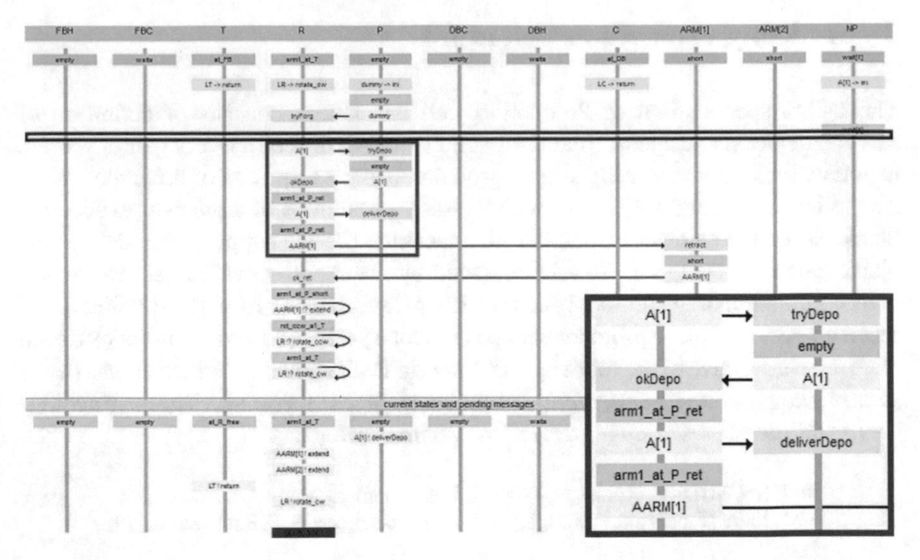

Fig. 2 The counterexample of a deadlock found in the server view of a 1-blank Production Cell

```
<i=1..N>{A[i].R.tryIn, R.arm1_at_T} -> {A[i].R.wait_a1extIn, R.arm1_at_T_ext},
        {AARM[1].R.extend, R.arm1_at_T_ext} -> {AARM[1].ARM[1].extend, R.arm1_at_T_ext},
        {AARM[1].R.ok_ext, R.arm1_at_T_ext} -> {AARM[1].R.retract, R.arm1_at_T_long},
<i=1..N>{A[i].R.wait_a1extIn, R.arm1_at_T_long} -> {A[i].T.ok, R.arm1_at_T_long},
■<i=1..N>{A[i].R.deliverIn, R.arm1_at_T_long} -> {A[i].R.wait_a1retIn, R.arm1_at_T_ret},
        {AARM[1].R.retract, R.arm1_at_T_ret} -> {AARM[1].ARM[1].retract, R.arm1_at_T_ret},
        {AARM[1].R.ok_ret, R.arm1_at_T_ret} -> {AARM[1].R.extend, R.arm1_at_T_short},
<i=1..N>{A[i].R.wait_a1retIn, R.arm1_at_T_short} -> {A[i].R.rotate_ccw_P, R.rot_ccw_a2_P},
...
```

The complete model of the traveling crane (C) is as follows:

```
server: C(agents A[N],LC; servers DBH,FBH),
states {at_FB_occ,at_FB_free,mov_to_DB,at_DB,mov_to_FB,waits},
services {deliver,go_to_FB,go_to_DB,ok,return,ok_ready},
actions{
<i=1..N>{A[i].C.deliver, C.at_DB} -> {A[i].C.go_to_FB, C.mov_to_FB},

<i=1..N>{A[i].C.go_to_FB, C.mov_to_FB} -> {A[i].FBH.try_C, C.at_FB_occ},
<i=1..N>{A[i].C.ok, C.at_FB_occ} -> {A[i].FBH.deliver_C, C.at_FB_free},
        {LC.C.return, C.at_FB_free} -> {LC.C.go_to_DB, C.mov_to_DB},
        {LC.C.go_to_DB, C.mov_to_DB} -> {LC.DBH.ready, C.waits},
        {LC.C.ok_ready, C.waits} -> {LC.C.return, T.at_DB},
}
```

4 Verification

The automatic verification is performed using the Dedan program [10]. The program contains internal CTL (Computation Tree Logic) temporal verifier TempoRG [11] (which uses explicit reachability space) for deadlock detection. It uses our own verification algorithm CBS (Checking By Spheres [41]).

A critical situation occurs when the robot exchanges the fresh blank with the forged one waiting inside the press. As the action of exchanging the forged blank with the incoming one is induced while the new blank is occurring, it is obvious that the last blank stays inside the press and cannot be taken out. If the circulation of blanks is applied in a closed loop, this requires at least two blanks to operate the system correctly. Indeed, the verification shows a deadlock in the press if only one blank is applied (Fig. 2). The figure shows the server view, including history of all servers. As the counterexample is rather long, most of it is suppressed between the two black lines. All servers are included in the picture, so the identifiers of states and messages are unreadable in the counterexample compressed to a paper's figure. We show it to illustrate a general shape of a trail. A region of the last actions leading to a deadlock is enlarged in a box at the right bottom corner. It consists of messages exchanged between the robot (`R`) and the press (`P`), and the states of these two servers: the robot asks the press if a blank may be passed to it (`tryDepo`), the press agrees (`okDepo`), then the bank is passed (`deliverDepo`). The blank is stuck in the press because the robot requires another blank to exchange it with a forged plate, but there is only one blank in the system. The last states of the servers and the last messages pending at them are shown under the trail.

In the server view of the trace, actions caused by various agents interleave in a history of every server (like `A[1]` and `AARM[1]` actions in Fig. 2). These actions are shown as a separate sequence of messages in a context of a given agent `A[1]` in the agent view. Figure 3 shows the same situation seen from the point of view of the agents. Servers history is on the left while agents history is on the right. Note that the picture is even less readable that one in Fig. 2, because both server timelines and agent timelines are included. The counterexample may be analyzed using enlarged fragments of the graphical file, or using counterexample-guided simulation in Dedan. Just like the server view, the counterexample is presented to show a general structure of it. Most of the trail is suppressed between the two black lines. Last actions (in a red box) leading to a deadlock in a context of the agent `A[1]` are enlarged in a box below. The agent timeline (on the right) shows to which server individual messages are sent (`P`-press or `R`-robot)

Note that the server view shows the perspective of cooperating controllers, and the deadlock concerns communication. The same system projected onto the agent view shows the perspective of blanks traveling through the cell, and it presents the history that leads a blank to be stuck inside the press.

The correctness of the model was indicated for 2 and 3 blanks. In the case of 4 blanks the model exceeds the available memory (8 GB). A computer with larger memory size might be used for further checking, but the experience shows that

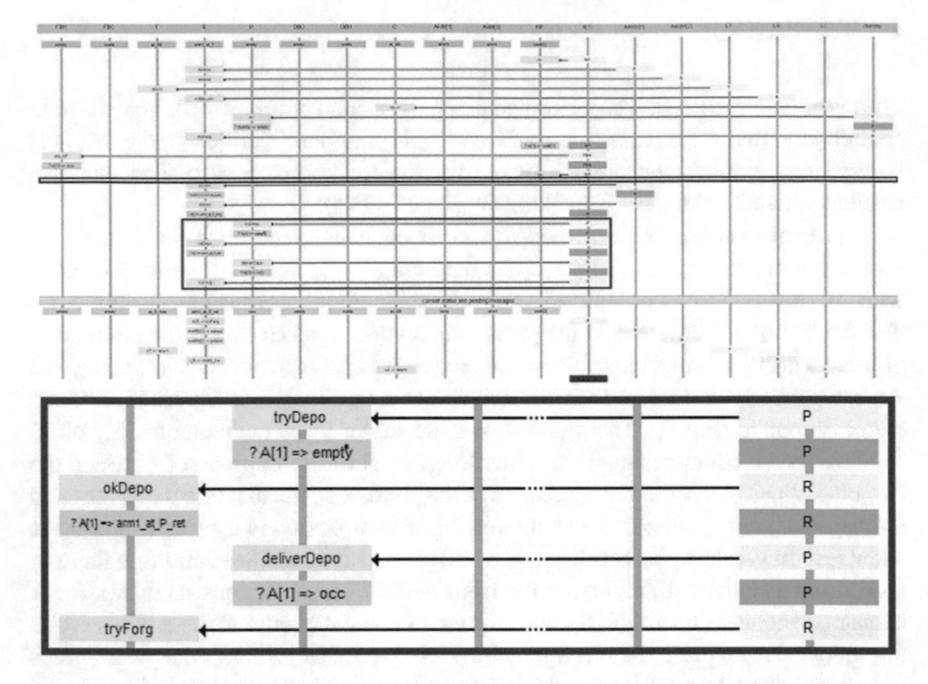

Fig. 3 The counterexample of a deadlock found in the agent view of a 1-blank Production Cell

expanding the memory even 20–30 times allows to add at most one blank because of model explosion problems in highly parallel system. Then, an external verifier Uppaal (with timeless modeling using *urgent states*, [12]) was used for 4 blanks and the verification showed the correctness. The model with 5 blanks is too big for Uppaal, therefore a simplified specification was used. The robot arms were not modeled as separate servers and their activity was included into the operation of the robot:

```
<i=1..N>{A[i].R.tryIn, R.arm1_at_T} -> {A[i].R.wait_a1extIn, R.arm1_at_T_long},
<i=1..N>{A[i].R.wait_a1extIn, R.arm1_at_T_long} -> {A[i].T.ok, R.arm1_at_T_long},
<i=1..N>{A[i].R.deliverIn, R.arm1_at_T_long} -> {A[i].R.wait_a1retIn, R.arm1_at_T_short},
<i=1..N>{A[i].R.wait_a1retIn, R.arm1_at_T_short} -> {A[i].R.rotate_ccw_P, R.rot_ccw_a2_P},
```

This model was successfully verified for 5 blanks, but for 6 blanks the problem returned. Therefore, the next step was modeling the Production Cell as an open system, in which new blanks appeared at the beginning of the feed belt and they disappeared at the end on the deposit belt (the traveling crane was not used). This allowed to verify the system with 6 blanks (with a deadlock concerning the last blank, which is expected).

5 Increasing Parallelism

When the robot reaches its target position (at the table, at the press or at the deposit belt), it extends one of its arms until a signal from an appropriate sensor stops its movement. Therefore, an arm cannot be extended in advance, before its direction points to the device. Analogous situation occurs after a blank is picked/released: the arm retracts to its rear position. It is modeled as a set of actions, starting from the point marked with ■ in the Sect. 3. When the robot reaches a state requiring retracting, a signal to the arm is issued, and after receiving of a confirmation the robot starts to rotate, leaded by the internal agent *LR*:

```
■<i=1..N>{A[i].R.deliverIn, R.arm1_at_T_long} -> {A[i].R.wait_a1retIn, R.arm1_at_T_ret},
        {AARM[1].R.retract, R.arm1_at_T_ret} -> {AARM[1].ARM[1].retract, R.arm1_at_T_ret},
        {AARM[1].R.ok_ret, R.arm1_at_T_ret} -> {AARM[1].R.extend, R.arm1_at_T_short},
<i=1..N>{A[i].R.wait_a1retIn, R.arm1_at_T_short} -> {A[i].R.rotate_ccw_P, R.rot_ccw_a2_P},

            {LR.R.rotate_ccw, R.rot_cw_a2_P} -> {LR.R.rotate_ccw, R.arm2_at_P},

<i=1..N>{A[i].R.rotate_ccw_P, R.arm1_at_P} -> {A[i].R.wait_a1extDepo, R.arm1_at_P_ext},
```

We noticed that the retracting differs from extending of an arm: it does not depend on any sensor external to the robot. Instead, an internal sensor of the robot is used. Therefore, the retracting of an arm may be executed in parallel with the rotation of the robot. In the model it may be achieved by simply replacing a state in which the arm position is expected, and adding a state `arm1_at_T_ret_goes` in which parallel actions occur:

```
■<i=1..N>{A[i].R.deliverIn, R.arm1_at_T_long} -> {A[i].R.wait_a1retIn, R.arm1_at_T_ret},
        {AARM[1].R.retract, R.arm1_at_T_ret} -> {AARM[1].ARM[1].retract, R.arm1_at_T_ret_goes},
<i=1..N>{A[i].R.wait_a1retIn, R.arm1_at_T_ret_goes} -> {A[i].R.rotate_ccw_P, R.rot_ccw_a2_P},

            {LR.R.rotate_ccw, R.rot_cw_a2_P} -> {LR.R.rotate_ccw, R.arm2_at_P},

<i=1..N>{A[i].R.rotate_ccw_P, R.arm1_at_P} -> {A[i].R.wait_a1extDepo, R.arm1_at_P_wait},
        {AARM[1].R.ok_ret, R.arm1_at_P_wait} -> {AARM[1].R. extend, R.rot_ccw_a2_P_ext},
```

6 Timed Verification

The verification with real time constraints is very important in modeling of controllers. In IMDS, adding time constraints is simple:

1. Equipping the actions with the minimal and maximal time of lasting (*x*,*y* in the example, open or closed bounds may be used):

```
{agent.server.service, server.value} -> <x,y>{agent.ohter_server.some_service, server.new_value}
```

2. Time delays of the channels (minimum and maximum time: *x,y*, open or closed bounds may be applied):

```
channels {FB->R(x,y)}
```

The internal verifier in Dedan program does not support real time constraints, therefore the external verifier Uppaal was used. However, the attempts to verify the Production Cell with time constraints failed due to insufficient memory (a PC with 8 GB was used). In the literature, some other approaches of real time verification, based on synchronous modeling, succeeded [26]. Our fail comes from explicit modeling of asynchronous channels built from pairs of synchronous channels, for every pair *(agent, target server)*. This largely expands the size of the reachability space for model checking.

Future implementation of our own real-time verifier will be free from such disadvantages because asynchrony is modeled directly in a succession of actions without any additional constructs (like asynchronous channels built from pairs of synchronous channels).

7 Conclusions

In the paper, modeling and verification of the Production Cell in the Dedan environment, using IMDS formalism is presented. Unlike in previous papers, the specification is asynchronous, which is rare among used formalisms. Also, none of the presented solutions join asynchrony with communication duality (message passing/resource sharing). Asynchrony is suitable for modeling of real distributed system features, like:

- locality of actions,
- autonomy of decisions,
- asynchrony of actions and communication.

Communication duality allows to observe the behavior of distributed systems in two perspectives: servers communicating by asynchronous messages versus agents traveling through the servers and sharing their variables. Such a duality allows to identify deadlocks in the Production Cell:

- in the agent view it is a malfunction of communication, seen from the perspective of cooperating subcontrollers;
- in the agent view it shows the sequence of actions concerning a single blank, leading to its stuck inside the Press.

Similar duality is presented in verification of Autonomous Vehicle Guidance System [42], in which the server view illustrates the cooperation of road segment controllers while the agent view shows how a vehicle travels through road segments. Deadlock is presented in both views. The idea of independent controllers cooperating by simple protocols to agree a coordinated behavior, presented in this paper and in [42], follows an IoT (Internet of Things) paradigm [43].

The timed model of verified system was built under Dedan. The timed verification was applied under external verifier Uppaal. The verification of the timed model of the Production Cell failed because of memory overrun. A probable source of the setback is that Uppaal is based on synchronous Timed Automata. Many additional automata must be used for modeling asynchronous channels, which greatly expands the model. In the future, internal verifier TempoRG [11] will be extended to cover real-time constraints. The asynchrony of channels is directly modeled in the LTS of IMDS specification, therefore no additional structures (expanding the reachability space) should be used.

References

1. Lewerentz, C., Lindner, T. (eds.): Formal Development of Reactive Systems. Springer, Berlin, Heidelberg (1995). https://doi.org/10.1007/3-540-58867-1
2. Rust, H.: A production cell with timing. In: Operational Semantics for Timed Systems, pp. 173–201. Springer, Berlin, Heidelberg (2005). https://doi.org/10.1007/978-3-540-32008-1_16
3. Flordal, H., Malik, R.: Modular nonblocking verification using conflict equivalence. In: 8th International Workshop on Discrete Event Systems, pp. 100–106. IEEE (2006). http://ieeexplore.ieee.org/document/1678415/
4. Larsen, P.G., Fitzgerald, J.S., Riddle, S.: Practice-oriented courses in formal methods using VDM++. Form. Asp. Comput. **21**(3), 245–257 (2009). https://link.springer.com/article/10.1007%2Fs00165-008-0068-5
5. El-Ansary, A., Elgazzar, M.M.: Real-time system using the behavioral patterns analysis (BPA). Int. J. Innov. Res. Adv. Eng. **1**(10), 233–245 (2014). http://www.ijirae.com/volumes/vol1/issue10/39.NVEC10091.pdf
6. Zimmermann, A.: Model-based design and control of a production cell. In: Stochastic Discrete Event Systems: Modeling, Evaluation, Applications, pp. 325–340. Springer, Berlin, Heidelberg (2008). https://doi.org/10.1007/978-3-540-74173-2_16
7. Lötzbeyer, A., Mühlfeld, R.: Task description of a flexible production cell with real time properties, FZI Technical Report, University of Karlsruhe (1996)
8. Chrobot, S., Daszczuk, W.B.: Communication dualism in distributed systems with Petri net interpretation. Theor. Appl. Inform. **18**(4), 261–278 (2006). https://taai.iitis.pl/taai/article/view/250/taai-vol.18-no.4-pp.261
9. Daszczuk, W.B.: Communication and resource deadlock analysis using IMDS formalism and model checking. Comput. J. **60**(5), 729–750 (2017). https://doi.org/10.1093/comjnl/bxw099
10. Dedan, http://staff.ii.pw.edu.pl/dedan/files/DedAn.zip
11. Daszczuk, W.B.: Verification of Temporal Properties in Concurrent Systems. PhD Thesis, Warsaw University of Technology (2003). https://repo.pw.edu.pl/docstore/download/WEiTI-0b7425b5-2375-417b-b0fa-b1f61aed0623/Daszczuk.pdf
12. Behrmann, G., David, A., Larsen, K.G.: A Tutorial on Uppaal 4.0. Aalborg University Report, Aalborg, Denmark (2006). http://www.it.uu.se/research/group/darts/papers/texts/new-tutorial.pdf

13. Heiner, M., Heisel, M.: Modeling safety-critical systems with Z and Petri nets. In: Felici, M., Kanoun, K., Pasquini, A. (eds.) SAFECOMP '99 Proceedings of the 18th International Conference on Computer Safety, Reliability and Security, Toulouse, France, 27–29 September 1999. LNCS, vol. 1698, pp. 361–374. Springer, Berlin, Heidelberg (1999). https://link.springer.com/chapter/10.1007%2F3-540-48249-0_31
14. Greenyer, J., Brenner, C., Cordy, M., Heymans, P., Gressi, E.: Incrementally synthesizing controllers from scenario-based product line specifications. In: Proceedings of the 2013 9th Joint Meeting on Foundations of Software Engineering—ESEC/FSE 2013, Sankt Petersburg, Russia, 18–26 August 2013, pp. 433–443. ACM Press, New York, NY (2013). https://doi.org/10.1145/2491411.2491445
15. Garavel, H., Serwe, W.: The unheralded value of the multiway rendezvous: illustration with the production cell benchmark. Electron. Proc. Theor. Comput. Sci. **244**, 230–270 (2017). https://doi.org/10.4204/EPTCS.244.10
16. Jacobs, J., Simpson, A.: A formal model of SysML blocks using CSP for assured systems engineering. In: Formal Techniques for Safety-Critical Systems, Third International Workshop, FTSCS 2014, Luxembourg, 6–7 November 2014. Communications in Computer and Information Science, vol. 476, pp. 127–141. Springer, Berlin, Heidelberg (2015). https://doi.org/10.1007/978-3-319-17581-2_9
17. Ma, C., Wonham, W.M.: The production cell example. Chapter 5. In: Nonblocking Supervisory Control of State Tree Structures. LNCIS, vol. 317, pp. 127–144. Springer, Berlin, Heidelberg (2005). https://doi.org/10.1007/11382119_5
18. Zorzo, A.F., Romanovsky, A., Xu, J., Randell, B., Stroud, R.J., Welch, I.S.: Using coordinated atomic actions to design safety-critical systems: a production cell case study. Softw. Pract. Exp. **29**(8), 677–697 (1999). https://doi.org/10.1002/(SICI)1097-024X(19990710)29:8<677::AID-SPE251>3.0.CO;2-Z
19. Sokolsky, O., Lee, I., Ben-Abdallah, H.: Specification and Analysis of Real-Time Systems with PARAGON (equivalence checking), Philadelphia, PA (1999). https://www.cis.upenn.edu/~sokolsky/ase99.pdf
20. Ramakrishnan, S., McGregor, J.: Modelling and testing OO distributed systems with temporal logic formalisms. In: 8th International IASTED Conference Applied Informatics' 2000, Innsbruck, Austria, 14–17 February 2000 (2000). https://research.monash.edu/en/publications/modelling-and-testing-oo-distributed-systems-with-temporal-logic-
21. Hoare, C.A.R.: Communicating sequential processes. Commun. ACM **21**(8), 666–677 (1978). https://doi.org/10.1145/359576.359585
22. Milner, R.: A Calculus of Communicating Systems. Springer, Berlin, Heidelberg (1984). ISBN 0387102353
23. Cassez, F., David, A., Fleury, E., Larsen, K.G., Lime, D.: Efficient on-the-fly algorithms for the analysis of timed games. In: 16th International Conference on Concurrency Theory (CONCUR'05), San Francisco, CA, 23–26 August 2005. LNCS, vol. 3653, pp. 66–80. Springer, Berlin, Heidelberg (2005). https://doi.org/10.1007/11539452_9
24. Dierks, H.: The production cell: a verified real-time system. In: 4th International Symposium on Formal Techniques in Real-Time and Fault-Tolerant Systems FTRTFT 1996: Uppsala, Sweden, 9–13 September 1996. LNCS, vol. 1135, pp. 208–227. Springer, Berlin, Heidelberg (1996). https://doi.org/10.1007/3-540-61648-9_42
25. Beyer, D., Lewerentz, C., Noack, A.: Rabbit: a tool for BDD-based verification of real-time systems. In: Computer Aided Verification, CAV 2003, Boulder, CO, 8–12 July 2003. LNCS, vol. 2725, pp. 122–125. Springer, Berlin, Heidelberg (2003). https://link.springer.com/chapter/10.1007%2F978-3-540-45069-6_13
26. Burns, A.: How to verify a safe real-time system—the application of model checking and timed automata to the production cell case study. Real-Time Syst. **24**(2), 135–151 (2003). https://doi.org/10.1023/A:1021758401878
27. Benghazi Akhlaki, K., Capel Tuñón, M.I., Holgado Terriza, J.A., Mendoza Morales, L.E.: A methodological approach to the formal specification of real-time systems by transformation of UML-RT design models. Sci. Comput. Program. **65**(1), 41–56 (2007). https://doi.org/10.1016/j.scico.2006.08.005

28. Barbey, S., Buchs, D., Péraire, C.: Modelling the Production Cell Case Study using the Fusion Method. Lausanne, Switzerland (1998). https://infoscience.epfl.ch/record/54618/files/Barbey98–298..ps.gz
29. Cattel, T.: Process control design using SPIN. In: Spin Workshop, 16 Oct 1995, Montreal, Canada (1995). http://spinroot.com/spin/Workshops/ws95/cattel.pdf
30. Schröter, C., Schwoon, S., Esparza, J.: The model-checking kit. In: 24th International Conference ICATPN 2003: Eindhoven, The Netherlands, 23–27 June 2003. LNCS, vol. 2697, pp. 463–472. Springer, Berlin, Heidelberg (2003). https://doi.org/10.1007/3-540-44919-1_29
31. Björnander, S., Seceleanu, C., Lundqvist, K., Pettersson, P.: ABV—a verifier for the architecture analysis and design language (AADL). In: 6th IEEE International Conference on Engineering of Complex Computer Systems, Las Vegas, USA, 27–29 April 2011, pp. 355–360. IEEE (2011). https://doi.org/10.1109/iceccs.2011.43
32. Capecchi, S., Giachino, E., Yoshida, N.: Global escape in multiparty sessions. Math. Struct. Comput. Sci. **26**(02), 156–205 (2016). https://doi.org/10.1017/S0960129514000164
33. Ruf, J., Kropf, T.: Modeling and checking networks of communicating real-time processes. In: CHARME 1999: Correct Hardware Design and Verification Methods, BadHerrenalb, Germany, 27–29 September 1999. LNCS, vol. 1704, pp. 267–279. Springer, Berlin, Heidelberg (1999). https://doi.org/10.1007/3-540-48153-2_20
34. Grosu, R., Broy, M., Selic, B., Stefănescu, G.: What is behind UML-RT? In: Kilov, H., Rumpe, B., Simmonds, I. (eds.) Behavioral Specifications of Businesses and Systems, pp. 75–90. Springer US, Boston, MA (1999). https://doi.org/10.1007/978-1-4615-5229-1_6
35. Žic, J.J.: Time-constrained buffer specifications in CSP + T and timed CSP. ACM Trans. Program. Lang. Syst. **16**(6), 1661–1674 (1994). https://doi.org/10.1145/197320.197322
36. Ehlers, R., Mattmüller, R., Peter, H.-J.: Combining symbolic representations for solving timed games. In: Chatterjee, K., Henzinger, T.A. (eds.) 8th International Conference on Formal Modeling and Analysis of Timed Systems, FORMATS 2010, Klosterneuburg, Austria, 8–10 September 2010. LNCS, vol. 6246, pp. 107–121. Springer, Berlin, Heidelberg (2010). https://doi.org/10.1007/978-3-642-15297-9_10
37. Ben-Abdallah, H., Lee, I.: A graphical language for specifying and analyzing real-time systems. Integr. Comput. Aided. Eng. **5**(4), 279–302 (1998). ftp://ftp.cis.upenn.edu/pub/rtg/Paper/Full_Postscript/icae97.pdf
38. Beyer, D., Rust, H.: Modeling a production cell as a distributed real-time system with cottbus timed automata. In: König, H., Langendörfer, P. (eds.) Formale Beschreibungstechniken für verteilte Systeme, 8. GI/ITG-Fachgespräch, Cottbus, 4–5 June 1998. Shaker Verlag, München, Germany (1998). https://www.sosy-lab.org/~dbeyer/Publications/1998-FBT.Modeling_a_Production_Cell_as_a_Distributed_Real-Time_System_with.Cottbus_Timed_Automata.pdf
39. Barbey, S., Buchs, D., Péraire, C.: A Case Study for Testing Object Oriented Software: A Production Cell. Swiss Federal Institute of Technology (1998)
40. Waeselynck, H., Thévenod-Fosse, P.: A case study in statistical testing of reusable concurrent objects. In: Third European Dependable Computing Conference Prague, Czech Republic, 15–17 September 1999, LNCS, vol. 1667, pp. 401–418. Springer, Berlin, Heidelberg (1999). https://doi.org/10.1007/3-540-48254-7_27
41. Daszczuk, W.B.: Evaluation of temporal formulas based on "checking by spheres." In: Proceedings Euromicro Symposium on Digital Systems Design, Warsaw, Poland, 4–6 September 2001, pp. 158–164. IEEE Computer Socity, New York, NY (2001). https://doi.org/10.1109/dsd.2001.952267
42. Czejdo, B., Bhattacharya, S., Baszun, M., Daszczuk, W.B.: Improving resilience of autonomous moving platforms by real-time analysis of their cooperation. Autobusy-TEST **17**(6), 1294–1301 (2016). http://www.autobusy-test.com.pl/images/stories/Do_pobrania/2016/nr%206/logistyka/10_l_czejdo_bhattacharya_baszun_daszczuk.pdf
43. Lee, G.M., Crespi, N., Choi, J.K., Boussard, M.: Internet of Things. In: Evolution of Telecommunication Services. LNCS, vol. 7768, pp. 257–282. Springer, Berlin Heidelberg (2013). https://doi.org/10.1007/978-3-642-41569-2_13

Implementing the Bus Protocol of a Microprocessor in a Software-Defined Computer

Julia Kosowska and Grzegorz Mazur

Abstract The paper describes a concept of a software-defined computer implemented using a classic 8-bit microprocessor and a modern microcontroller with ARM Cortex-M core for didactic and experimental purposes. Crucial to this design is the timing analysis and implementation of microprocessor's bus protocol using hardware and software resources of the microcontroller. The device described in the paper, SDC_One, is a proof-of-concept design, successfully demonstrating the software-defined computer idea and showing the possibility of implementing time-critical logic functions using a microcontroller. The project is also a complex exercise in real-time embedded system design, pushing the microcontroller to its operational limits by exploiting advanced capabilities of selected hardware peripherals and carefully crafted firmware. To achieve the required response times, the project uses advanced capabilities of microcontroller peripherals – timers and DMA controller. Event response times achieved with the microcontroller operating at 80 MHz clock frequency are below 200 ns and the interrupt frequency during the computer's operation exceeds 500 kHz.

Keywords Software-defined computer · Microprocessor · Bus protocol · Bus monitor · Microcontroller peripherals · Real-time system · Event response time

1 Software-Defined Computer

1.1 Concept

The concept of a software-defined computer is an offspring of work on laboratory hardware environment used for teaching the digital system design and microprocessor systems in particular. Contemporary microcontrollers used in embedded systems

J. Kosowska · G. Mazur (✉)
Institute of Computer Science, Warsaw University of Technology,
Nowowiejska 15/19, 00-665 Warsaw, Poland
e-mail: g.mazur@ii.pw.edu.pl

R. Bembenik et al. (eds.), *Intelligent Methods and Big Data in Industrial Applications*, Studies in Big Data 40, https://doi.org/10.1007/978-3-319-77604-0_10

have powerful hardware mechanisms supporting software debugging but the possibilities of monitoring and presenting their hardware operation at the level of bus transactions are almost non-existent. It is generally not possible to view the bus operation or the process of execution of an instruction using a contemporary microcontroller. In this aspect, 8-bit microprocessors from 1980s appear to be useful for didactic purposes but their typical hardware environments – accompanying peripheral controllers and glue logic – are technologically outdated. A common trend to use software simulators to illustrate hardware operation has one major drawback – students are unable to get any hardware-related skills and they stop realizing that a computer is implemented as a real electronic device.

The goal of a software-defined computer project was to make it possible for students to see and understand the operation of a microprocessor at bus transaction level, using hardware utilizing both vintage and contemporary microprocessor technology. The idea was to design a fully functional and usable computer with hardware monitoring capabilities, using a classic microprocessor as its CPU and implementing all the other components of a computer with a modern single-chip microcontroller.

The design uses a real microprocessor rather than a CPU core implemented in an FPGA. Also, since the device was to be used for microcomputer-, not programmable logic-related course, we have decided to implement the functions of a computer and bus monitor using a microcontroller rather than an FPGA. This way the design may be easily reproduced and modified by students using a popular, inexpensive microcontroller development kit connected to the target microprocessor placed on a small PCB or even on a breadboard. Such approach has a side effect of familiarizing the students with advanced design techniques employed to implement the robust and time-critical functionality of computer hardware using microcontroller's peripherals and firmware.

1.2 Design Overview

Our software-defined computer, named SDC_One (Fig. 1), is a fully usable microcomputer with the capabilities typical to 8-bit microprocessor-based computers designed in the late 1970s/early 1980s. It is capable of running a standard operating system such as CP/M. It is also equipped with a hardware bus monitor, described in Sect. 4.

SDC_One connects to a PC using single USB cable used for powering the device and for communication. On the PC side it is visible as a composite device with two virtual serial ports and one mass storage device. One of the virtual serial ports is used for hardware monitor console; the other for target computer terminal.

Hardware components. The SDC_One design was implemented in four versions, using four different target microprocessors: Zilog Z80 CPU, Motorola MC68008, Intel 8085 and WDC W65C02S. Support for a few other processors is under development. All the versions use the same host platform for implementing the computer's memory, peripherals, glue logic and bus monitor. We have chosen ST Microelectron-

Fig. 1 An SDC_One, Z80 CPU version. Target CPU board connects directly to the ST Microelectronics NUCLEO-L476RG board

ics STM32L476 microcontroller [1, 2], mounted on popular Nucleo-64 development board because of its popularity, rich on-chip resources and low price.

Target computer resources. SDC_One provides the target microprocessor with the following resources:

- 64 KiB of read-write memory;
- pushbutton and 4 LEDs with on/off and PWM control;
- terminal interface;
- interval timer;
- two non-volatile mass storage devices (virtual diskette drives) implemented with microcontroller's onboard Flash memory.

1.3 Similar Projects

During the implementation of our design, two similar designs were published on the web. They are software-defined computers based on the WDC W65C02S processor with the Parallax Propeller microcontroller emulating some other parts of the computer designed by Jac Goudsmit. The Parallax Propeller is a multicore microcontroller with 8 cores called cogs and 2 KiB of RAM per core. Apart from that, it also has 32 KiB of shared ROM, 32 KiB of shared RAM called hub memory, and 32 I/O lines. Except for some hobby projects, it is not widely used.

The first of the aforementioned designs, named Propeddle [3], uses a separate SRAM chip connected directly to the target processor, as well as some glue logic chips, such as latches and buffers, to compensate for the low number of I/O lines available in the Propeller. The Propeller microcontroller generates clock and control signals for the 65C02, activates the SRAM chip at the proper times and emulates I/O devices for the target processor.

The second design, named L-Star [4], is a simplified version of the Propeddle, in which the SRAM and glue logic chips were removed leaving only the 65C02 with its address, data and R/-W lines connected directly to the Propeller. The target processor's RAM is emulated using Propeller's 32 KiB hub memory. Since that size of memory may be insufficient for some applications, an extended version called L-Star Plus contains an SRAM chip connected to the 65C02 and controlled by the Propeller much like in the Propeddle.

There are 3 main advantages of SDC_One over these designs:

- SDC_One is based on a popular, standard microcontroller, not a niche device.
- It does not require external memories nor any glue logic.
- It provides comprehensive facilities for monitoring and controlling the operation of the target processor.

2 Implementing the Bus Protocol

2.1 Overview of a Transfer Cycle

Sequence of events. Emulating the microprocessor's environment with a microcontroller requires in-depth analysis of its bus protocol. To implement a bus monitor and a computer's functionality in microcontroller's firmware, the microcontroller must be able to extend the transfer cycle initiated by a target processor. Most microprocessors implement such mechanism with the help of an input signal controlling transfer completion or extension. When the signal is active, the processor keeps the control signal states unchanged, allowing the external hardware to finalize the transfer at a proper time. The signal is named differently by various CPU manufacturers. In the Z80 CPU it is called -WAIT and its active (low) state signals a bus cycle extension request.

A typical bus protocol of a classic microprocessor, like the Z80 CPU, includes the following sequence of events:

- presenting the transfer address and signaling a data transfer request, its direction and type,
- in case of a write transfer, setting the value of data on the data bus,

- checking the state of the -WAIT signal and possibly extending the cycle at the request of a peripheral module,
- finishing the cycle upon detecting -WAIT signal deactivation which includes deactivating transfer request signals and possibly other control signals as well as latching the data in case of a read transfer.

Operation of software-defined computer. In our design the -WAIT signal is kept active most of the time to ensure that no transfer cycle will be missed by the microcontroller. It gets activated for a short time when the transfer request is serviced and is to be finished. It must get deactivated again before the processor probes it in the next transfer cycle. In order to emulate the operation of memory and peripherals, the microcontroller must detect the start of a new cycle, check its type and direction and according to that decide on subsequent actions. If the processor is trying to read data, the microcontroller must retrieve the data from its own memory or poll an input/output device it interacts with, output the data onto the data lines and activate the data output buffers. If the processor it writing data, the microcontroller should save the data, interpret it and possibly take some further actions and/or notify appropriate input/output device. Once the transfer is serviced, -WAIT signal should be deactivated to let the processor finish the cycle.
Time-critical operations. Upon recognizing the end of transfer, -WAIT signal may be activated, which must happen before the processor checks its state in the next cycle. If the microcontroller drove the data bus, it must also turn off its data output buffers in time to avoid a bus conflict in case the next cycle is a write cycle and the processor drives the bus.

Activating the -WAIT signal in time to ensure recognition of bus cycle extension request and turning off data output buffers at the end of a read cycle are time critical actions that must be performed within a given time period to ensure the correct operation of a software-defined computer. One approach would be to use the edges of control signals activated upon starting a new cycle by the target processor to trigger an interrupt and performing the necessary actions in the interrupt service routine. However, with an estimated microcontroller's externally-triggered interrupt response time together with necessary firmware overhead of about 250 ns, it would be impossible to perform the necessary actions in time. Thus, a different faster solution was needed to correctly implement the time-critical bus operations.

2.2 *Implementation of Time-Critical Pieces of Bus Protocol Using DMA*

Rather than relying entirely on firmware, we decided to use hardware modules of STM32 microcontroller to change the state of critical signals. The bus signals of a target processor are connected to GPIO ports of a microcontroller. The state of the GPIO lines may be changed by means of DMA transfers to port control registers, which requires notably less time than firmware-based reaction. Unfortunately, in

STM32 microcontrollers, a change of GPIO input state cannot trigger a DMA transfer. DMA transfers may, however, be requested by other peripheral modules, including timers, and a timer event may be caused by a GPIO's input state change. In our design we use a timer operating in capture mode to generate capture events triggering DMA transfers upon detecting a change of relevant control signals. The timer counter value is ignored.

For most types of CPUs used in SDC_One it is necessary to provide hardware-controlled -WAIT signal activation and data buffer turn-off. Both actions should occur as soon as the target CPU signals the end of a transfer cycle. They must be implemented using two distinct DMA transfers; end-of-cycle (-WAIT) signal state is changed by a transfer to GPIO BSRR register while the data bus buffers are turned off by means of writing GPIO MODER register (setting the data bus lines as inputs).

Multiple trigger inputs and multiple DMA transfers. While with the MC68008 microprocessor the end of transfer may be detected based on the state of a single signal, -DS (data strobe), many other CPUs, including the Z80, use one of a few signals to identify different transfer cycles. In such a case, the DMA must be triggered based on a combinatorial function of several inputs. Fortunately, the STM32 microcontrollers' timers have some features that helped to keep our design simple. The capture event may be triggered by a XOR of up to 3 input signals. Moreover, the timer's channels form pairs and it is possible to feed both channels in a pair with the same input. Each channel can issue a separate DMA request. Therefore, a change in input control signal may easily trigger the two required transfers using two DMA channels.

2.3 *Interrupt-Driven Data Transfers*

If -WAIT activation is timely performed, the target processor begins a machine cycle by setting all relevant bus signals and keeps them unchanged until the -WAIT signal gets deactivated. Since meeting all critical bus timing requirements is ensured by hardware means, the remaining actions of the microcontroller may be dealt with in firmware by means of interrupt service routines. The change of the target processor's output control signals at the beginning of a new machine cycle triggers an interrupt request. In the interrupt service routine the microcontroller needs to check the control signals, determine the type of the cycle and therefore the actions necessary to emulate the processor's environment. The correct operation of the software-defined computer requires that every machine cycle gets serviced exactly once. Therefore, it is necessary to ensure that the beginning of every cycle triggers an interrupt by choosing an appropriate set of control signals so that at least one of them is changed when a certain type of cycle begins. If at the beginning of some type of cycle more than one of the monitored control signals changes, multiple interrupt requests may be generated and we must ensure that any spurious interrupts are ignored. Before determining the type of transfer, all the relevant signal states should be fixed.

2.4 *CPU-Specific Aspects of Bus Protocol*

The sources for the generation of DMA and interrupt requests needed to implement the bus protocol depend on the set of bus control signals implemented in a given CPU. To choose the right set of signals, the target's processor bus protocol should be thoroughly analyzed. Moreover, some additional DMA transfers or actions performed in the interrupt service routine may be necessary in case of some processors.

In Figs. 2, 3 and 4, the edges marked in green are used to issue DMA requests that both activate the -WAIT signal and put the data bus in a high-impendence state while the ones marked in blue are used to signal an interrupt. The critical time parameters are marked in red.

Z80 CPU. Figure 2 shows the memory access cycle of the Zilog Z80 CPU processor. In the memory access cycle, the processor first activates the -MREQ signal to inform that it wants to access memory (as opposed to I/O space) and then the -RD or -WR signal to inform about the direction of the transfer. The data being transferred to memory in a write cycle is valid before the -WR signal gets activated. The signals used as a trigger for the DMA transfer include -MREQ and -IORQ since one of them gets activated in every type of cycle with its rising edge occurring at the end of the cycle. The signals used as a trigger for start-of-cycle interrupt, beside -RD and -WR, also include -IORQ to ensure the correct recognition of interrupt acknowledge cycle, in which neither -RD nor -WR gets activated. The -MREQ and -IORQ are not

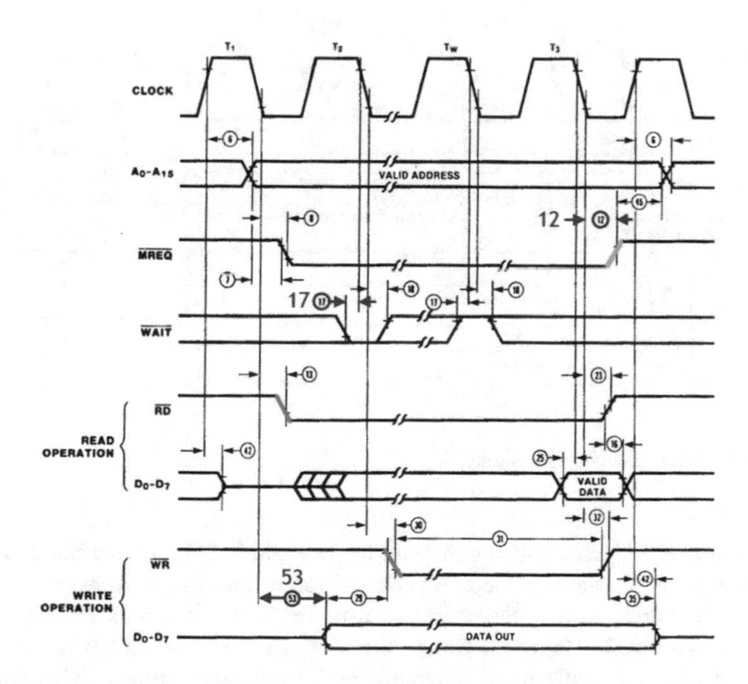

Fig. 2 Z80 CPU bus timing taken from [5] with critical sequences identified

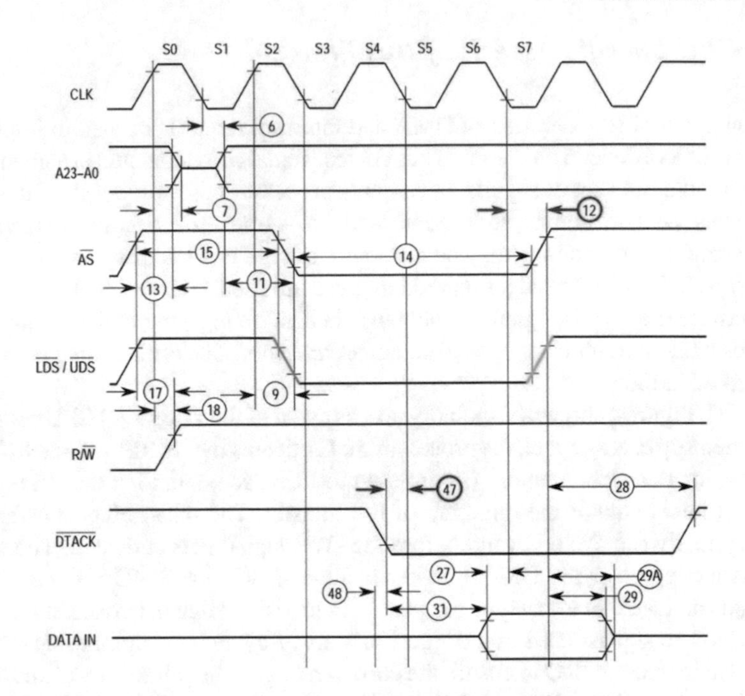

Fig. 3 MC68008 bus timing taken from [6] with critical sequences identified

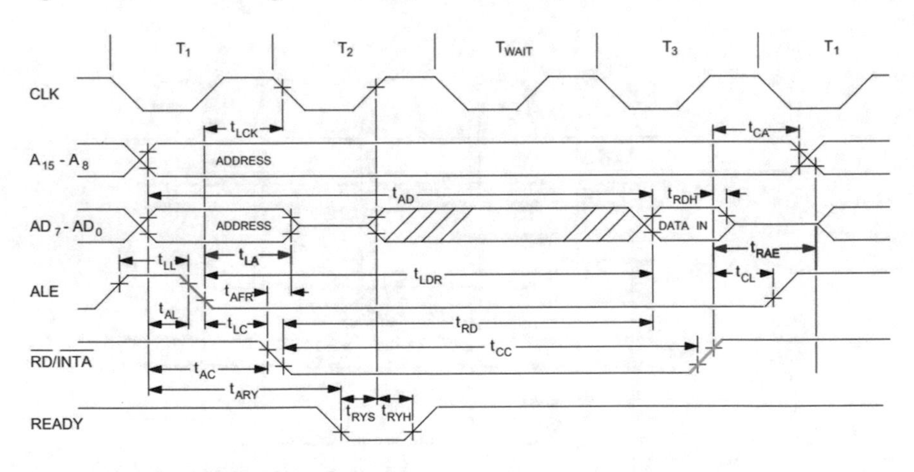

Fig. 4 OKI MSM80C85AH bus timing [8]

used because they get activated before all the relevant bus signals are fixed, making it more difficult to ensure the correct recognition of the type of transfer cycle and validity of data retrieved by the processor from the data bus in write cycles.

Motorola MC68008. Figure 3 shows the read cycle of the Motorola MC68008 processor. In case of MC68008 the leading edge of the data strobe (-DS) signal that occurs every time the processor request a transfer of a piece of data is used as a trigger for start-of-cycle interrupt and its trailing edge is used as a trigger for the DMA transfer to activate the bus cycle extension request (put -DTACK low) and to tristate the data bus.

Intel 8085/OKI 80C85AH. The Intel 8085 microprocessor [7] and its derivatives differ from the CPUs described above by a multiplexed data and address bus (Fig. 4). The least significant byte of address is transferred using the data bus at the beginning of each transfer cycle. Address transfer is signaled with the accompanying ALE signal strobe; address transfer time depends on the CPU clock frequency and it cannot be extended. The SDC_One design must guarantee correct latching of address with proper relation to ALE strobe timing. The presence of ALE strobe is also an indicator of bus cycle start. The critical times to be observed are t_{LA} and t_{RAE} – the address must be recorded no later than t_{LA} from the trailing edge of ALE and t_{RAE} is the required bus turnaround time.

The end of a transfer cycle is signaled with the trailing edge of -RD, -WR or -INTA signal. Two timer channels with XOR input function using -RD, -WR and -INTA inputs are used to trigger two DMA transfer; one for deactivating the bus buffer, the other for activating the READY line, which causes the microprocessor to extend the next bus cycle. ALE input is captured by another timer, triggering the DMA transfer to read the address from a multiplexed data/address bus. The same event also triggers the interrupt used to service the data transfer cycle.

3 Timing Analysis

To confirm the validity of our draft design, it was necessary to determine the exact values of the critical time parameters and to ensure that all of them are met. Such an evaluation was needed for every target processor used in our design. Some of the important time parameters are marked in red in Fig. 2 for the Zilog Z80 CPU and in Fig. 3 for the Motorola MC68008. To ensure that a software-defined computer operates correctly we had to calculate the maximum time it may take for the time critical actions to be performed after the DMA trigger in the worst-case scenario. The time is expressed in terms of the target processor clock cycle time and is used to calculate the maximum theoretical frequency of the target processor clock. If the obtained value is greater than or equal to the minimum clock frequency specified in the processor's datasheet, then it is possible to build a correctly operating software-defined computer with a given microprocessor using the scheme presented. We present the results of the timing constraints evaluation for the Zilog Z80 CPU [5], Motorola MC68008 [6] and OKI MSM80C85AH [8]. In case of the aforementioned processors, both the deactivation of the data output buffers and activation of the cycle extension signal can be performed as soon as the current transfer is finished. Denoting clock cycle time as T_{CLK} and using the time parameters from the processor's datasheets, we obtained the following results:

- Zilog Z80:

– data bus turnoff maximum time from the trailing edge of the -RD strobe to putting the data lines into high impendence state is T_{CLK} – 85 ns,
– the maximum time from the trailing edge of the -RD or -WR strobe to activating the WAIT signal is $2 \cdot T_{CLK} - 155$ ns.

- Motorola MC68008:

– data bus turnoff time from the trailing edge of the -DS strobe to putting the data lines into high impendence state is T_{CLK} – 30 ns,
– the maximum time from the trailing edge of the -DS strobe to activating the WAIT signal is $2.5 \cdot T_{CLK} - 60$ ns.

- OKI80C85AH:

– maximum data bus turnoff time from the trailing edge of the -RD or -INTA strobe to putting the data lines into high impendence state (t_{RAE}) is $T_{CLK}/2$ – 10 ns,
– A0..7 valid delay from ALE trailing edge (t_{LA}) is $T_{CLK}/2$–50 ns,
– the maximum time from the trailing edge of the -RD, -WR or -INTA strobe to READY sampling exceeds T_{CLK} and is not critical.

The DMA response time of a microcontroller depends on many factors and is not specified in the microcontroller's datasheet. Therefore, we had to measure the time from the change of input port signal issuing a DMA request till the change of output port signal altered by the DMA transfer in case of 2 different priority DMA requests issued simultaneously which is the case when both tristating of the data bus and activation of WAIT signal are triggered by the same event (which is a typical case). To measure the time a microcontroller's timer was programmed to generate a PWM waveform with a frequency of a few hundred kHz. Two channels of another timer were fed with the PWM waveform and programmed to capture the edges of the PWM waveform and trigger two outputs accordingly thereby mirroring the PWM waveform on two port outputs. Using an oscilloscope we measured the time between the relevant edges of the original and mirrored PWM waveforms obtaining 141 ns for the higher priority DMA transfer and 172.5 ns for the lower priority transfer on STM32L476RGT microcontroller running at 80 MHz. After performing the necessary calculations using the aforementioned timing constraints, we ascertained the minimum clock cycle time as 226 ns for the Z80 CPU, 171 ns for the MC68008 and 382 ns for the 80C85 and therefore the maximum clock frequency as 4.4 MHz for the Zilog Z80, 5.8 MHz for the MC68008 and 2.6 MHz for the 80C85. The calculated frequencies are significantly higher than the minimum operating frequencies required by non-static versions of the processors, which proves the possibility of constructing a correctly operating software-defined computer using the presented approach.

4 Hardware Monitor

Our design features comprehensive hardware monitoring facilities allowing to observe and control the operation of the processor at machine cycles level with the following capabilities:

- setting the frequency of the target microprocessor's clock signal;
- initializing the target (RESET signal generation);
- loading programs to the target computer's memory;
- display and modification of memory and I/O port content;
- control of target processor's operation: single-stepping, breakpoints, trace collection and fast free run mode;
- controlling the operation of the target's mass storage.

Execution modes. By default the computer operates in machine cycle stepping mode which allows the closest supervision over the operation of the processor. In this mode, the processor is stopped in every cycle with the cycle extension (-WAIT) signal kept active until a command to finish the current cycle is issued via the monitor console. Information about the current cycle including its type, transfer address and data is displayed on the monitor console. A disassembled instruction is displayed after the cycle information in the first cycle of an instruction (in most implementations), which facilitates easier software debugging and monitoring of the processor operation. In the instruction stepping mode the processor is stopped only in the first cycle of an instruction. In continuous run mode machine cycles are finished as soon as they are serviced unless they match one of the defined breakpoints, in which case the processor is stopped. Information about executed machine cycles is gathered in a circular buffer and may be displayed once the processor is stopped. In continuous run mode the event that cause the processor to be stopped may be specified via breakpoints, which are defined by giving transfer address and a bit mask of types of cycles in which the breakpoint is to be active. In fast run mode both trace collection and breakpoints checking are disabled and only the actions essential for the emulation of the computer environment are performed in the start-of-cycle interrupt service routine, which facilitates the fastest operation of the computer. The mode is suitable for running the vintage OS and application software for demonstrative and hobbyist's purposes.

The hardware monitor features include commands for displaying and modifying the target computer's environment, including target clock frequency, memory and I/O ports content display and modification and interrupt injection. When the processor is stopped in a transfer cycle, any changes get immediately reflected in the information about current cycle and the disassembled instruction. Observing how certain conditions affect the actions of the processor allows very low-level in-depth insight into the operation of the computer.

5 Conclusion

At the time of writing, versions of SDC_One with four significantly different processors – the Zilog Z80 CPU, Intel 8085/OKI 80C85, Motorola MC68008 and WDC W65C02S – have been implemented and tested. The design have already proven useful for demonstrating the operation of a processor to students. With the SDC_One we can easily show the phases of instruction's execution and data addressing modes using real hardware. The low level insight into the actions of the CPU our design enables lets us notice and demonstrate to students some particularities of processors that are only observable at the machine cycle level, such as an optimization of a non-taken conditional branch execution in 8085 microprocessor or the operation of a prefetch mechanism in MC68008.

The effective performance of SDC_One has been checked by measuring the number of machine cycles executed per second. In case of the Z80 CPU running at 2.5 MHz and executing a stream of NOPs the result was 357 thousand machine cycles per second (7 clock cycles per transfer), which is 1.75 times slower than best-case performance. Three versions of SDC_One run CP/M – a popular operating system from the 1980s, which supports Intel 8080 series, Zilog Z80 and Motorola 68 k family of CPUs. Some popular CP/M applications such as WordStar text processor and ZASM assembler have been installed on SDC_One and used to create and assemble a correctly operating program thereby proving SDC_One to be a fully functional computer.

References

1. STMicroelectronics: RM0351 STM32L4x5 and STM32L4x6 advanced ARM®-based 32-bit MCUs Reference Manual (2016)
2. STMicroelectronics: STM32L476xx Datasheet (2015)
3. Goudsmit, J.: Propeddle – Software Defined 6502 Computer (2012) https://propeddle.com
4. Goudsmit, J.: L-Star: Software-Defined 6502 Computer (2014) https://hackaday.io/project/3620
5. Zilog: PS0178 Z8400/Z84C00 Z80 CPU Product Specification (2002)
6. Motorola: MC68000UM M68000 8-16-32-Bit Microprocessors User's Manual (REV 9.1) (2006)
7. Intel: Component Data Catalog 1982, Intel Corporation Santa Clara, CA, USA, pp. 7-10–7-25 (1982)
8. OKI Semiconductor: MSM80C85AHRS/GS/JS 8-Bit CMOS MICROPROCESSOR (2003)

Part III
Data Mining

ISMIS 2017 Data Mining Competition: Trading Based on Recommendations - XGBoost Approach with Feature Engineering

Katarzyna Baraniak

Abstract This paper presents an approach to predict trading based on recommendations of experts using XGBoost model, created during ISMIS 2017 Data Mining Competition: Trading Based on Recommendations. We present a method to manually engineer features from sequential data and how to evaluate its relevance. We provide a summary of feature engineering, feature selection, and evaluation based on experts recommendations of stock return.

1 Introduction

Data mining competitions are constantly popular among researchers as a way of learning and testing different models. Recently, some of the approaches are used more often due to its efficiency and possible application to the variety of problems. XGBoost is one of the most popular models among data mining competition so it is valuable to make a study if it is appropriate for a problem of stock market prediction based on expert recommendations [1].

2 Data

Data used for experiments in this paper was provided by Tipranks in ISMIS 2017 Data Mining Competition: Trading Based on Recommendations. Training data contains 12234 records. Data has a tabular format. Every record corresponds to decision at different time point. The first column contains stock id, second - various number of

---

The work was partially done when author was working at National Information Processing Institute.

K. Baraniak (✉)
Polish-Japanese Academy of Information Technology, Warsaw, Poland
e-mail: katarzyna.baraniak1@pjwstk.edu.pl

K. Baraniak
National Information Processing Institute, Warsaw, Poland

R. Bembenik et al. (eds.), *Intelligent Methods and Big Data in Industrial Applications*, Studies in Big Data 40, https://doi.org/10.1007/978-3-319-77604-0_11

Table 1 Data summary

Value	Training data set		Test data set
	Recommendations	Decisions	Recommendations
Buy	34306	5722	34166
Hold	18388	3403	18751
Sell	2564	3109	2287

recommendations, third and the last one - final Decision. Each recommendation from the second column is put in a '{}' bracket and contains following information: id of expert, a number of days before the decision when the recommendation was made. Training data come from a period of two months. Test set contains 7555 records in the same format except for 'Decision' column. Test set was gathered in the time of two months following training data set.

2.1 Preliminary Data Exploration

Table 1 presents number of particular decisions and recommendation. Values in 'Decision' and 'Recommendation' column are not equally frequent. The most frequently recommended class is Buy in both training and test set. The least frequent value 'Sell' is about 13 times less frequent than 'Buy' in both sets.

2.2 Data Preparation

To improve accuracy we used an oversampling method to train the model on a similar number of classes. XGBoost is a regression model so all data has to be a numeric value. Every recommendation and decision replaced by a numeric value as follows:

- Sell - 0
- Hold - 1
- Buy - 2

As it is stock recommendation we simply assumed that values are Sell$<$ Hold$<$ Buy. We do not choose one hot encoding because then our decision classes would be independent.

3 Solution

3.1 Gradient Tree Boosting

Our approach to this problem was to build extreme gradient boosting trees model with manually engineered features. The solution was implemented in python using pandas, numpy and XGBoost packages. Gradient Tree Boosting is a tree ensemble model. Extended description of extreme gradient boosting trees can be found here [2]. According to this article, we briefly explain the model as follows.

We can denote a set of n examples and m features as $D = (x_i, y_i)(|D| = n, x \in R^m, y \in R)$. The predicted value will be:

$$\hat{y}_i = \phi(x_i) = \sum_{k=1}^{K} f_k(x_i),\ f_k \in F, \tag{1}$$

where K is the number of trees and f is a function from space F of regression trees. In order to optimise the parameters of functions we need to minimise the following objective:

$$Obj = \sum_{i=1}^{n} l(y_i, \hat{y}_i) + \sum_{k=1}^{K} \Omega(f_k), \tag{2}$$

where

$$\Omega(f) = \gamma T + \lambda ||\omega||^2 \tag{3}$$

and $l(y_i, \hat{y}_i)$ is a loss function measuring the difference between the prediction$\hat{y}_i$ and the target $y.T$ is a number of leaves in tree and w is a score sum of corresponding leaves.

Optimal objective function:

$$\tilde{L}^{(t)}(g) = -\frac{1}{2} \sum_{j=1}^{T} \frac{\left(\sum_{i \in I_j} g_i\right)^2}{\sum_{i \in I_j} h_i + \lambda} + \gamma T, \tag{4}$$

where $g_i = \partial_{\hat{y}^{(t-1)}} l(y_i, \hat{y}^{(t-1)})$ and $h_i = \partial^2_{\hat{y}^{(t-1)}} l(y_i, \hat{y}^{(t-1)})$.

As it is impossible to enumerate all possible tree structures, a greedy algorithm is used instead of this. The following loss reduction formula is used to evaluate split candidates:

$$L_{split} = \frac{1}{2} \left[\frac{\left(\sum_{i \in I_L} g_i\right)^2}{\sum_{i \in I_L} h_i + \lambda} + \frac{\left(\sum_{i \in I_R} g_i\right)^2}{\sum_{i \in I_R} h_i + \lambda} - \frac{\left(\sum_{i \in I} g_i\right)^2}{\sum_{i \in I} h_i + \lambda} \right] - \gamma, \tag{5}$$

where I_L is left tree node I_R is right tree node and $I = I_L \cup I_R$.

Table 2 Features selected for model

Feature	Description
Weighted sum of buy recommendations	$\sum_{i=1}^{n} \frac{1}{d_i} x$, where d is a number of a day before decision when recommendation was given and $x \in \{0, 1\}$ where 1 means recommendation has buy value
Weighted sum of hold recommendation	$\sum_{i=0}^{n} \frac{1}{d_i}$, where d is number of day before decision and $x \in \{0, 1\}$ where 1 means recommendation has hold value
Double exponential smoothing	Predicted value using double exponential smoothing: for $d = 1$: $s_1 = x_1, b_1 = x_1 - x_0$, and for $d > 2$: $s_d = \alpha x_d + (1 - \alpha)(s_{d-1} + b_{d-1})$, $b_d = \beta(s_d - s_{d-1}) + (1 - \beta)b_{d-1}$, $F_{d+1} = s_d + b_d$ where $\alpha = 0.9, \beta = 0.9$ s are parameters b_d is a trend at a day d and s_d is a level at day d, F_{d+1} is the smoothed trend value for day $d + 1$
Last day and expert accuracy	Intersection of mean last day recommendation and expert accuracy of these recommendations

3.2 Feature Engineering

We have tried to create various features however most of them has little impact on the final score. Features used in final solution are presented in Table 2. The rest of engineered features which was excluded from final solution is presented in Table 3.

3.3 Feature Selection

Following evaluation of features was made to select proper ones to built model: evaluating model on a training set, feature importance and feature correlation. Additionally, features were evaluated based on a score on a leaderboard, although this score cannot be trusted because it contains just part of test examples and was collected in a different period of time than training set.

Based on cross-validation we were able to reject the weakest features which do not improve prediction or make it even worse. It was also checked with leaderboard score as it was easy to overfit the training data.

XGBoost enables to compute and plot feature importance. If feature gain low score it was removed from a model. Importance of all computed features is presented on Fig. 1. F-score is a metric which sums how many times each feature is split on. A

Table 3 Rest of evaluated features (excluded from final solution)

Feature	Description
Value	Value of stock change from experts recommendations
Weighted sum of sell recommendations	$\sum_{i=0}^{n} \frac{1}{d_i} x_i$, where d is number of day before decision and $x \in \{0, 1\}$ where 1 means recommendation has sell value
Number of recommendations	Number of all recommendations based on all recommendations of experts
Number of sell recommendations	Number of sell recommendations based on all recommendations of experts
Number of hold recommendations	Number of hold recommendations based on all recommendations of experts
Number of buy recommendations	Number of buy recommendations based on all recommendations of experts
Mean last 3 days recommendations	Mean of recommendations from last 3 days before decision counted on numeric values
Mean last week recommendations	Mean of recommendations from last week before decision counted on numeric values
Mean second week before recommendations	Mean of recommendations from second week before decision counted on numeric values
Last recommendation	Mean of last day recommendations counted on numeric values
Recommendation of company with the best accuracy	Last recommendation of highest accuracy company
Recommendation of expert with the best accuracy	Last recommendation of highest accuracy expert
Exponential smoothing	for $d = 0$: $s_0 = x_0, b_1 = x_1 - x_0$, and for $d > 0$: $s_d = \alpha x_d + (1 - \alpha) s_{d-1}$, where d means number of day
Mean accuracy of companies from last day	Accuracy of all companies that their experts made a recommendation last day
Mean accuracy of experts from last day	Accuracy of all experts that made a recommendation last day
Best company recommendation of buy	If a company with the highest accuracy made buy recommendation
Best company recommendation of hold	If a company with the highest accuracy made hold recommendation
Best company recommendation of sell	If a company with the highest accuracy made sell recommendation
Best expert recommendation of buy	If an expert with the highest accuracy made buy recommendation
Best expert recommendation of hold	If an expert with the highest accuracy made hold recommendation
Best expert recommendation of sell	If an expert with the highest accuracy made sell recommendation

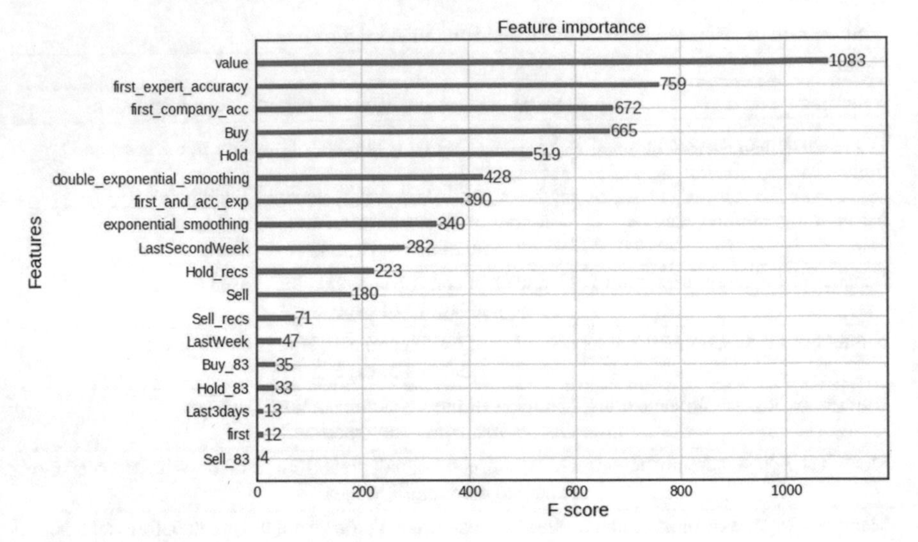

Fig. 1 Importance of all features

detailed description is provided in [3] For K-class problem it is calculated as follows. For each class $k = 1, 2, \ldots, K$ we denote sum of M trees as

$$f_k(x) = \sum_{m=1}^{M} T_{km}(x), \tag{6}$$

where T_{km} is a single tree. Then importance measure is

$$I_{lk}^2 = \frac{1}{M} \sum_{m=1}^{M} I_l^2(T_{km}). \tag{7}$$

I_l relevance measure of each predictor variable X_l, denoted as

$$I_l^2(T) = \sum_{t=1}^{J-1} i_t^2 I(v(t) = l) \tag{8}$$

where (t) is a node and the sum is over J-1 internal nodes of the tree.

XGBoost is not sensitive to correlated features same as other tree-based algorithms but it is unnecessary to keep highly correlated features as they do not provide new information. Because of that, we plotted correlations of features to select these with high correlation. We were able to do that as every feature was converted to numeric values as explained before. Figure 2 presents correlations in data from training set and Fig. 3. Both are quite similar. The highest correlation is between features describing

Fig. 2 Correlation of all features on training set

recommendations from last days and weeks before a decision and also between numbers of recommendations.

3.4 Parameter Tuning, Training and Evaluation

The performance of the model was checked with different values changed manually. Parameters chosen for final model are presented in Table 4 The rest of parameters was set to default.

The model was trained and evaluated using 10-fold cross-validation. Mean error was used as a metric in cross-validation. We achieved following results based on cross-validation

- train mean error: 0.4263
- test mean error: 0.5490

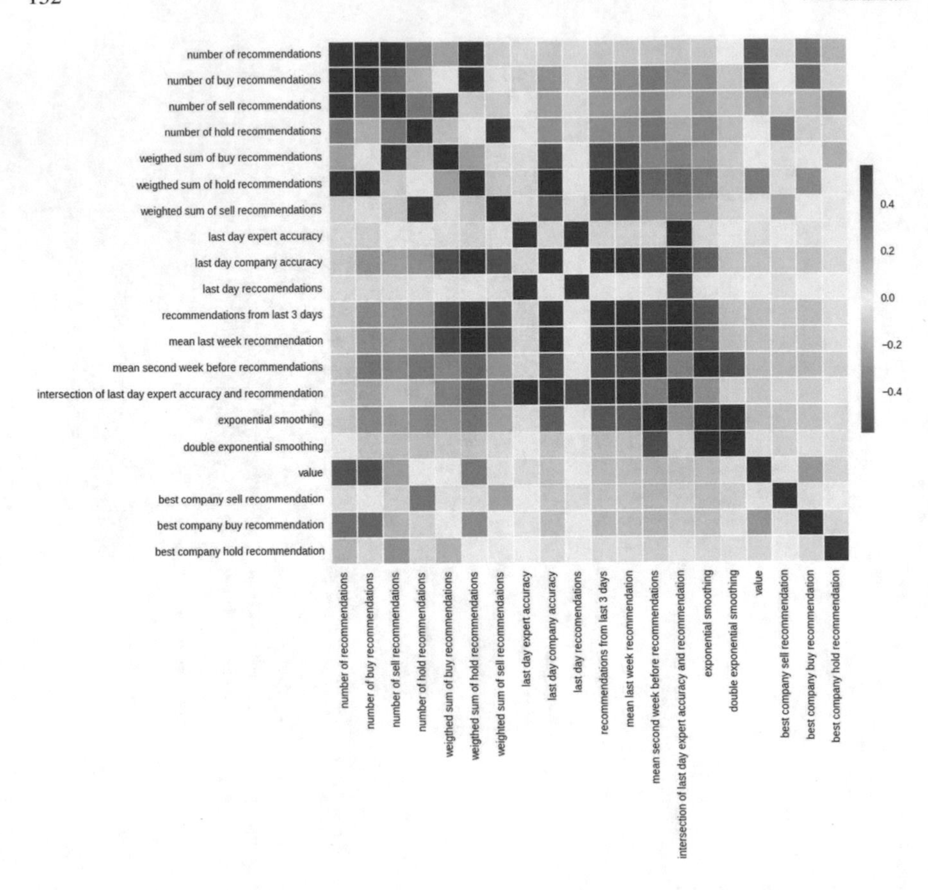

Fig. 3 Correlation of all features on test set

Table 4 Parameters used in final model

Parameter	Value
Number of boosting iterations	50
Maximum tree depth for base learners	6
Boosting learning rate	0.2
Objective function	multi:softmax
Class number	3

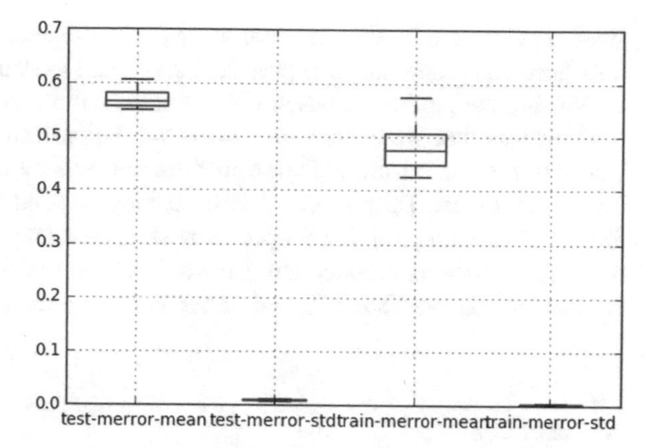

Fig. 4 Training and test data sets error

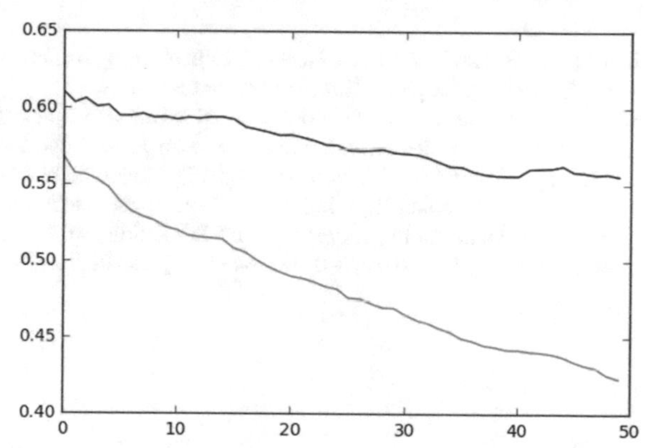

Fig. 5 Model evaluation on split data set: 20% test (blue line) and 80% training set (green line)

We were able to achieve lower error but we had to be careful about overfitting. Presented results give better score on the leaderboard which was also taken into consideration during training. Figure 4 presents test and training errors. In both standard deviation is low, close to zero. The mean error is higher for the test set and slightly lower for the training set.

Additionally, we evaluated model simply splitting training data o to data set: 80% of the data was in training data set and 20% of the data was in the test set. Then we plot train versus test error presented on Fig. 5. Training error decreased quickly but test error decreases slower so that means we have still high variance.

4 Summary and Future Improvements

Presented method gain 0.4451 accuracy on the leader board, and 0.39741139 accuracy on the final board. The final result was lower than baseline solution of 0.43958408 accuracy which could be achieved by setting 'Sell' label on every record.

Because classes of decisions were not equally frequent it would be appropriate to introduce other measures like precision and recall to evaluate the model.

Making the prediction based on recommendations is a difficult task. The main difficult was that we cannot trust cross-validation results on training set because test and training set are gathered in different periods of time and it was easy to overfit the model. Despite this XGBoost may be used to make stock predictions. Period of sampling training set probably should be longer to catch dependencies in the recommendations. Extended data set from the various period of time would be helpful to make predictions based on recommendations.

References

1. Dey, S., Kumar, Y., Saha, S., Basak, S.: Forecasting to classification: predicting the direction of stock market price using Xtreme Gradient Boosting
2. Chen, T., Guestrin, C.: XGBoost: a scalable tree boosting system. In: Proceedings of the 22nd ACM SIGKDD International Conference on Knowledge Discovery and Data Mining (KDD '16), pp. 785–794. ACM, New York (2016). https://doi.org/10.1145/2939672.2939785
3. Hastie, T., Tibshirani, R., Friedman, J.: Boosting and additive trees. The Elements of Statistical Learning: Data Mining, Inference, and Prediction, pp. 337–387. Springer, New York (2009). https://doi.org/10.1007/978-0-387-84858-7_10. (ISBN 978-0-387-84858-7)

Fast Discovery of Generalized Sequential Patterns

Marzena Kryszkiewicz and Łukasz Skonieczny

Abstract Knowledge in the form of generalized sequential patterns finds many applications. In this paper, we focus on optimizing GSP, which is a well-known algorithm for discovering such patterns. Our optimization consists in more selective identification of nodes to be visited while traversing a hash tree with candidates for generalized sequential patterns. It is based on the fact that elements of candidate sequences are stored as ordered sets of items. In order to reduce the number of visited nodes in the hash tree, we also propose to use not only parameters *windowSize* and *maxGap* as in original GSP, but also parameter *minGap*. As a result of our optimization, the number of candidates that require final time-consuming verification may be considerably decreased. In the experiments we have carried out, our optimized variant of GSP was several times faster than standard GSP.

Keywords Data mining · Sequential patterns · Generalized sequential patterns GSP

1 Introduction

Discovery of sequential patterns belongs to important tasks of the data mining area. Sequential patterns were introduced in [1] as frequent patterns representing buying behavior of customers over time. A generalized notion of a sequential pattern was introduced in [11]. The latter notion differed from the former one in that it incorporated additional constraints on a *minimal time gap* between consecutive elements of a sequence, *maximal time gap* within which each pair of consecutive elements of the sequence should occur, and *window size* specifying a time period within which all

M. Kryszkiewicz (✉) · Ł. Skonieczny
Institute of Computer Science, Warsaw University of Technology, Nowowiejska 15/19, 00-665 Warsaw, Poland
e-mail: mkr@ii.pw.edu.pl

Ł. Skonieczny
e-mail: lskoniec@ii.pw.edu.pl

R. Bembenik et al. (eds.), *Intelligent Methods and Big Data in Industrial Applications*, Studies in Big Data 40, https://doi.org/10.1007/978-3-319-77604-0_12

purchased items belonging to one element of the sequence should occur (so called *sliding window constraint*).

Sequential patterns as well as their generalized version are useful in many applications. One may search such patterns not only in sales transaction data, but also in event or genetic data sets etc. Generalized sequential patterns with time gaps constraints and sliding window constraint turn out very useful for text analysis [10].

A variety of algorithms have been proposed for discovering sequential patterns or their representatives called closed sequential patterns (please, see e.g. [1, 2, 4, 5, 7, 8, 12–14]). The discovery of generalized sequential patterns is more complex. The GSP algorithm offered in [11] is capable of discovering generalized sequential patterns, as well as simple sequential patterns as a special case of the former ones. GSP can be easily adapted to discovering (generalized) sequential patterns with even additional constraints of different kind, for example constraints on size and duration of a sequence as a whole. The SPIRIT algorithm, which is an adaptation of GSP, allows discovering generalized sequential patterns that meet user-defined regular expression constraints [4]. The problem of pushing various constraints into sequential pattern mining was presented, for example, in [3, 9].

In this paper, we propose an optimization of the GSP algorithm, which consists in more selective identification of nodes to be visited while traversing a hash tree with candidates for generalized sequential patterns. It is based on the fact that elements of candidate sequences are stored as ordered sets of items. In order to reduce the number of visited nodes in the hash tree, we also propose to use not only parameters *windowSize* and *maxGap* as in original GSP, but also parameter *minGap*. As a result of our optimization, the number of candidates that require time-consuming verification may be considerably decreased. We have experimentally verified the usefulness of our approach on a synthetic data set, which was generated by IBM Quest Synthetic Data Generator [6].

The layout of this paper is as follows. In Sect. 2, we recall the notions of a sequential pattern and generalized sequential pattern. Section 3 outlines the GSP algorithm and presents its aspects being subject to optimization in more detail. In Sect. 4, we present our optimization of GSP. In Sect. 5, we report the performance of GSP and our optimized variant of GSP. The obtained results are concluded in Sect. 6.

2 Basic Notions

Let $I = \{i_1, i_2, \ldots, i_m\}$, $I \neq \emptyset$, be a set of distinct literals, called *items*. In the case of a transactional database, an item corresponds to a product purchased by a customer. In general, any set of items is called an *itemset*. In the remainder of the paper, we assume that the items in each itemset X, where $X \subseteq I$, are ordered.

Definition 1 (*a transaction*) A *transaction T* is a triplet <*dsid*, *tid*, *itemset*>, where *dsid* is a data sequence identifier and *tid* is a transaction identifier. □

Table 1 Example data sequence D [11] and 3 example transactions (highlighted) supporting respective elements of sequence S = <(4) (3) (2,4)>

dsid	*tid*	*items*
1	10	1, 2
1	25	4, 6
1	45	3
1	50	1, 2
1	65	3
1	90	2, 4
1	95	6

In the case of a transactional data set, *dsid* may be understood as the identifier of a customer, and *tid* as a time stamp of the sales transaction registered for this customer.

Definition 2 (*a data sequence*) A *data sequence* is a list of *transactions* with the same *dsid* that are ordered by *tid* increasingly. □

Table 1 presents an example data sequence providing information about purchases of customer 1. In the following, $\mathcal{D}$ will denote a data set consisting of data sequences. One may be interested in discovering sequential patterns (or frequent sequences) based on available data sequences.

Definition 3 (*a sequence*) A *sequence* is a list of *elements* that are itemsets. The *size* of the sequence equals the sum of the cardinalities of the elements of the sequence. □

Example 1 Let sequence S = <(1,2) (2,5,6) (4,5) (1)>. Then S consists of 4 elements and its size equals 8. Item 5 occurs in the second and third element of S. □

Definition 4 (*a subsequence, a supersequence*) A sequence S is a *subsequence* of sequence S' if S can be obtained from S' by removing any items from any elements in S'. A sequence S' is a *supersequence* of a sequence S if S is a subsequence of S'. □

Example 2 Let sequence S' = <(1,2) (2,5,6) (4,5) (1)> and sequence S = <(2) (2,6) (5)>. Then S is a subsequence of S' (and S' is a supersequence of S). □

Definition 5 (*support of a sequence*) A *transaction T supports a sequence element elt* if $elt \subseteq T.itemset$. *A data sequence D* supports a *sequence* $S = <(elt_1) \dots (elt_n)>$ if there are n transactions $T_1, \dots, T_n$ in D such that $T_1.tid < \cdots < T_n.tid$ and elt_j is supported by $T_j, j = 1..n$. *A support* of a sequence S is the number of data sequences in $\mathcal{D}$ supporting S. □

Example 3 Let D be a data sequence of customer 1 which is presented in Table 1. D supports sequence S = <(4) (3) (2,4)> (example 3 transactions supporting its elements are highlighted), but does not support sequence S' = <(1,4) (3) (2,4)> (since no transaction in D supports element (1,4) of sequence S'). □

Definition 6 (*a sequential pattern*) A sequence *S* is a *sequential pattern* if its support is not less than a user-specified threshold *minSup*. □

The notion of a *generalized sequential pattern* can be introduced by means of a *generalized support*, which allows a sequence element to be supported by a transaction window of limited size (*windowSize*) rather than only by a single transaction, as well as imposes additional constraints on *tid* values of consecutive elements in a sequence (*minGap* and *maxGap*).

Definition 7 (*a transaction window in a data sequence D*) $(D, t_{\text{start}}, t_{\text{end}})$-*transaction window*, where *D* is a data sequence and $t_{\text{start}} \leq t_{\text{end}}$, is denoted by *TransWindow*$(D, t_{\text{start}}, t_{\text{end}})$ and is defined as the set of transactions in *D* with *tid* values belonging to the interval $[t_{\text{start}}, t_{\text{end}}]$; that is:

$$TransWindow\,(D, t_{\text{start}}, t_{\text{end}}) = \{T \in D | T.tid \in [t_{\text{start}}, t_{\text{end}}]\}.$$

□

Definition 8 (*supporting a sequence element by a transaction window*) *TransWindow*$(D, t_{\text{start}}, t_{\text{end}})$ *supports a sequence element elt* if *elt* is contained in the union of itemsets of transactions in *TransWindow*$(D, t_{\text{start}}, t_{\text{end}})$; that is, if:

$$elt \subseteq \bigcup \{T.itemset | T \in TransWindow\,(D, t_{\text{start}}, t_{\text{end}})\}.$$

□

Definition 9 (*generalized support of a sequence*) Given user-specified *windowSize*, *minGap* and *maxGap, a data sequence D supports a sequence* $S = <(elt_1) \ldots (elt_n)>$ *in a generalized way* if there are *n* transaction windows *TransWindow*$(D, t_{\text{start}_1}, t_{\text{end}_1})$, …, *TransWindow*$(D, t_{\text{start}_n}, t_{\text{end}_n})$ in *D* such that $t_{\text{start}_1} \leq t_{\text{end}_1} < t_{\text{start}_2} \leq t_{\text{end}_2} \ldots < t_{\text{start}_n} \leq t_{\text{end}_n}$ and the following conditions are fulfilled:

- $\forall_{j=1 \ldots n}$ *TransWindow*$(D, t_{\text{start}_j}, t_{\text{end}_j})$ supports elt_j,
- $\forall_{j=1 \ldots n}$ $(t_{\text{end}_j} - t_{\text{start}_j} \leq windowSize)$,
- $\forall_{j=2 \ldots n}$ $(t_{\text{start}_j} - t_{\text{end}_{j-1}} > minGap)$,
- $\forall_{j=2 \ldots n}$ $(t_{\text{end}_j} - t_{\text{start}_{j-1}} \leq maxGap)$.

A generalized support of a sequence *S* is the number of data sequences in $\mathcal{D}$ that support *S* in a generalized way. □

Example 4 Let *D* be a data sequence of customer 1 which is presented in Table 2. We will show that *D* supports sequence *S* = <(1,4) (2,3) (4,6)> in a generalized way with respect to *windowSize* = 20, *minGap* = 19, *maxGap* = 50. Let us consider the following three transaction windows in *D* (highlighted in Table 2):

- TW_1 = *TransWindow*$(D, t_{\text{start}_1}, t_{\text{end}_1})$, where $t_{\text{start}_1} = 10$ and $t_{\text{end}_1} = 25$,
- TW_2 = *TransWindow*$(D, t_{\text{start}_2}, t_{\text{end}_2})$, where $t_{\text{start}_2} = 45$ and $t_{\text{end}_2} = 50$,

Table 2 Example data sequence *D* [11] and 3 example transaction windows (highlighted) supporting respective elements of sequence *S* = <(1,4) (2,3) (4,6)> in a generalized way with respect to *windowSize* = 20, *minGap* = 19, *maxGap* = 50

dsid	*tid*	*items*
1	10	1, 2
1	25	4, 6
1	45	3
1	50	1, 2
1	65	3
1	90	2, 4
1	95	6

- $TW_3 = TransWindow(D,\ t_{start_3},\ t_{end_3})$, where $t_{start_3} = 90$ and $t_{end_3} = 95$.

Clearly, $t_{start_1} \leq t_{end_1} < t_{start_2} \leq t_{end_2} < t_{start_3} \leq t_{end_3}$. The first (second and third) element of sequence *S* is supported by TW_1 (TW_2, and TW_3, respectively). In addition, the start and end times $t_{start_1}, t_{end_1}, t_{start_2}, t_{end_2}, t_{start_3}, t_{end_3}$ of these transaction windows fulfill *windowSize*, *minGap* and *maxGap* constraints as specified in Definition 9. Thus, data sequence *D*, which contains transaction windows TW_1, TW_2 and TW_3, supports sequence *S* in a generalized way with respect to given values of *windowSize*, *minGap* and *maxGap*. □

Observation 1 Let *D* be a data sequence in $\mathcal{D}$ and *S* = <(elt_1) … (elt_n)>.

(a) If *D* supports sequence *S* in a generalized way, then *D* supports each its one-element subsequence <(elt_j)>, $j = 1 \ldots n$, in a generalized way.
(b) If *D* supports sequence *S* in a generalized way, then *D* supports each its two-element subsequence <(elt_j) (elt_{j+1})>, $j = 1 \ldots n - 1$, in a generalized way.
(c) If there is a one-element subsequence <(elt_j)>, $j = 1 \ldots n$, of sequence *S* that is not supported by *D* in a generalized way, then *D* does not support *S* in a generalized way.
(d) If there is a two-element subsequence <(elt_j) (elt_{j+1})>, $j = 1 \ldots n - 1$, of sequence *S* that is not supported by *D* in a generalized way, then *D* does not support *S* in a generalized way.

Definition 10 (*a generalized sequential pattern*) The sequence *S* is a *generalized sequential pattern* if its generalized support is not less than a user-specified threshold *minSup*. □

3 Discovering Generalized Sequential Patterns with GSP

In this section, we first recall a general idea of the GSP algorithm [11], which discovers all generalized sequential patterns from a data set $\mathcal{D}$. Then, in more detail, we describe: (1) creation of a hash tree whose leaves store candidates for generalized sequential patterns and (2) traversing this tree for initial identification of leaves storing candidates that have a chance to be supported in a generalized way by a data sequence from $\mathcal{D}$. The latter task will be subject to our optimization in Sect. 4; namely, we will propose a method for making this identification more efficient and reducing the number of candidates that will have to undergo final verification whether they are really supported in a generalized way by a data sequence.

3.1 General Idea of GSP

The GSP algorithm finds generalized sequential patterns in an iterative way. At first, one-item generalized sequential patterns are determined. Then, within each iteration j, where $j \geq 2$, GSP:

- creates candidate sequences for generalized sequential patterns of size j based on the found generalized sequential patterns of size $j - 1$ (the newly created candidate sequences of size j are stored in leaves of a hash tree),
- calculates generalized supports for the candidate sequences of size j,
- identifies the candidate sequences of size j with generalized supports not less than *minSup* as generalized sequential patterns.

The calculation of generalized supports of candidates of size j consist in determining for each data sequence D in $\mathcal{D}$ a subset of those candidates that are supported in a generalized way with respect to given threshold values of *windowSize*, *minGap* and *maxGap*, and incrementing their support counters. The process of determining which of candidate sequences of size j are supported by a data sequence D in a generalized way is carried out in the following way:

1. the hash tree storing these candidate sequences is traversed in order to reach all leaves that have a chance to contain candidate sequences of size j that are supported by D in a generalized way;
2. each candidate in the reached leaves is finally verified if it is really supported by D in a generalized way (please see [11] for details).

GSP guarantees that the candidate sequences of size j stored in the reached leaves include all candidate sequences of size j that are supported by D in a generalized way. Nevertheless, it does not guarantee that each reached leaf contains at least one candidate sequence of size j supported by D in a generalized way. Since the final verification of candidate sequences is a very time-consuming operation, the efficiency of the algorithm strongly depends on the number of the reached leaves.

The goal of our optimization, which will be presented in Sect. 4, is more selective identification of nodes (in particular, leaves) to be visited while traversing a hash tree with candidates for generalized sequential patterns.

3.2 Creating a Hash Tree for Storing Candidate Sequences

In this subsection, we will describe briefly the way a hash tree for storing candidate sequences is built. To this end, we will first introduce auxiliary notions of a *flattened sequence* and a *hash function for an item.*

Definition 11 (*a flattened sequence*) *Flattened*(S), where S is a sequence, is a multi-list of items obtained by concatenating all elements of S. □

Example 5 Let sequence S = <(1,4) (2,3) (4,6)>. Then, *Flattened*(S) = (1,4,2,3,4,6) and |*Flattened*(S)| = 6; that is, equals the size of S. □

Definition 12 (*a hash function for an item*) A *hash function for an item i* is denoted by $h(i)$ and is defined as $h(i) = i \bmod k$, where k is a given natural number. □

Now, we are ready to present the way a hash tree of candidate sequences is built. It is assumed that the root of the hash tree is at level 1 and the level of each other node N in the tree equals the number of nodes on the path from the root of the tree to node N. It is also assumed that at most some m candidate sequences can be stored in a leaf of the hash tree.[1]

Initially, the root of the hash tree is a leaf node. Hence, first m candidate sequences can be stored in the root. An attempt to store one more candidate sequence in the root causes its split. In such a case, the root is transformed from a leaf into an internal node and becomes a parent of new k child leaf nodes. Each candidate sequence S that originally was stored in the root is placed in the root's child node determined based on the first item, say i, in *Flattened*(S); that is, sequence S is placed in $h(i)$-th child node of the root.

Now, let us consider the case when a candidate sequence S is to be stored in the hash tree whose root, say N_1, is not a leaf. Then, $h(i)$-th child node, say N_2, of N_1, where i is the first item in *Flattened*(S), is determined. If N_2 is a leaf node, but is not completely full, then S is placed in it. On the other hand, if N_2 is a leaf node, but is completely full, then N_2 is transformed into an internal node and all candidate sequences stored in it so far as well as sequence S will be distributed among N_2's new k child leaf nodes; namely, each such a sequence, say S', will be moved to $h(j)$-th child node of N_2, where j is the second item in *Flattened*(S'). Otherwise; that is, if N_2 is an internal node, then, its $h(j)$-th child node, say N_3, where j is the second item in *Flattened*(S), is determined and the described procedure is repeated for N_3, and eventually its descendants, in an analogous way as in the case of node N_2.

Please note that the hash tree may be imbalanced.

---

[1]In our implementation of the hash tree, this restriction is kept for leaves at levels not exceeding |*Flattened*(S)|, but leaves at level |*Flattened*(S)| + 1 are allowed to store more than m candidate sequences.

3.3 Searching Leaves Possibly Containing Candidate Sequences Supported by a Data Sequence in a Generalized Way

In this subsection, we recall the method of traversing the hash tree in the GSP algorithm within an iteration j in order to identify leaves that are likely to contain candidate sequences of a size j that are supported by a given data sequence D in a generalized way. The method guarantees that no leaf with a candidate sequence of size j that is really supported by D in a generalized way is skipped.

The procedure of traversing the hash tree depends on the type of the visited internal node. It is different for the root and other internal nodes. Beneath we consider these two cases:

An internal node *N* is a root. The value of the hash function $h(i)$ is calculated for each item i in data sequence D. Let $HS = \{h(i)|\ T \in D \wedge i \in T.itemset\}$. Then each j-th child node of the root, where $j \in HS$, will be also visited.

An internal node *N* is not a root. Let $V_N = \{(i, t)|$ i is an item such that node N was reached by hashing on i and t is the value of *tid* of a transaction T in D such that $i \in T.itemset\}$. Then, for each (i, t) in V_N, the hash function is applied to all occurrences of items in D whose values of *tid* belong to $[t{-}windowSize, t + max(windowSize, maxGap)]$. The obtained values of the hash function indicate child nodes of node N that will be visited next.

For each path being traversed, the procedure continues until a leaf node is reached.

Example 6 Let $windowSize = 15$, $minGap = 30$, $maxGap = 40$, D be a data sequence from Table 2, N be an internal node and $(4, 25) \in V_N$. Then, the hash function will be applied to all occurrences of items in D whose values of *tid* belong to $[25 - 15, 25 + \max(15, 40)] = [10, 65]$. Hence, the hash function will be applied to all occurrences of items in *TransWindow*(D, 10, 65); that is, to 8 item occurrences:

- the occurrences of items 1 and 2 in the transaction with *tid* = 10,
- the occurrences of items 4 and 6 in the transaction with *tid* = 25,
- the occurrence of item 3 in the transaction with *tid* = 45,
- the occurrences of items 1 and 2 in the transaction with *tid* = 50,
- the occurrence of item 3 in the transaction with *tid* = 65. □

Please note that the GSP method of searching leaves possibly containing candidate sequences that are supported by a data sequence in a generalized way does not make use of the *minGap* parameter. In the next section, we offer an optimization of looking for promising leaves that uses all three GSP parameters: *windowSize*, *maxGap* and *minGap*, and takes into account the fact that elements of candidate sequences are ordered itemsets.

4 Optimizing GSP by Reducing the Number of Visited Nodes in a Hash Tree

The optimization we propose consists in reduction of the number of visited nodes of a hash tree. More precisely, for each internal node which is not a root, we propose a new method of determining the set of items to which hashing should be applied. The new method is more selective than the one offered in [11], which we recalled in Sect. 3.3. Our optimization is based on a relation we observed between a candidate sequence and its location in the hash tree.

Let N_1, ..., N_n be a path of nodes in the hash tree, where N_1 denotes the root of the hash tree and N_n—a leaf. Let S be a candidate sequence stored in leaf N_n. Then, S is reachable after applying hashing to first $n - 1$ items in *Flattened*(S); namely, j-th item in *Flattened*(S) is used for hashing in node N_j, $j = 1,\ldots, n - 1$. The result of hashing in N_j indicates a child node of N_j that should be visited next.

Let us consider an l-th item in *Flattened*(S), say item i, and $(l + 1)$-th item in *Flattened*(S), say item i', that were used for hashing. Clearly, item i directly precedes item i' in *Flattened*(S). In consequence, if i and i' are in the same element of S, then i directly precedes item i' in this element of S; otherwise, i is the last item of an element of S, while i' is the first item of the next element of S. Thus, sequence S has either the form of $<\_\left({}^*i, i'^*\right)\_>$ or $<\_({}^*i)\left({}^*i, i'^*\right)\_>$, where "*" stands for any items or no item, and "_" stands for any elements or no element.

Case $S = <\_({}^*i,i'^*)\_>$. Items i and i' belong to the same element of sequence S. Since items in each sequence element are ordered, then $i' > i$. In addition, a necessary condition for supporting sequence $<\_({}^*i,i'^*)\_>$ by a data sequence D in a generalized way is existence of transactions T and T' in D such that $i \in T.itemset$, $i' \in T'.itemset$ and $T'.tid \in [T.tid - windowSize, T.tid + windowSize]$.

Case $S = <\_({}^*i)\ (i'^*)\_>$. Items i and i' belong to consecutive elements of sequence S. Hence, a necessary condition for supporting $<\_({}^*i)\ (i'^*)\_>$ by a data sequence D in a generalized way is existence of transactions T and T' in D such that $i \in T.itemset$, $i' \in T'.itemset$ and $T'.tid \in (T.tid + minGap, T.tid + maxGap]$.

These observations allow us to formulate Proposition 1.

Proposition 1 *Let $V_N = \{(i, t) \mid i$ is an item such that node N was reached by hashing on i and t is the value of tid of a transaction T in data sequence D such that $i \in T.itemset\}$. Then, in order to find all candidate sequences in a hash tree that are supported by data sequence D in a generalized way, it is sufficient to apply the hash function to:*

1. *each occurrence of any item $i' > i$ in D whose tid belongs to $[t–windowSize, t + windowSize]$, and*
2. *each occurrence of any item i' in D whose tid belongs to $(t + minGap, t + maxGap]$.* □

Example 7 Let $windowSize = 15$, $minGap = 30$, $maxGap = 40$, D be a data sequence from Table 2, N be an internal node and $(4, 25) \in V_N$. In Example 6, we showed that the original GSP method of identifying promising leaves requires applying hashing on 8 occurrences of items in *D*. However, according to our proposed method, it suffices to apply the hash function:

- to all occurrences of items $i' > 4$ in *D* whose values of *tid* belong to [25 – 15, 25 + 15] = [10, 40] (that is, to the occurrence of item 6 in the transaction with *tid* = 25),
- to all occurrences of items in *D* whose values of *tid* belong to (25 + 30, 25 + 40] = (55, 65] (that is, to the occurrence of item 3 in the transaction with *tid* = 65).

Thus, our method requires applying the hash function to 2 occurrences of items in *D*, which is 4 times less than in the case of the original GSP method. □

Please note that for *minGap* ≥ *maxGap*, the second interval in Proposition 1 becomes an empty set, so Proposition 1 reduces to Proposition 2.

Proposition 2 *Let minGap ≥ maxGap and* V_N *= {(i, t)| i is an item such that node N was reached by hashing on i and t is the value of tid of a transaction T in data sequence D such that i ∈ T.itemset}. Then:*

(a) in order to find all candidate sequences in a hash tree that are supported by data sequence D in a generalized way, it is sufficient to apply the hash function to each occurrence of any item i′ > i in D whose tid belongs to [t–windowSize, t + windowSize],
(b) each generalized sequential pattern contains no more than one element. □

If, nevertheless, *minGap* < *maxGap*, then intervals [*t* – *windowSize*, *t* + *windowSize*] and (*t* + *minGap*, *t* + *maxGap*], which occur in Proposition 1, are not guaranteed to be disjoint. Actually, if *windowSize* ≤ *minGap* < *maxGap*, then the two intervals are disjoint, indeed (as shown in Example 7). Otherwise, if *minGap* < *windowSize* ≤ *maxGap* or *minGap* < *maxGap* ≤ *windowSize*, they are not disjoint. Thus, incautious application of Proposition 1 in such cases could result in repeated hashing on occurrences of items from transactions indicated by common parts of both *tid* intervals. Our Proposition 3 eliminates overlapping of the *tid* intervals, but still guarantees that the hash function will be applied to the same occurrences of items as in the case of Proposition 1.

Proposition 3 *Let minGap < maxGap and* V_N *= {(i, t)| i is an item such that node N was reached by hashing on i and t is the value of tid of a transaction T in data sequence D such that i ∈ T.itemset}. Then, in order to find all candidate sequences in a hash tree that are supported by data sequence D in a generalized way, it is sufficient to apply the hash function to:*

1. *each occurrence of any item i′ > i in D whose tid belongs to [t–windowSize, t + min(windowSize, minGap)] ∪ (t + min(maxGap, windowSize), t + windowSize], and*
2. *each occurrence of any item i′ in D whose tid belongs to (t + minGap, t + maxGap].* □

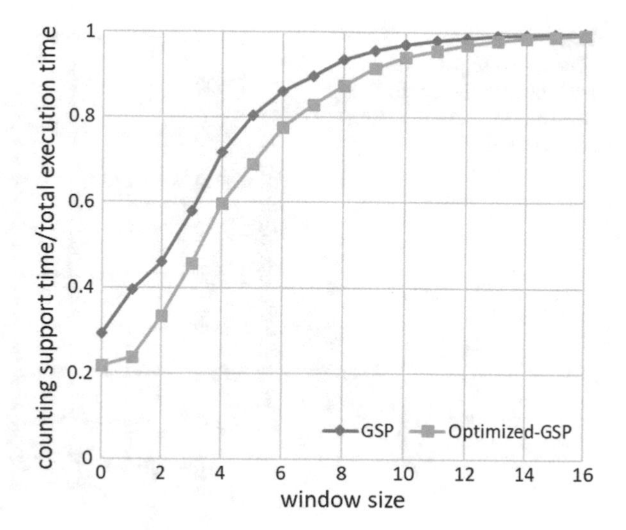

Fig. 1 The ratio of the counting support time to the total execution time with respect to *windowSize*; *minSup* = 200, *minGap* = 30, *maxGap* = 45

Please note that:

- for $windowSize \leq minGap < maxGap$, the first interval in Proposition 3 shortens to the original first interval from Proposition 1 $[t\text{–}windowSize, t + windowSize]$,
- for $minGap < windowSize \leq maxGap$, it shortens to $[t\text{–}windowSize, t + minGap]$,
- for $minGap < maxGap \leq windowSize$, it shortens to $[t\text{–}windowSize, t + minGap] \cup (t + maxGap, t + windowSize]$.

As follows from Propositions 1–3, our method of determining items for hashing uses *minGap* in addition to *windowSize* and *maxGap*, and is more selective than the method applied originally in GSP. Thus, the application of our method in the GSP algorithm may result in decreasing the number of visited nodes in a hash tree and, in particular, in decreasing the number of visited leaves, whose candidate sequences have to undergo final verification, which is a time-consuming operation. The usefulness of applying our optimization was examined experimentally. The obtained results are presented in the next section.

5 Experimental Results

To examine practical relevance of our proposal, we have performed an experimental evaluation of the GSP algorithm and our optimization of GSP, which we will call Optimized-GSP. Both algorithms were implemented by us. They differ only in the counting support function in the part responsible for identifying leaves in the hash tree that have a chance to contain candidate sequences supported by a given data sequence in a generalized way. In the remainder of the paper, the term *counting support time* embraces both the search of leaves with promising candidate sequences in a hash tree, and the final verification of such candidates.

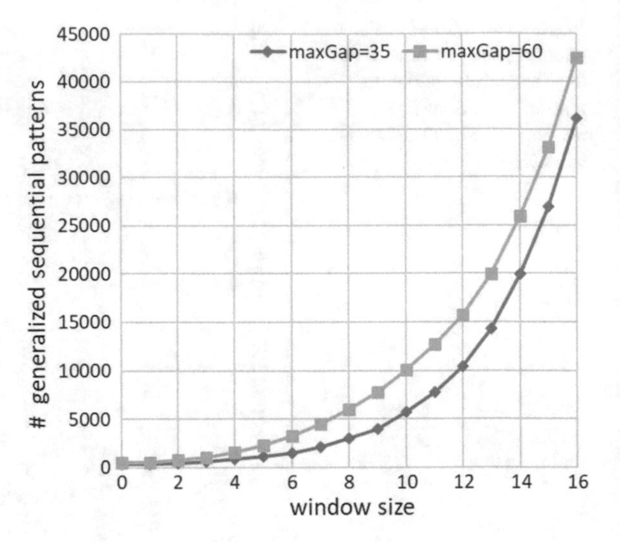

Fig. 2 Number of generalized sequential patterns with respect to *windowSize*; *minSup* = 200, *minGap* = 30

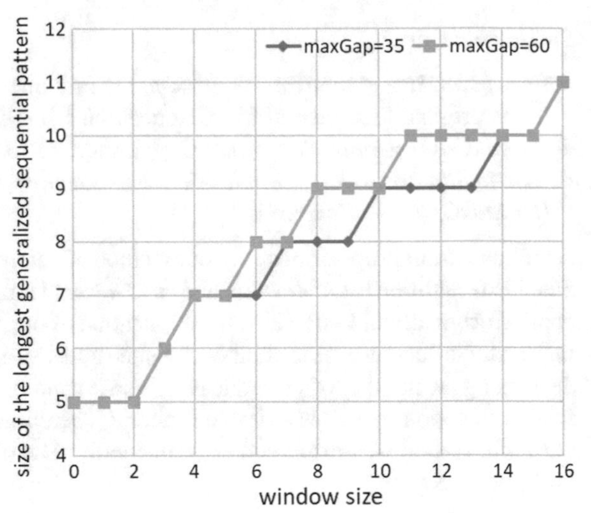

Fig. 3 Size of the longest generalized sequential pattern with respect to *windowSize*; *minSup* = 200, *minGap* = 30

We used a synthetic data set generated by IBM Quest Synthetic Data Generator [6]. The created data set contained 1000 customer sequences, with 50 transactions per customer on average, and 4 items per transaction on average. The total number of different items was set to 1000.

Figure 1 shows the ratio of the counting support time to the total execution time for GSP and Optimized-GSP. The results show that the phase of counting support is indeed a very time consuming phase of the GSP algorithm, so optimizing it makes sense. It can be noticed that the mentioned ratio comes asymptotically to 1 when *windowSize* increases. In fact, it can be claimed that the more patterns in the data sets and the greater size of these patterns, the greater the mentioned ratio. Figures 2

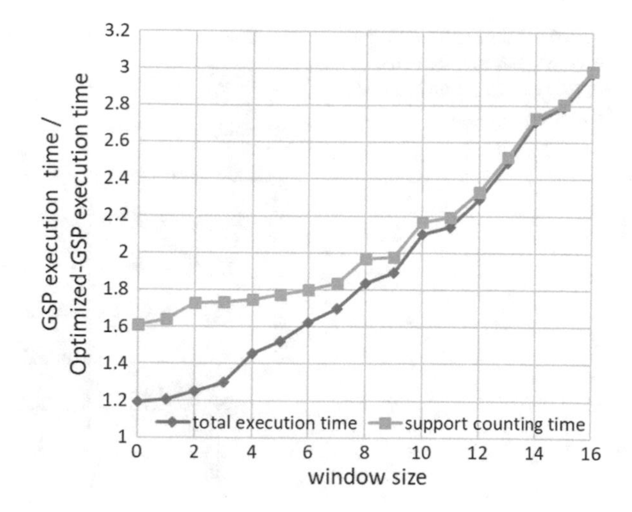

Fig. 4 Optimization performance with respect to *windowSize*; *minSup* = 200, *minGap* = 30, *maxGap* = 45

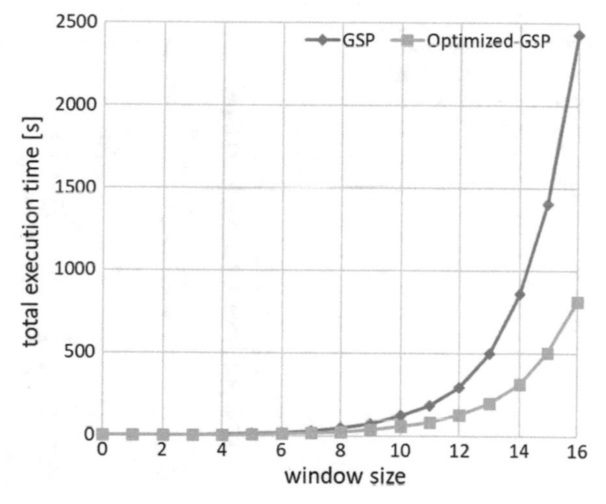

Fig. 5 Total execution time with respect to *windowSize*; *minSup* = 200, *minGap* = 30, *maxGap* = 45

and 3 show that an increase in value of *windowSize* results in an increase in both the total number of generalized sequential patterns and their size.

The Optimized-GSP algorithm turns out 2–4 times faster than GSP for reasonable values of parameters in the carried out experiments. Figure 4 suggests that the advantage from the optimization grows in an unlimited way with an increase in value of *windowSize*, while the total execution time (Fig. 5) grows exponentially due to exponential explosion of generalized sequential patterns (Fig. 2).

The remaining figures show the influence of other GSP parameters. An increase in value of *maxGap* (Figs. 6 and 7) results in an increase in the number of patterns and time of their discovery until the point when *maxGap* exceeds maximal intervals between transaction identifiers of data sequences in the data set. According to Figs. 6 and 7, Optimized-GSP is several times faster than GSP. It can be seen there that the

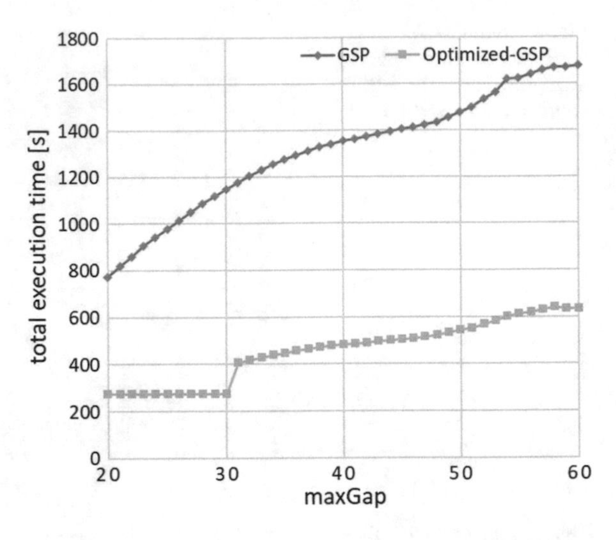

Fig. 6 Total execution time with respect to *maxGap*; *minSup* = 200, *minGap* = 30, *windowSize* = 15

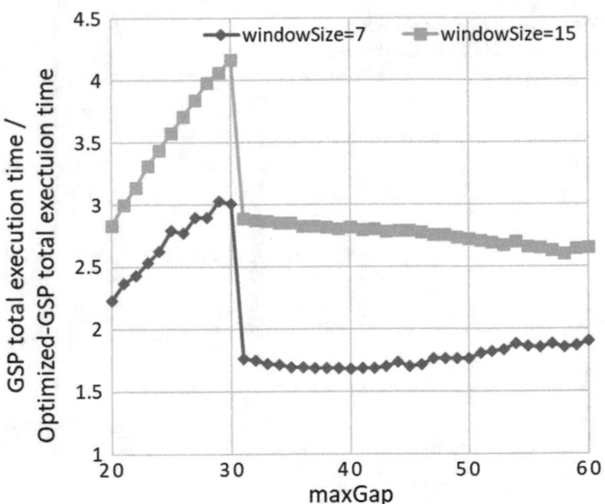

Fig. 7 Optimization performance with respect to *maxGap*; *minSup* = 200, *minGap* = 30

improvement in the speedup obtained by Optimized-GSP is particularly high in the case when *minGap* >= *maxGap*, which is the effect of applying Proposition 2. Figure 8 shows that an increase in value of *minGap*, which is the reason of a decrease in the number of generalized sequential patterns, results in a decrease in the execution time. Finally, Fig. 9 shows that Optimized-GSP visits up to 4 times less hash tree nodes than GSP, which confirms that the proposed optimization reduces unnecessary node visits while traversing the hash tree with candidate sequences.

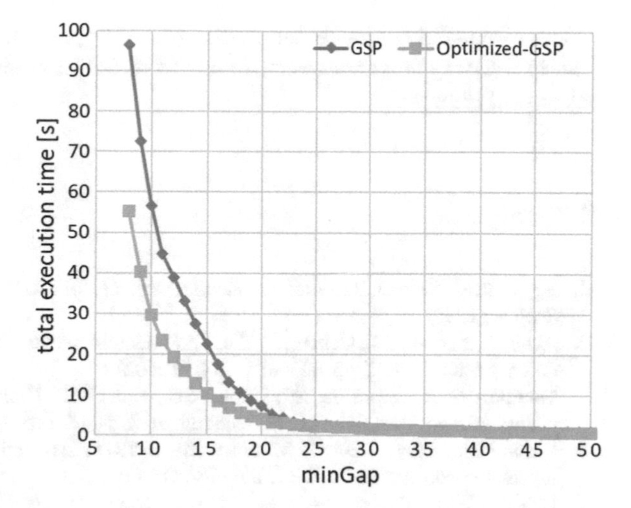

Fig. 8 Total execution time with respect to *minGap*; *minSup* = 200, *maxGap* = 40, *windowSize* = 0

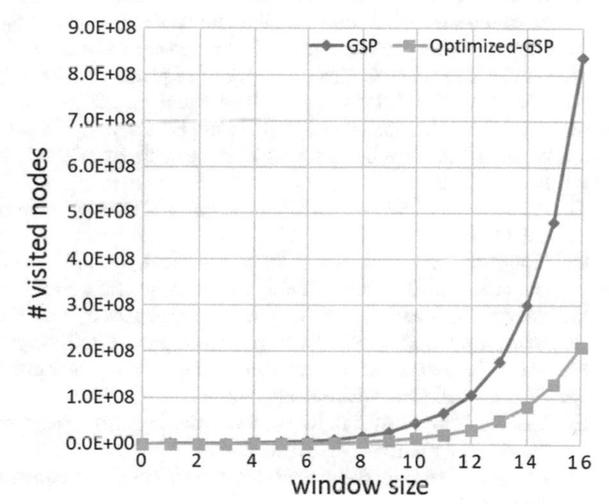

Fig. 9 Number of visited nodes with respect to *windowSize*; *minSup* = 200, *minGap* = 30, *maxGap* = 45

6 Conclusions

In this paper, we have offered an optimization of the GSP algorithm, which reduces unnecessary node visits while traversing the hash tree with candidate sequences. Our optimization is based on the fact that elements of candidate sequences are stored as ordered sets of items. In order to reduce the number of visited nodes in the hash tree, we have also proposed to use not only parameters *windowSize* and *maxGap* as in original GSP, but also parameter *minGap*. As a result of our optimization, the number of candidates that require final time-consuming verification may be considerably decreased. In the experiments we have carried out, our optimized variant of GSP was several times faster than standard GSP.

Our proposed optimization can be easily applied in any GSP-like algorithms (e.g. in SPIRIT) that impose constraints of different kinds on generalized sequential patterns of interest.

References

1. Agrawal, R., Srikant, R.: Mining sequential patterns. In: ICDE 1995, pp. 3–14. IEEE Computer Society (1995)
2. Ayres, J., Flannick, J., Gehrke, J., Yiu, T.: Sequential pattern mining using a bitmap representation. In: KDD 2002, pp. 429–435. ACM (2002)
3. Fournier-Viger, P., Lin, J.C.W., Kiran, R.U., Koh, Y.S., Thomas, R.: A survey of sequential pattern mining. Data Sci. Pattern Recognit. **1**(1), 54–77 (2017)
4. Garofalakis, M.N., Rastogi, R., Shim, K., SPIRIT: sequential pattern mining with regular expression constraints. VLDB J. 223–234 (1999)
5. Han, J., Pei, J., Mortazavi-Asl, B., Chen, Q., Dayal, U., Hsu, M.: FreeSpan: frequent pattern-projected sequential pattern mining. In: KDD 2000, pp. 355–359. ACM (2000)
6. IBM Almaden Quest Research Group, Quest Synthetic Data Generator
7. Lin, M.Y., Lee, S.Y.: Fast discovery of sequential patterns by memory indexing. In: DaWaK 2002. LNCS, vol. 2454, pp. 150–160. Springer (2002)
8. Pei, J., Han, J., Mortazavi-Asl, B., Pinto, H., Chen, Q., Dayal, U., Hsu, M.: PrefixSpan: mining sequential patterns by prefix-projected growth. In: ICDE 2001, pp. 215–224. IEEE Computer Society (2001)
9. Pei, J., Han, J., Wang, W.: Mining sequential patterns with constraints in large databases. In: CIKM 2002, pp. 18–25. ACM (2002)
10. Protaziuk, G., Kryszkiewicz, M., Rybinski, H., Delteil, A.: Discovering compound and proper nouns. In: RSEISP 2007. LNCS, vol. 4585, pp. 505–515. Springer (2007)
11. Srikant, R., Agrawal, R.: Mining sequential patterns: generalizations and performance improvements. In: EDBT 1996. LNCS, vol. 1057, pp. 3–17. Springer (1996)
12. Wang, J., Han, J.: BIDE: efficient mining of frequent closed sequences. In: ICDE 2004, pp. 79–90. IEEE Computer Society (2004)
13. Yan, X., Han, J., Afshar, R.: CloSpan: mining closed sequential patterns in large datasets. In: SDM 2003, pp. 166–177. SIAM (2003)
14. Zaki, M.J.: SPADE: an efficient algorithm for mining frequent sequences. Mach. Learn. **42**(1/2), 31–60 (2001)

Seismic Attributes Similarity in Facies Classification

Marcin Lewandowski and Łukasz Słonka

Abstract Seismic attributes are one of the component of reflection seismology. Formerly the advances in computer technology have led to an increase in number of seismic attributes and thus better geological interpretation. Nowadays, the overwhelming number and variety of seismic attributes make the interpretation less unequivocal and can lead to slow performance. Using the correlation coefficients, similarities and hierarchical grouping the analysis of seismic attributes was carried out on several real datasets. We try to identify key seismic attributes (also the weak ones) that help the most with machine learning seismic attribute analysis and test the selection with Random Forest algorithm. Obtained quantitative factors help with the overall look at the data. Initial tests have shown some regularities in the correlations between seismic attributes. Some attributes are unique and potentially very helpful with information retrieval while others form non-diverse groups. These encouraging results have the potential for transferring the work to practical geological interpretation.

Keywords Seismic attributes · Geophysics · Correlation · Similarity · Machine learning

1 Introduction

1.1 Geoscience Objective

One of the main task of geoscience is the recognition of the Earth's interior. For this purpose, geophysicists use various methods (e.g. seismology, gravity and magnetics methods, electrical sounding etc.) to extract subsurface geological information. One of those methods is reflection seismology [1], usually called seismic reflec-

M. Lewandowski (✉)
Institute of Computer Science, Warsaw University of Technology, Warszawa, Poland
e-mail: M.Lewandowski.3@elka.pw.edu.pl

Ł. Słonka
Institute of Geological Sciences, Polish Academy of Sciences, Warsaw, Poland
e-mail: lukasz.slonka@twarda.pan.pl

R. Bembenik et al. (eds.), *Intelligent Methods and Big Data in Industrial Applications*, Studies in Big Data 40, https://doi.org/10.1007/978-3-319-77604-0_13

tion method, used particularly in the study of sedimentary basins. This involves stimulating Earth's subsurface by a controlled seismic source of energy and then recording seismic signal reflected from different geological boundaries characterized by contrasting physical properties (such as velocity of seismic waves and bulk density). General processing of such signals is well-developed and highly advanced, and involves various denoising, stacking and filtering procedures.

Interpretation of seismic data includes detection and mapping of various geological structures (structural interpretation) and detection and mapping of lateral and vertical changes of sedimentary successions (seismic stratigraphic interpretation, seismic facies analysis etc.).

In order to facilitate interpretation procedure, seismic attribute has been proposed [2]. Generally, seismic attribute can be described as a quantity extracted from seismic data that can enhance information retrieval from a traditional seismic image.

But the vast number and variety of seismic attributes [3] is almost overwhelming and there are many distinct categorization styles.[1] However, sometimes, attributes contain nearly the same information. It has been shown in [4] that amplitude based attributes analysis reveals fairly linear or quadratic relationships between them.

1.2 Machine Learning

Interpretation is one of the stages that machine learning can aid human work. Seismic attribute value can by natural feature for particular observation (physical coordinates with multiple attributes assigned to it). There are expert systems that help interpreting data, see e.g. [5]. One can find approaches based on classification or clustering solutions, both semi-supervised [6] or unsupervised [7]. Recently, a new feature selection algorithm (called *Boruta*) has been designed [8, 9].

The main goal of this paper is to identify the most relevant seismic attributes. Proper selection of the attributes among a large number of them can drastically improve seismic stratigraphic interpretation.

2 Materials and Methods

2.1 Seismic Data

In the study seven seismic profiles from two distinct areas were used. The first area is located between Kłodawa salt diapir and Ponętów–Wartkowice salt structure in central Poland [10], another one is the so-called Nida Trough, located in South Poland.

---

[1] Attributes revisited http://www.rocksolidimages.com/attributes-revisited.

2.2 Calculation of Seismic Attributes

We have used the Paradigm ® 15.0[2] software to organize and extract attributes from raw time migrated sections. The following attributes have been extracted:

- signal envelope - commonly known in signal-processing; the envelope of a signal can be described as boundary within which the signal is contained, when viewed in the time domain;
- thin-bed indicator - geophysics specific; can extract a locations where instantaneous frequencies jump or go in the reverse direction; defined as the absolute value of the instantaneous frequency minus weighted average instantaneous frequency;
- instantaneous frequency - the time derivative of the instantaneous phase.

No cutoff filters were used while extracting the data. For the frequency based attributes the range of 10–60 Hz was used instead of default 10–50 Hz. This decision was dictated by non negligible signal strength at 50–60 Hz. The rest parameters for attribute extraction were set to default. The number of extracted attributes was 74. Adding three plain ones (time migrated section, depth and offset) gave the total number of 77 potential features. The data was exported via standard SEG-Y format[3] and imported to dedicated software for analysis and interpretation. Some linear normalization and other minor attribute-specific processing was applied to the data.

2.3 Applied Methods

In the paper the following common correlation coefficients were used:

- Pearson product-moment,
- Spearman's rank.

In order to compare distance between two different attributes (see Sect. 3.3) we had to change correlation coefficient to dissimilarity value using the following formula:

$$dissimilarity = 1.0 - abs(correlationCoefficient).$$

The formula ignores a sign of correlation value (negative correlation coefficient is still correlation). This way dissimilarities values are within the range [0, 1]. These values ware input for hierarchical grouping. Single-link algorithm [11] was used for grouping. In this method, the distance between two groups is the minimum of the distances between all pairs of attribute taken from the two clusters.

---

[2]Paradigm - E&P Subsurface Software Solutions, http://www.pdgm.com/.

[3]SEG Y rev1 Data Exchange Format, http://www.seg.org/documents/10161/77915/seg_y_rev1.pdf.

2.4 Classification Problem

For experimental comparison of the attribute selection impact, the classification problem was specified. Each coordinate of particular seismic trace and time position was considered as an observation. Each observation consisted of calculated attribute value and was treated as a vector.

Taking the particular seismic profile, the vertical fragment of data was used as training set. The fragment had width of 7 seismic traces and simulated well data driven approach. Classes were labeled by time/depth of particular 7 observations and different facies combinations were taken into consideration. The seismic facies analysis was focused mainly on the Upper Jurassic interval. All seismic horizons for interval identification were interpreted by the expert.

The Random Forest algorithm was utilized because it is considered as a good first choice for classifying lithology [12]. It is resilient method and it computes naturally attribute importances. The following parameters values were used:

- number of estimators: 300;
- function to measure the quality of a split: *gini* index;
- number of features to consider when looking for the best split: 9 (square root of attributes count);
- minimum number of samples required to split an internal node: 10;
- minimum number of samples required to be at a leaf node: 5;
- maximum depth of the tree: unlimited;
- no maximum number of leaf nodes;
- samples drawn with replacement;
- node impurity threshold: 1e-7.

For the time being, we did not have an objective method for the algorithm results verification. However, for initial experimental attribute selection comparison it was not necessary. For each possible attribute, we trained the Random Forest without this single attribute and compared the classification results with full attribute version of the Random Forest performance (averaged over 100 independent runs).

For comparing results we used Euclidean distance applied to each observation prediction. The sum of differences was obtained for each independent run.

The number of estimators was determined experimentally. Lower values (below 150) gave highly unstable classification results. Except the number of estimators, no profound algorithm parameters tuning was applied. For each problem, 100 independent runs of the algorithm were performed.

3 Results and Discussion

3.1 Distribution

Figure 1 represents arbitrary chosen attribute values histograms. Charts are representative as they show that there are no significant differences between two chosen areas (Kłodawa and Nida). Of course there are some minor differences in attribute distribution between each of seven profiles.

It should be noted that similar distribution can be the result of similar depth of study data (and frequencies used at signal acquisition). There is possibility that the change of frequency of signal put into the ground and received later can affect significantly actual distribution.

3.2 Correlation

Figure 2 shows a couple of cross plots of attributes pairs. The correlation varies: from high value represented by Fig. 2a to no correlation shown at Fig. 2c. Worth noting is Fig. 2b that indicates a potential problem with outstanding values: in this case RMS Frequency had, quite often, value of zero. This fact should be considered when utilizing machine learning. Such a faulty values could be, for example, treated as missing data and properly imputed.

When calculating correlation coefficients on such big sample sets (average seismic profile consists of about 6 000 000 samples) all calculation methods gave practically the same results. Generally P-value was very low (with some exceptions, when correlation was insignificantly low: below 0.001).

Figure 3 presents correlation for all attributes for Kłodawa 04 profile. For other profiles, correlations were practically the same. Because of large number of attributes, only part (interesting one) of them is presented. One can identify couple of groups that consist of similar (correlated) groups.

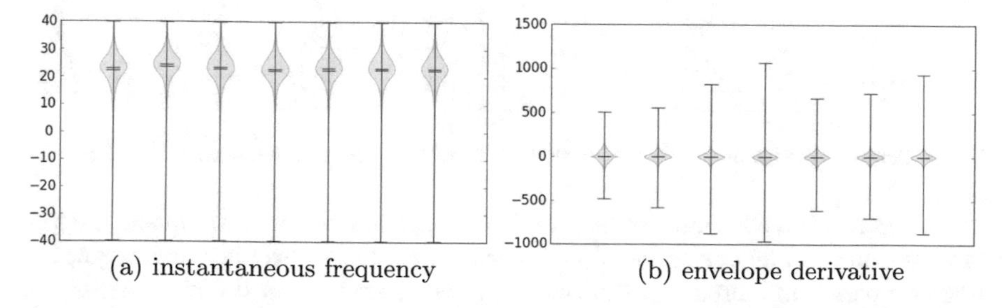

(a) instantaneous frequency (b) envelope derivative

Fig. 1 Representative attribute values distribution. Y axis specifies attribute value while X axis represents selected seismic profiles: Kłodawa03-07, then 5-5-92K(Nida) and 10-5-92K(Nida). There are marked extrema, mean and median on each plot

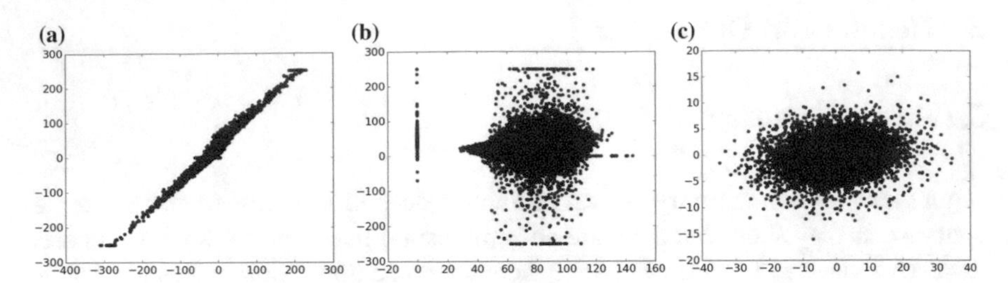

Fig. 2 Common correlation types in the data extracted from Kłodawa 05 profile. **a** Thin Bed Indicator versus Instantaneous Frequency, **b** RMS Frequency versus Instantaneous Frequency, **c** Relative Acoustic Impedance versus Time Migrated Section

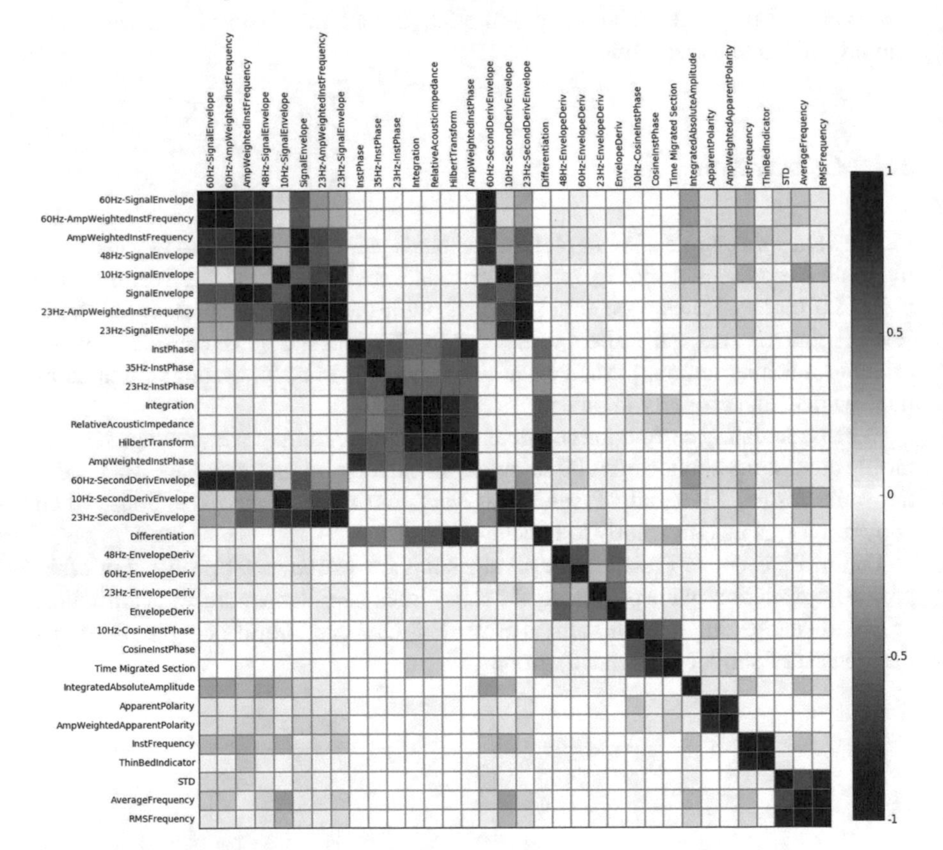

Fig. 3 Pearson correlation matrix for selected attributes obtained from Kłodawa 04 profile

First group was composed of *amplitude weighted instantaneous frequency*, *signal envelope* and *second derivative signal envelope*. All these attributes had frequency-bounded parts. Unfiltered *signal envelope* was correlated the most with *35*-Hz *signal envelope*, thus the great part of information is within spectrum near 35-Hz. This fact was also confirmed in time migrated section frequency spectrum (not shown in the paper).

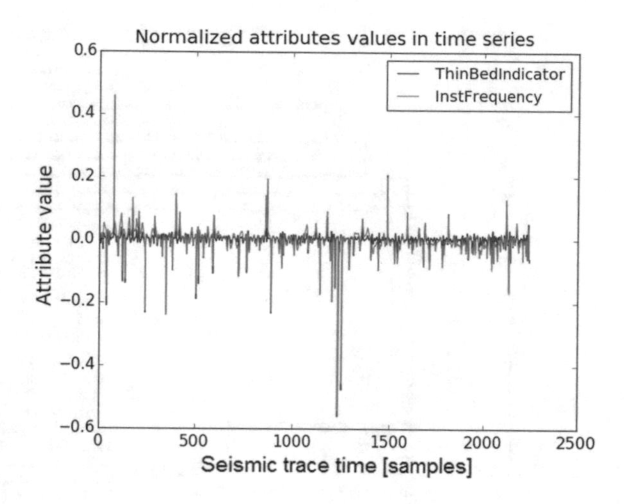

Fig. 4 High correlation on plot for two selected attributes. Plot presents single traces in time domain selected from whole seismic profile Values are normalized in order to show correspondence.

Second group consisted of frequency fragmented *instantaneous phase*, *Hilbert transform*, *relative acoustic impedance* and *integration*. There was no other attribute that was correlated with this group (except negatively correlated *differentiation*).

Next smaller or weaker-linked groups were: *cosine instantaneous phase* with *signal amplitude*, *envelope derivative* and *average frequency*.

Figure 4 presents example of one pair of attribute precise comparison. Only one seismic trace from each attribute is shown. Attributes had high correlation coefficient (0.895659, p-value 0.000000) and it can be observed that they generally overlap. Peaks for both series had the same position and amplitude.

3.3 Grouping

We identified a couple of groups within attributes basing on dissimilarity measure. Figure 5 presents results of hierarchical grouping in a form of dendrogram. It reflects a portion of the information included at Fig. 3. Groups can be easily perceived, but with less information within each of the group.

3.4 Classification Performance

For experimental comparison of the attribute selection impact, we performed some tests. For interpreted seismic interval in selected profile the Random Forest classifier was utilized. Figure 6 shows the correspondence between classifier results and attribute selection. *Maximum correlation value* is the maximum (absolute) value of correlation coefficient for chosen attribute and each from the other attributes set. The lower this value is, the more unique attribute values are.

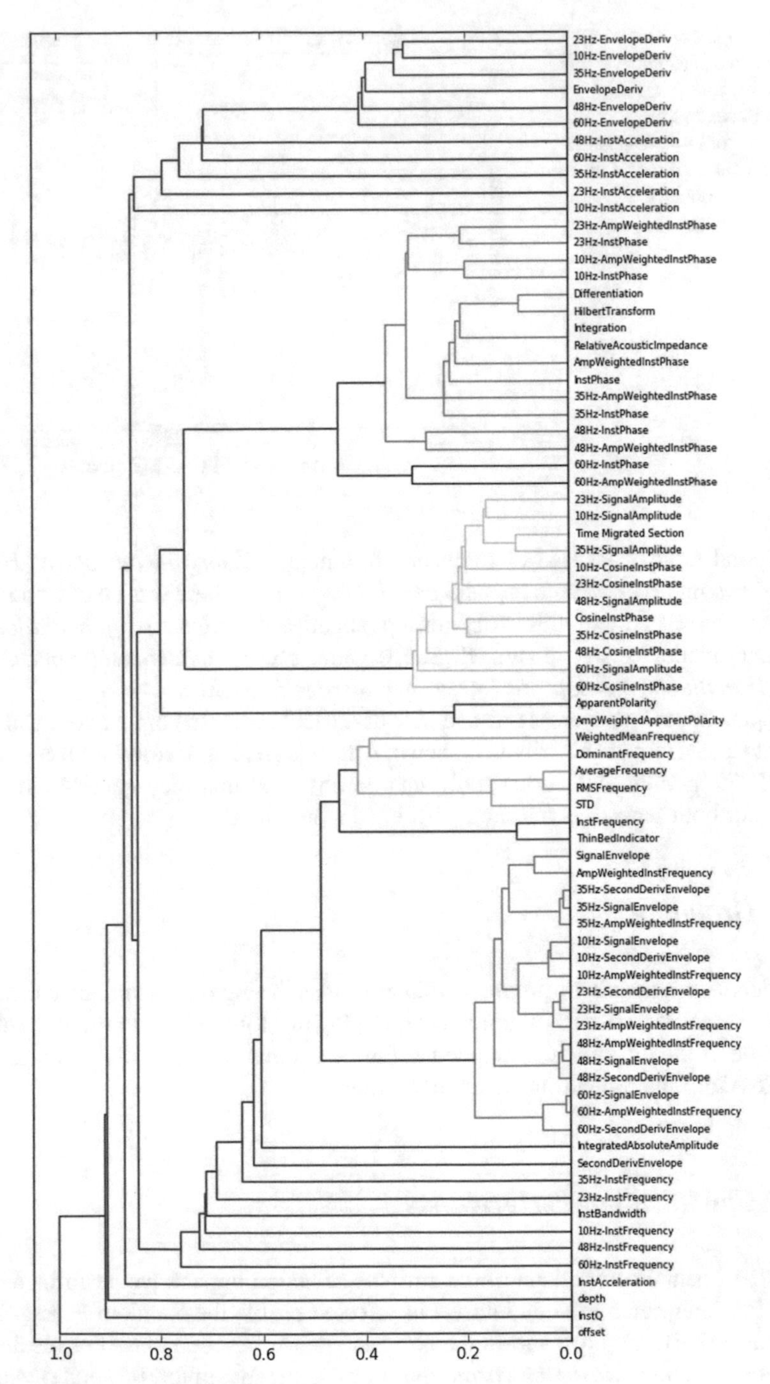

Fig. 5 Dendrogram representing attributes grouping for Kłodawa05 profile. The differences for other profiles are negligible. Coloring threshold was arbitrary chosen to value of 0.38

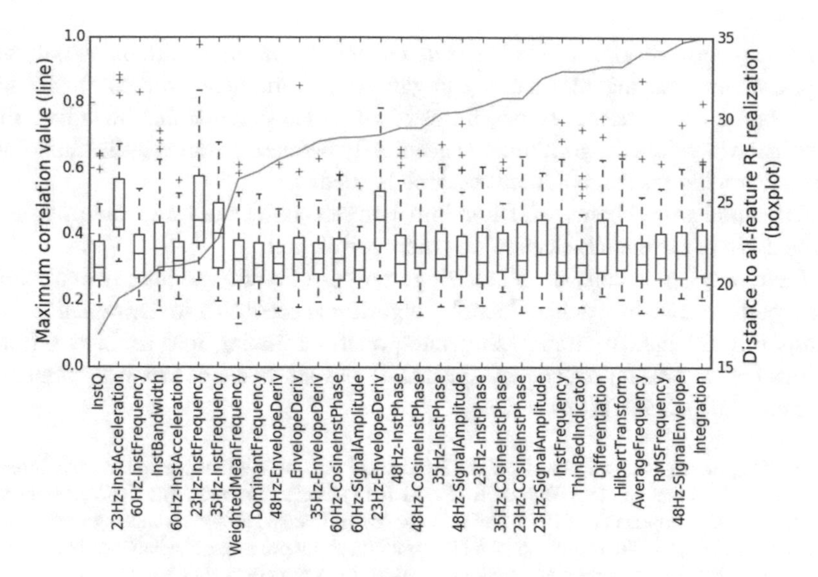

Fig. 6 Comparison of selection attributes impact on Random Forest performance. Each boxplot represents difference between full-attribute Random Forest realizations for 100 runs of algorithm. Values represented by line correspond to each attribute maximum correlation with all other attributes

The lower *distance to all-feature RF realization* is, the less difference for standard results is. Higher values inform that lack of the attribute gives different classification results. It can be interpreted as how much impact particular attribute has on overall model.

We observed no solid relationship. Some of attributes (i.e. 23 Hz Instantaneous Acceleration) appeared to have higher impact on overall results, but there was no general tendency when taking into account correlation between attributes.

Explanation for such results can be the fact that Random Forest gave various results in independent runs for taking quite small training set. Probably considering more subsets of attributes sets can lead to further results. Anyway, more research is needed in order to make clear if one can make use of the correlation fact.

4 Conclusions

Some problems within area were identified, but not all of them were thoroughly investigated. Overwhelming number of various numeric attributes can make the interpretation not easy and can lead to slow performance.

Taking into account different regions (as considered in the paper), the distribution of seismic attributes was similar, thus some steps of machine learning can potentially be region independent.

Some unique attributes (like *integrated absolute amplitude*) are potentially very helpful when retrieving information using machine learning algorithms. On the other hand, strong correlation of some groups of attributes suggests that including all of them in the mentioned algorithms can needlessly increase resource utilization. These are encouraging results, but more research is needed.

The initial experiments with Random Forest classifier did not show particular relationship of attributes selection and their correlation.

Future work plan consists of studying how selection of attributes actually influences performance of machine learning algorithms according to their quality. Final results can aid human stratigraphic interpretation. Taking into account different regions (other working frequency spectrum) and attributes common in other software could be advantageous.

Acknowledgements Seismic data from the Kłodawa-Ponetów–Wartkowice area was acquired for the BlueGas-JuraShale project (BG1/JURASHALE/13) funded by the Polish National Centre for Research and Development (NCBiR). San Leon Energy is thanked providing access to seismic data from the Nida Trough. We would also thank to Paradigm ® for providing academic software license for seismic attributes extraction and to the authors and developers of Python, NumPy, Matplotlib and ObsPy.

References

1. Zbigniew Kasina. Metodyka badan sejsmicznych. Polish. Wydaw. Instytutu GSMiE PAN [Gospodarki Surowcami Mineralnymi i Energia Polskiej Akademii Nauk], 1998. 289 pp. isbn: 83-87854-20-4
2. Chopra, S., Marfurt, K.J.: Seismic attributes - a historical perspective. Geophysics **70**(5), 3SO–28SO (2005). https://doi.org/10.1190/1.2098670
3. Chopra, S., Marfurt, K.J.: Seismic attributes for prospect identification and reservoir characterization. Soc. Explor. Geophys. (2007). https://doi.org/10.1190/1.9781560801900
4. Barnes, A.: Too many seismic attributes? vol. 31 (Landmark Graphics Corporation, Colorado, USA, 2006)
5. Barnes, A.E.: Attributes for automating seismic facies analysis. SEG Tech. Progr. Expand. Abstr 2000. Soc. Explor. Geophys. **1**, 553–556 (2000). https://doi.org/10.1190/1.1816121
6. Qi, J., et al.: Semisupervised multiattribute seismic facies analysis. Interpretation **4**(1), SB91–SB106 (2016). https://doi.org/10.1190/INT-2015-0098.1
7. Coléou, T., Poupon, M., Azbel, Kostia: Unsupervised seismic facies classification: a review and comparison of techniques and implementation. Lead. Edge **22**(10), 942–953 (2003). https://doi.org/10.1190/1.1623635
8. Kursa, M.B., Rudnicki, W.R.: Feature Selection with the- BorutaPackag. J. Stat. Softw. **36**(11) (2010). https://doi.org/10.18637/jss.v036.i11
9. Kursa, M.B., Jankowski, A., Rudnicki, W.R.: Boruta – a system for feature selection. Fundam. Inf. **101**(4), 271–285 (2010). https://doi.org/10.3233/FI-2010-288. ISSN: 0169-2968
10. Krzywiec, P., et al. Control of salt tectonics on mesozoic unconventional petroleum system of the central mid-polish trough. In: AAPG Annual Convention and Exhibition 2015, Denver, USA (2015)
11. Jain, A.K., Murty, M.N., Flynn, P.J.: Data clustering: a review. ACM Comput. Surv. **31**(3), 264–323 (1999). https://doi.org/10.1145/331499.33150

12. Cracknell, Matthew J., Reading, Anya M.: Geological mapping using remote sensing data: A comparison of five machine learning algorithms, their response to variations in the spatial distribution of training data and the use of explicit spatial information. Comput. Geosci. **63**, 22–33 (2014). https://doi.org/10.1016/j.cageo.2013.10.008

Efficient Discovery of Sequential Patterns from Event-Based Spatio-Temporal Data by Applying Microclustering Approach

Piotr S. Maciąg

Abstract Discovering various types of frequent patterns in spatiotemporal data is gaining attention of researchers nowadays. We consider spatiotemporal data represented in the form of events, each associated with location, type and occurrence time. The problem is to discover all significant sequential patterns denoting spatial and temporal relations between event types. In the paper, we adapted a microclustering approach and use it to effectively and efficiently discover sequential patterns and to reduce size of dataset of instances. Appropriate indexing structure has been proposed and notions already defined in the literature have been reformulated. We modify algorithms already defined in literature and propose an algorithm called Micro-ST-Miner for discovering sequential patterns in event-based spatiotemporal data.

1 Introduction

Discovering frequent patterns is among the most important fields of data mining. The problem has been originally proposed for databases containing transactions of purchased products (items). A set of items occurring together with a minimal predefined frequency is defined as a frequent pattern. Discovered frequent patterns are also used in several more complex data mining tasks: association rules mining [1] or associative classification [13]. Also, several types of extensions of frequent patterns have been considered: discovering sequential patterns, time-constrained sequential patterns, etc. [12].

Among the most important algorithms proposed for frequent patterns discovery problem are: Apriori-Gen [1], FP-Growth [5], DECLAT [14], SPADE [15] or GSP [12]. The above algorithms often use additional structures for improving computations efficiency or for patterns storage and representation.

P. S. Maciąg (✉)
Institute of Computer Science, Warsaw University of Technology,
Nowowiejska 15/19, 00-665 Warsaw, Poland
e-mail: pmaciag@ii.pw.edu.pl
URL: http://www.ii.pw.edu.pl/ii_eng

R. Bembenik et al. (eds.), *Intelligent Methods and Big Data in Industrial Applications*, Studies in Big Data 40, https://doi.org/10.1007/978-3-319-77604-0_14

The problem of discovering frequent patterns is also applied to more specific, application-dependent domains. In particular, the usefulness of discovering frequent patterns in spatiotemporal data has been shown in the following areas: analysis of trajectories of movements of objects [3, 7], detection of convoys of moving objects [7], crime incidents analysis [10], weather prediction and climate change analysis [11].

In the paper, we consider the problem of discovering sequential patterns from spatiotemporal event-based datasets. The event-based spatiotemporal dataset contains a set of instances denoting occurrences of predefined event types, i.e. each instance is associated with an event type, occurrence location and occurrence time. We assume that instances of some event types may precede or follow instances of other types in terms of both their spatial and temporal neighborhoods. For example: seismic activity in an area may be followed by a gas hazard in the neighboring coal mine and thus, by potential disaster. If the above pattern occurs frequently in the set of registered events, then it may provide useful information for a variety of systems and services.

The problem of finding sequential patterns for spatiotemporal data is still subjects to development. In the paper, we propose an indexing structure and the so-called microclustering approach for reducing and transforming the size of original dataset of instances. The proposed index allows us to efficiently discover sequential patterns while maintaining appropriate level of effectiveness and correctness of discovered patterns. To appropriately adjust already proposed solutions for the problem, we reformulate definitions already proposed in literature.

The layout of the paper is as follows: in Sect. 2 we describe related works and current research state in the area. In Sect. 3 we introduce elementary notions. In Sect. 4 we define the problem considered in the paper and describe proposed methods, discussing their computational properties and implementation. We give results of conducted experiments on generated datasets and consider possible results applicable to real datasets in Sect. 5. Proposed Micro-ST-Miner algorithm has been implemented in C++ language and compared with algorithms proposed in the literature.

2 Related Works

In the paper, we consider the problem originally raised in publication [6]. We partially adopt the notation introduced there. For a set of event types $F = \{f_1, f_2, \ldots, f_n\}$ and for a dataset D storing events instances, where each instance $e \in D$ is associated with a quadruple *eventID, location, occurrence time* and *event type* $\in F$, the problem is to discover all significant sequential patterns in the form $\overrightarrow{s} = f_{i_1} \rightarrow f_{i_2} \rightarrow \cdots \rightarrow f_{i_k}$, where $f_{i_1}, f_{i_2}, \ldots, f_{i_k} \in F$. Informally we say that any two event types constitute a sequential pattern $f_{i_1} \rightarrow f_{i_2}$, if occurrences of instances of type f_{i_1} attract in their spatial proximity and, in temporal context afterwards, occurrences of instances of event type f_{i_2}. In other words, *attract* means that the number of occurrences of instances of type f_{i_2} in some spatial proximities and within some time intervals afterwards of occurrences of instances of event type f_{i_1} is greater than the number of occurrences of instances of type f_{i_2} in a whole considered spatiotemporal space. For a given instance $e \in D$, some its spatial proximity and some time interval afterwards an occurrence time of e, defines a spatiotemporal neighborhood of instance e. The notion of spatiotemporal neighborhood space is formally given in Definition 1.

Informally, the notation $\overrightarrow{s} = f_{i_1} \rightarrow f_{i_2} \rightarrow \cdots \rightarrow f_{i_k}$ denotes the fact, that instances of event type f_{i_1} attract in their predefined spatiotemporal neighborhood spaces occurrences of instances of event type f_{i_2}. Those instances of event type f_{i_2}, attract occurrences of instances of the next event type in $\overrightarrow{s}$ and so forth. The last type of instances participating in the sequential pattern $\overrightarrow{s}$ is f_{i_k}. Please note that in a given pattern $\overrightarrow{s} = f_{i_1} \rightarrow f_{i_2} \rightarrow \cdots \rightarrow f_{i_k}$, we do not assume any relation between nonconsecutive event types participating in $\overrightarrow{s}$ (e.g. for a pattern $f_{i_1} \rightarrow f_{i_2} \rightarrow f_{i_3}$, we cannot obtain any knowledge from this pattern that instances of event type f_{i_1} attract in their spatiotemporal neighborhoods occurrences of instances of event type f_{i_3}). We use notions *sequential pattern* and *sequence* interchangeably.

The above described notion of sequential patterns has been extended in several ways. In publication [10], the authors propose a new type of patterns: *cascading patterns* assuming that occurrences of instances of a given type may be attracted by occurrences of instances of more than one other type (e.g. occurrences of instances of event type f_{i_3} may be attracted by occurrences of instances of event types f_{i_1} and f_{i_2} only if they occur together in some spatial and temporal proximity). The methods presented in [10] may produce exact and approximate results.

The topic related to this described in our paper is *collocation patterns* mining. In [4], the authors provide an overview of the most recent methods proposed in the area. For a dataset of spatial event types and their instances, the problem is to find these event types, which instances occur in spatial proximity. Other authors consider the problem of discovering frequent patterns and periodical patterns in spatiotemporal trajectories. In [9], a method which discovers frequently visited areas based on the set of given movements of an object is given. The article assumes that each trajectory is represented in the form of a sequence of locations and lengths of trajectories are equal (the length is called a period). The problem of discovering frequent patterns in trajectories with non-equal periods or lengths is raised in [8].

3 Elementary Notions

Let us consider the spatiotemporal event-based dataset given in Fig. 1. The dataset contains $F = \{A, B, C, D\}$ event types and $D = \{a1, a2, \ldots, d9\}$ events instances. The dataset is embedded in the spatiotemporal space V, by $|V|$ we denote the volume of that space, calculated as the multiplication of spatial area and size of time dimension. Spatial and temporal sizes of spatiotemporal space are usually specified by the domain of considered task (e.g. for a weather dataset which may be used for discovering patterns between climate event types, the spatial area is specified by geographical coordinates of registered instances and size of time dimension is given by times of occurrences of instances). On the other hand, by $V_{N(e)}$ we denote the

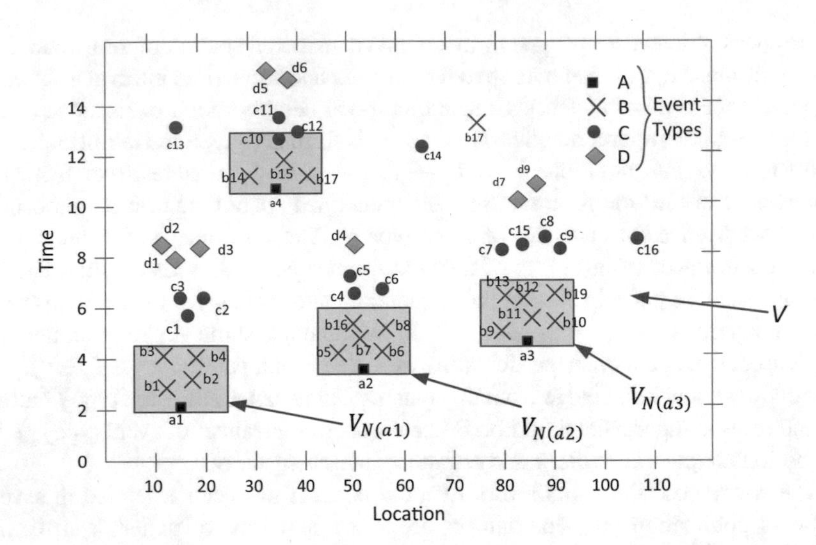

Fig. 1 An example of the dataset containing occurrences of instances of several types and possible neighborhoods [6]

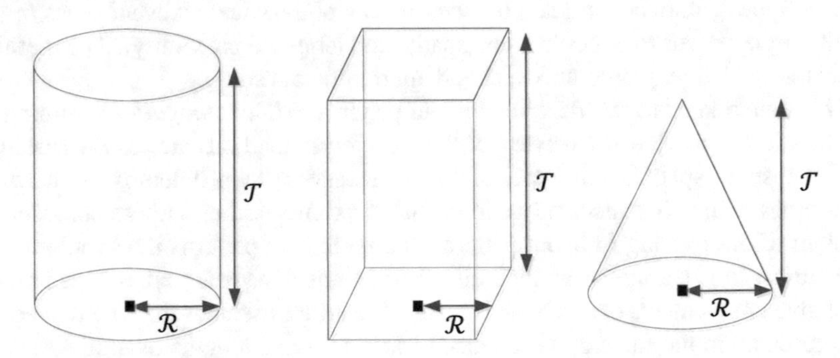

Fig. 2 Examples of types of neighborhood spaces $V_{N(e)}$ (figure adapted from [6])

spatiotemporal neighborhood space of instance e (the formal statement of $V_{N(e)}$ is given in Definition 1). Sizes and shape of $V_{N(e)}$ are usually given by an expert.

Definition 1 (*Neighborhood space*) By $V_{N(e)}$ we denote the neighborhood space of instance e. $\mathcal{R}$ denotes spatial radius and $\mathcal{T}$ temporal interval of that space (and, due to that, $V_{N(e)}$ has a cylindrical shape).

Let us comment Definition 1 and let consider Fig. 2 where we denote possible shapes of neighborhood spaces. It is possible to select other type of neighborhood space than provided in Definition 1. For example, it may be appropriate to select conical neighborhood space for some applications.

Definition 2 (*Neighborhood definition* [6]) For a given occurrence of event instance e, the neighborhood of e is defined as follows:

$$N(e) = \{p | p \in D \wedge distance(e.location, p.location) \leq \mathcal{R} \\ \wedge (p.time - e.time) \leq \mathcal{T} \wedge p.type = e.type\} \quad (1)$$

where $\mathcal{R}$ denotes spatial radius and $\mathcal{T}$ temporal interval of the neighborhood space $V_{N(e)}$.

The neighborhood $N(e)$ of instance e is the set of instances of the same type as e contained inside the neighborhood space $V_{N(e)}$.

Definition 3 (*Density* [6]) For a given spatiotemporal space V, event type f and its events instances in D, density is defined as follows:

$$Density(f, V) = \frac{|\{e | e.type = f \wedge e \text{ is inside } V\}|}{|V|} \quad (2)$$

density is the number of instances of type f occurring inside some space V divided by the volume of space V.

Definition 4 (*Density ratio* [6]) Density ratio for two event types f_{i_1}, f_{i_2} is defined as follows:

$$DensityRatio(f_{i_1} \rightarrow f_{i_2}) = \frac{avg_{(e.type=f_{i_1})}(Density(f_{i_2}, V_{N(e)}))}{Density(f_{i_2}, V)} \quad (3)$$

$avg_{(e.type=f_{i_1})}(Density(f_{i_2}, V_{N(e)}))$ specifies the average density of instances of type f_{i_2} occurring inside the neighborhood spaces $V_{N(e)}$ defined for instances, whose type is f_{i_1}. V denotes the whole considered spatiotemporal space and $Density(f_{i_2}, V)$ specifies density of instances of type f_{i_2} in space V.

Density ratio for the pattern $f_{i_1} \rightarrow f_{i_2}$ is defined as the ratio of the average density of instances of type f_{i_2} inside the neighborhood spaces $V_{N(e)}$ defined for each instance $e \in f_{i_1}$ and the overall density of instances of type f_{i_2} in the whole spatiotemporal space V.

Density ratio expresses the strength of the attraction relation $f_{i_1} \rightarrow f_{i_2}$ between any two event types. If its value is greater than one, then occurrences of instances of type f_{i_1} attract in their spatiotemporal neighborhood spaces occurrences of instances of type f_{i_2}. If its value is equal to one, then there is no correlation between these two event types. The value below one indicates negative relation (instances of type f_{i_1} repel in their spatiotemporal neighborhood spaces occurrences of instances of type f_{i_2}). However, it may be very difficult to provide its appropriate value taking into account usually arbitrarily chosen V and $V_{N(e)}$. Careful study of experimental results provided in [6] show some deficiencies in that subject.

Definition 5 (*Sequence* $\overrightarrow{s}$ *and tailEventSet*($\overrightarrow{s}$) [6]) $\overrightarrow{s}$ denotes a k-length sequence of event types: $s[1] \rightarrow s[2] \rightarrow \cdots \rightarrow s[k-1] \rightarrow s[k]$, where $s[i] \in F$.

tailEventSet($\overrightarrow{s}$) is the set of instances of type $\overrightarrow{s}[k]$ participating in the sequence $\overrightarrow{s}$.

If a k-length sequence $\vec{s}$ is considered to be expanded, then the neighborhood spaces $V_{N(e)}$ and neighborhoods $N(e)$ are calculated for each $e \in$ tailEventSet($\vec{s}$).

Definition 6 (*Sequence index* [6]) For a given k-length sequence $\vec{s}$, sequence index is denoted as follows:

1. If $k = 2$ then:

$$SeqIndex(\vec{s}) = DensityRatio(\vec{s}[1] \to \vec{s}[2]) \tag{4}$$

2. If $k > 2$ then:

$$SeqIndex(\vec{s}) = \min \begin{cases} SeqIndex(\vec{s}[1:k-1]), \\ DensityRatio(\vec{s}[k-1] \to \vec{s}[k]) \end{cases} \tag{5}$$

Definition 7 Sequence (sequential pattern) $\vec{s}$ is significant, if $SeqIndex(\vec{s}) \geq \theta$, where θ is significance threshold given by the user.

The value of $SeqIndex(\vec{s})$ for a k-length sequence $\vec{s}$ denotes the minimal strength of the $\to$ relation between any two consecutive event types participating in $\vec{s}$. In overall, sequence index provides means for discovering sequences for which its value is greater than the threshold given by the user. Possible significant sequences for the dataset shown in Fig. 1 are: $\vec{s_1} = (A \to B \to C \to D)$, $\vec{s_2} = (B \to C \to D)$ and $\vec{s_3} = (C \to D)$.

To illustrate the above presented notions and give an outline of the algorithm discovering sequential patterns, please consider the sequence $\vec{s_1} = A \to B \to C \to D$. Initially, the strength of the following relation $A \to B$ will be considered. The number of instances of type B is significantly greater in the neighborhood spaces created for instances of type A than the number of instances of type B in the whole space V. So we assume significant value of density ratio for the relation $A \to B$ and create a sequence $\vec{s} = A \to B$. The tail event set of $\vec{s}$ will be $tailEventSet(\vec{s}) = \{b1, b2, \ldots, b18\}$. The sequence $\vec{s}$ may be then expanded with event type C, because the average density of instances of type C is significantly high in the neighborhood spaces defined for the instances of type B contained in $tailEventSet(\vec{s})$. Similarly to the above, $\vec{s}$ will be expanded with event type D to create $\vec{s_1}$.

Discovering all significant sequences is the aim of the algorithm ST-Miner [6]. The algorithm expands all significant sequences starting with 1-length sequences (singular event types). Each sequence is expanded in a dept-first manner by possible appending other event types. If the actual value of the sequence index is below a predefined threshold, then the sequence is not expanded any more. The crucial step in the expanding process is to compute $N(e)$ and density for each instance e in the tail event set of the processed sequence. Computing $N(e)$ may be seen as performing spatial join and for this purpose we use the *plane sweep algorithm* proposed in [2]. The complexity of the algorithm is $O(\mathcal{N}\sqrt{(\mathcal{N})})$ where $\mathcal{N}$ is the number of instances in D.

4 Reducing Dataset Size by Indexing Instances Using Microclustering Approach

Searching for the neighborhoods $N(e)$ by means of the plane sweep algorithm is a costly operation. Initially, for each event type and its instances e in D, a neighborhood $N(e)$ has to be computed. In the next steps, the number of computed neighborhoods will depend on the cardinalities of tail events sets of already discovered sequences. To resolve the mentioned problem we propose to use a *microclustering* approach. The aim of the approach is to reduce the number of instances of different types by grouping neighboring instances of the same type into one microcluster. Sequential patterns discovery may then proceed using the modified dataset. In Fig. 3 we show a conception of microclustering.

Microcluster e_c is a quintuple consisting of *microcluster identifier (CID), event type* of contained instances, *a set of instances* contained in the microcluster, *number of instances contained in the microcluster* and *representative location* (for spatial and temporal) aspects. Representative location in both spatial and temporal dimensions is computed as a mean from respectively spatial coordinates and timestamps of instances contained in the microcluster.

Microclustering index MC_D created for dataset D, is a structure containing set of created microclusters and it may be used for appropriate computation of density ratios and sequence indexes on the reduced dataset. According to the proposed method, tail event sets of sequences contain microclusters rather than the original events instances.

Now we will propose our algorithm which will be based on the above introduced concept of microclustering. We need to appropriately modify definitions of density ratio and define facilitating indexing structures. We provide our modifications of the ST-Miner algorithm.

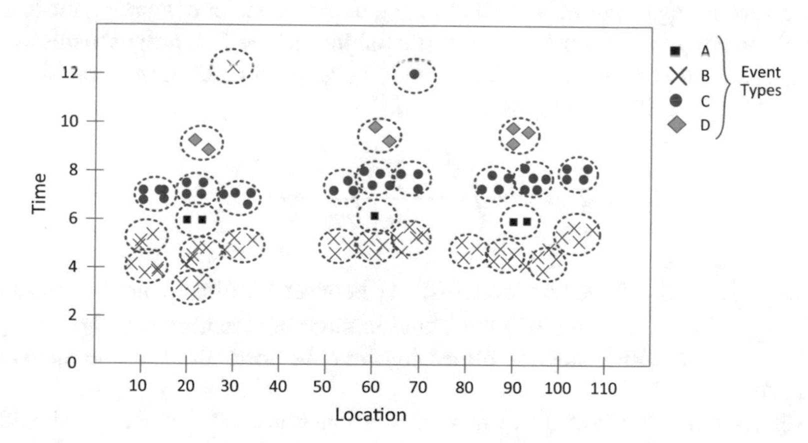

Fig. 3 A microclustering approach applied to the problem of reducing the size of a dataset

Table 1 An example of the microclustering index created by means of Algorithm 1 applied for the dataset shown in Fig. 1

CID	Event type	Contained instances	Num. of instances	Location (spatial, temporal)
1	A	$\{a1\}$	1	(15, 2)
2	A	$\{a2\}$	1	(51, 4)
⋮	⋮	⋮	⋮	⋮
5	B	$\{b1, b2, b3, b4\}$	4	(15, 3)
6	B	$\{b5, b6, b7, b8, b16\}$	5	(51, 5)
⋮	⋮	⋮	⋮	⋮
10	C	$\{c1, c2, c3\}$	3	(16, 6)
11	C	$\{c4, c5, c6\}$	3	(55, 6)
⋮	⋮	⋮	⋮	⋮
17	D	$\{d1, d2, d3\}$	3	(15, 8)
18	D	$\{d4\}$	1	(51, 8)
⋮	⋮	⋮	⋮	⋮

Let us consider the dataset given in Fig. 1. The corresponding microclustering index is shown in Table 1. For the microclustering step one may apply similar approach as proposed in [16]. The size of a microcluster may be limited by specifying its diameter threshold. In particular, for a given microcluster e_c and its contained instances $\{e_1, e_2, \ldots, e_n\}$, as the size of a microcluster we propose to use its diameter defined according to Formula (6). By $\overrightarrow{e_i}$ we denote a vector containing location in spatial and temporal dimensions of a particular instance e_i. The proposed microclustering approach is presented in Algorithm 1. Algorithm 1 has been inspired by the phase of building CF tree proposed in [16].

$$\mathcal{D}(e_c) = \sqrt{\frac{\sum_{i=1}^{m}\sum_{j=1}^{m}(\overrightarrow{e_i} - \overrightarrow{e_j})^2}{m*(m-1)}} \tag{6}$$

where $\sqrt{(\overrightarrow{e_i} - \overrightarrow{e_j})^2}$ is the Euclidean distance between locations of any two instances e_i and e_j, $i, j = 1 \ldots m$ and m is the actual number of instances contained in e_c. If needed, the spatial and temporal dimensions may be normalized for the microclustering step.

Algorithm 1 is illustrated in Fig. 4. Definitions introduced in Sect. 3 should be appropriately modified according to the above proposed concept of microclustering.

Algorithm 1 Algorithm for microclustering dataset D

1: **for** each event type $f \in F$ **do**
2: Take first instance of type f from D and initialize first microcluster of type f by inserting this instance to that microcluster.
3: **while** There are instances in dataset D of type f **do**
4: Take next instance of typee f from dataset D and insert it to the nearest microcluster e_c of type f.
5: **if** Diameter of e_c is greater than the predefined threshold **then**
6: Find a pair of farthest instances in e_c and mark them as seeds of two new microclusters.
7: Redistribute the rest of instances in e_c assigning them to the nearest from the two new microclusters.
8: **end if**
9: **end while**
10: **end for**

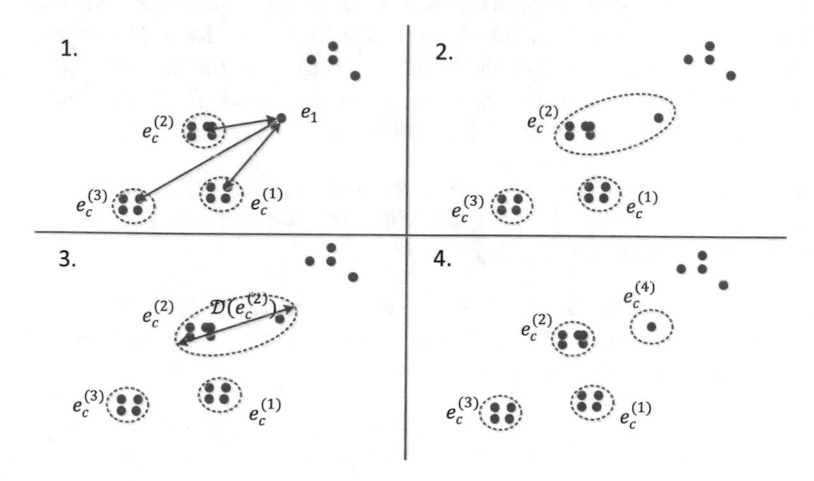

Fig. 4 The four phases of Algorithm 1

By MC_D we denote a microclustering index created for a dataset D. For any entry $e_c \in MC_D$ we propose two definitions of neighborhoods $n(e_c)$ and $N(e_c)$.

Definition 8 For a microcluster $e_c^{'} \in MC_D$, the neighborhood $n(e_c^{'})$ of $e_c^{'}$ is defined as follows:

$$n(e_c^{'}) = \{e_c | e_c \in MC_D \wedge distance(e_c^{'}.location, e_c.location) \leq \mathcal{R} \wedge (e_c^{'}.time - e_c.time) \leq \mathcal{T} \wedge e_c^{'}.type = e_c.type\} \tag{7}$$

that is, the neighborhood $n(e_c^{'})$ of a given microcluster $e_c^{'}$ is defined as the set of those microclusters, whose centers are contained inside space $V_{N(e_c^{'})}$ specified with respect to parameters $\mathcal{R}$ and $\mathcal{T}$. $e_c^{'}.location$ and $e_c.location$ refer respective to representative spatial locations of $e_c^{'}$ and e_c.

Definition 9 For a microcluster e_c' and for its neighborhood $n(e_c')$, the set of instances covered by entries from $n(e_c')$ is defined as follows:

$$N(e_c') = \{p | p \in D \wedge p \text{ is inside } e_c \wedge e_c \in n(e_c') \wedge e_c'.type = e_c.type\} \quad (8)$$

that is, as the neighborhood $N(e_c')$ of a given microcluster e_c' we define a set of those instances from D, that each of which is contained in the microclusters from $n(e_c')$.

The above introduced definitions $n(e_c)$ and $N(e_c)$ are useful in reformulating definitions of density and density ratio introduced in Sect. 3. By $MC_D(f)$ we define a set of entries in MC_D of event type f.

In Definitions 8 and 9 we do not refer to any particular event type, which microclusters or instances occur respectively in $n(e_c)$ or $N(e_c)$. However, it may be noticed from already provided notions, that density ratio considered for any two event types always take into account only neighborhoods of those particular event types. For example, considering the pattern $f_{i_1} \rightarrow f_{i_2}$, for any $e_c \in MC_D(f_{i_1})$, $n(e_c)$ and $N(e_c)$ contain respectively only microclusters and instances of type f_{i_2}.

Remark 1 For a microcluster $e_c \in MC_D$, the number of instances contained in this microcluster is denoted by $|e_c|$. For example, the number of instances contained in microcluster $e_c^{(5)}$ from Table 1 is $|e_c^{(5)}| = 4$.

Definition 10 Modified density. For a given spatiotemporal space V, event type f and its set of microclusters $MC_D(f) = \{e_c^{(1)}, e_c^{(2)}, \ldots, e_c^{(n)}\}$ contained inside V, the modified density is defined as follows:

$$density_{MC_D}(f, V) = \frac{|e_c^{(1)}| + |e_c^{(2)}| + \cdots + |e_c^{(n)}|}{|V|}, \quad \text{where } \{e_c^{(1)}, e_c^{(2)}, \ldots, e_c^{(n)}\} \text{ inside } V \quad (9)$$

that is, we define modified density as the quotient of the number of instances contained in $e_c^{(i)} \in MC_D(f)$ and the volume of space V. We say that a microcluster e_c is inside space V, if its representative location is inside V.

Definition 11 Modified density ratio. For two event types f_{i_1}, f_{i_2} and their entries in microclustering index MC_D, the modified density ratio is defined as follows:

$$DensityRatio_{MC_D}(f_{i_1} \rightarrow f_{i_2}) = \frac{\sum_{e_c \in f_{i_1}} \left(|e_c| * density_{MC_D}(f_{i_2}, V_{N(e_c)})\right)}{\left(\sum_{e_c \in f_{i_1}} |e_c|\right) * density_{MC_D}(f_{i_2}, V)} \quad (10)$$

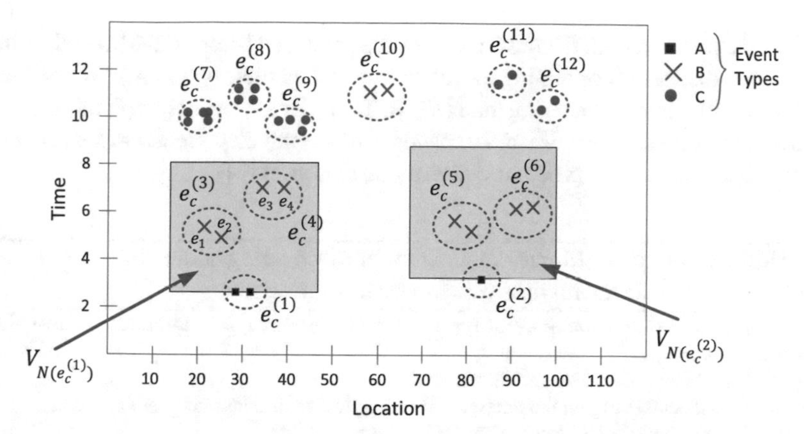

Fig. 5 An example of the dataset containing microclusters and their neighborhoods

comparing to Definition 4, the above definition takes into account proposed microclustering index and its entries.

The idea of Definition 11 is to calculate density ratio for the relation $f_{i_1} \rightarrow f_{i_2}$ using entries in microclustering index MC_D. Taking a microcluster $e_c \in MC_D(f_{i_1})$ and its neighborhood space $V_{n(e_c)}$, $density_{MC_D}(f_{i_2}, V_{N(e_c)})$ is weighted by the number $|e_c|$, that is by the number of instances contained in e_c. That is, the term:

$$\frac{\sum_{e_c \in f_{i_1}} \left(|e_c| * density_{MC_D}(f_{i_2}, V_{N(e_c)})\right)}{\left(\sum_{e_c \in f_{i_1}} |e_c|\right)}$$

specifies the weighted average density of instances of type f_{i_2} occurring in the neighborhood spaces $V_{N(e_c)}$ defined for microclusters $e_c \in MC_D(f_{i_1})$ with weights $|e_c|$. Before we proceed with an algorithm description we need to introduce one more definition.

Definition 12 For a sequence $\overrightarrow{s} \rightarrow f$ of length $k+1$, that f follows event type $\overrightarrow{s}[k]$. tailEventSet($\overrightarrow{s} \rightarrow f$) is equal to the the set of microclusters of type f in MC_D following the sequence $\overrightarrow{s}$. This fact is denoted by the formula: tailEventSet($\overrightarrow{s} \rightarrow f$) $= \{n(e_c^{(1)}) \cup n(e_c^{(2)}) \cup \cdots \cup n(e_c^{(n)})\}$ where tailEventSet($\overrightarrow{s}$) = $\{e_c^{(1)}, e_c^{(2)}, \ldots, e_c^{(n)}\}$.

To illustrate above notions, please consider Fig. 5 and assume that we investigate the relation $A \rightarrow B$. Assume that at the beginning $\overrightarrow{s} = A$ and $\overrightarrow{s}$ should be expanded with B, that is consider $\overrightarrow{s} \rightarrow B$. $tailEventSet(\overrightarrow{s}) = \{e_c^{(1)}, e_c^{(2)}\}$ and $tailEventSet(\overrightarrow{s} \rightarrow B) = n(e_c^{(1)}) \cup n(e_c^{(2)}) = \{e_c^{(3)}, e_c^{(4)}, e_c^{(5)}, e_c^{(6)}\}$. Additionally: $n(e_c^{(1)} = \{e_c^{(3)}, e_c^{(4)}\})$ and $N(e_c^{(1)}) = \{e_1, e_2, e_3, e_4\}$.

Now we proceed with the description of the proposed Micro-ST-Miner algorithm. For a given dataset D containing events instances and their types, Algorithm 2 starts with computing microclustering index MC_D. Then, for any event type f, the algorithm creates 1-length sequence $\overrightarrow{s}$ and expands it using the *ExpandSequence* procedure (our modification of the procedure already proposed in [6]).

Algorithm 2 Micro-ST-Miner - algorithm for discovering sequential patterns from event-based spatiotemporal data using microclustering index

Input: D - a dataset containing event types and their instances, θ - significance threshold for sequence index.
Output: A set of significant sequential patterns.
1: Perform microclustering on the dataset D resulting in a microclustering index MC_D.
2: **for** each event type f and its entries in $MC_D(f)$ **do**
3: Create 1-length sequence $\overrightarrow{s}$ from f.
4: tailEventSet($\overrightarrow{s}$) := $MC_D(f)$.
5: *ExpandSequence*($\overrightarrow{s}$, MC_D, θ).
6: **end for**

Algorithm 3 ExpandSequence($\overrightarrow{s}$, MC_D, θ) procedure

Input: $\overrightarrow{s}$ - a sequence of event types, MC_D - microclustering index created for D, θ - threshold for sequence index.
1: **for** each event type f and its entries in MC_D **do**
2: Calculate SequenceIndex($\overrightarrow{s} \rightarrow f$).
3: **if** value of the sequence index $\geq \theta$ **then**
4: Remember $\overrightarrow{s} \rightarrow f$ as significant.
5: tailEventSet($\overrightarrow{s} \rightarrow f$) := $\{n(e_c^{(1)}) \cup n(e_c^{(2)}) \cup \cdots \cup n(e_c^{(n)})\}$ where tailEventSet($\overrightarrow{s}$) = $\{e_c^{(1)}, e_c^{(2)}, \ldots, e_c^{(n)}\}$.
6: ExpandSequence(($\overrightarrow{s} \rightarrow f$), MC_D, θ).
7: **end if**
8: **end for**

Algorithm 4 Calculate SequenceIndex($\overrightarrow{s} \rightarrow f$) procedure

Input: $\overrightarrow{s} \rightarrow f$ - a sequence of event types; $k+1$ - the length of the sequence $\overrightarrow{s} \rightarrow f$
1: return min(SequenceIndex($\overrightarrow{s}$), DensityRatio($\overrightarrow{s}[k] \rightarrow f$)).

Table 2 Numbers of instances in generated datasets

Ps = 5, Pn = 10, Nf = 20, $\mathcal{R}$ = 10, $\mathcal{T}$ = 10, *DSize* = 1000, *TSize* = 1200

Ni	100	110	120	130	140	150	160
Avg. dataset size	10000	11000	12000	13000	14000	15000	16000

5 Experimental Results

In this section, we provide results of experiments. In particular, we describe proposed data generator, experiments on generated data and review future experiments on real data.

5.1 Data Generation

We implemented a generator similar to proposed in [1, 6]. The main parameters of the generator are described in [6].

- *Ps* denotes the average length (the number of participating event types) of generated sequences.
- *Pn* defines the average number of maximal sequences in a given dataset. Both *Pn* and *Ps* are used as the averages for Poisson distributions (that is, we generate the length of a particular sequence and the number of sequences in a dataset using Poisson distributions with the averages equal to the above parameters *Ps* and *Pn*).
- *Ni* the average number of events sequences participating in a pattern.
- *Dsize* and *TSize* denote sizes of spatial and temporal dimensions of considered spatiotemporal embedding space V.
- *Nf* is the total number of event types available in data generator. The actual number of event types in a sequence is generated using Poisson distribution with average *Ps*.

$\mathcal{R}$ and $\mathcal{T}$ define sizes of neighborhood spaces $V_{N(e)}$. In our experiments, we use cuboid neighborhood spaces (the reader may refer to Fig. 2). For each *Ni* value we generate a series of 10 datasets. The average number of instances in a dataset is calculated as follows $Ti = 2 * Ni * Ps * Pn$. We refer reader to [6] for the precise description of data generation method (Table 2).

Table 3 Average dataset size, averages of reduced dataset size and Compression Ratio after applying microclustering algorithm to datasets with different event sequences values

ps = 5, pn = 10, Nf = 20, $\mathcal{R}$ = 10, $\mathcal{T}$ = 10, *DSize* = 1000, *TSize* = 1200

Ni	Avg. dataset size		Microcluster's diameter threshold			
			40	60	80	100
100	10000	Microclustering index size	5845	5269	4497	3679
		Compression Ratio	1.46518	1.62536	1.90438	2.32781
110	11000	Microclustering index size	6373	5702	4806	3899
		Compression Ratio	1.47011	1.64311	1.94944	2.40292
120	12000	Microclustering index size	8245	7103	5740	4490
		Compression Ratio	1.50625	1.74842	2.16359	2.76592
130	13000	Microclustering index size	7345	6467	5380	4297
		Compression Ratio	1.49081	1.69321	2.03532	2.54829
140	14000	Microclustering index size	8523	7357	5957	4664
		Compression Ratio	8523	7357	2.13933	2.73242
150	15000	Microclustering index size	8608	7476	6085	4766
		Compression Ratio	1.4849	1.70974	2.10058	2.68191
160	16000	Microclustering index size	8483	7324	5923	4629
		Compression Ratio	1.49322	1.72952	2.13861	2.73644

5.2 Experiments Conducted on Generated Data

We have to comment conducted experiments. In the first step, we adjust microcluster's diameter by performing microclustering on the generated datasets. The averages are shown in Table 3. For the reduced datasets, their size is denoted by the number of entires in a microclustering index.

Compression Ration in Table 3 is calculated as the ratio of original dataset size and microclustering index size. In our experiments we use basic sequence threshold equal to 1 (Table 4).

Table 4 Average computation time (in milliseconds) for the microclustering algorithm applied to datasets with different event sequences values

ps = 5, pn = 10, Nf = 20, $\mathcal{R} = 10$, $\mathcal{T} = 10$, $DSize = 1000$, $TSize = 1200$

Ni	Avg. dataset size		Microcluster's diameter threshold			
			40	60	80	100
100	10000	Average computation time (in milliseconds)	1112.3	995	847.3	714.8
110	11000		1373.2	1211.9	1024.1	847.4
120	12000		2425.5	2099.7	1661	1420.4
130	13000		1823.8	1581.7	1304.9	1074.4
140	14000		2525.1	2141.3	1732.8	1391.9
150	15000		2420	2105	1695.5	1373.3
160	16000		2573.5	2152.2	1745.9	1395.7

Table 5 Computation time (in milliseconds) for the ST-Miner algorithm and for our Micro-ST-Miner algorithm applied to reduced datasets

ps = 5, pn = 10, Nf = 20, $\mathcal{R} = 10$, $\mathcal{T} = 10$, $DSize = 1000$, $TSize = 1200$

Ni		Microcluster's diameter threshold				
		Original	40	60	80	100
100	Computation time (in milliseconds)	64321.4	9690.7	5885.7	3427.3	1926.1
110		282505	10580.5	6647.9	3675.4	2148.2
120		474749	19435.4	12481.9	5053	2813.4
130		289212	15220.8	9167.9	5184.4	2533
140		577154	25342.6	14191.5	6384.1	3163.2
150		444738	21404	12877.8	6630.2	3046.9
160		381737	17064.3	11035.4	5376.5	2847.3

In Table 5 we show execution times for the original ST-Miner algorithm proposed in [6] and for our improvements. For each *Ni* value we select from generated dataset one dataset, which number of instances is similar to the average shown in Table 2. Results presented in Table 6 have been generated in the similar way, but focusing on the number of discovered patterns for each dataset and method. This approach has been motivated by reducing computation time and time constraints.

Presented approach may be adapted for a variety of real situations. In particular, in our future studies we will investigate the usefulness of proposed approach using the weather dataset presented in [6] or datasets used in [10].

Table 6 Number of discovered patterns for the ST-Miner algorithm and for our Micro-ST-Miner algorithm applied to reduced datasets

ps = 5, pn = 10, Nf = 20, $\mathcal{R}$ = 10, $\mathcal{T}$ = 10, *DSize* = 1000, *TSize* = 1200

Ni		Microcluster's diameter threshold				
		Original	40	60	80	100
100	Number of discovered patterns	2635.3	130.3	80.2	54.6	36.7
110		7710	79.4	70.4	50.1	38.8
120		9438.3	91.6	74.3	39.5	29.7
130		7224.5	102.3	69.1	50.4	33.5
140		8012.3	114.7	91.4	53.2	35
150		9810.1	99.4	69.8	51.6	30.1
160		5218.8	87.3	72.8	46.1	30.9

6 Conclusions

In the paper, we proposed modification of recently proposed notions and algorithms for the problem of discovering sequential patterns from event-based spatiotemporal data. The proposed method starts with computing microclustering index for a given dataset and appropriately redefining notions of neighborhood, density and density ratio. We offered our algorithm for the microclustering step and algorithms for computing significant sequential patterns using the reduced dataset. In summary, we have the following most important conclusions:

1. We plan to focus on presenting usage of proposed approach to real datasets. At the moment there is still lack of useful and good quality datasets of event-based spatiotemporal data.
2. Presented experiments show the usefulness of the proposed algorithm for generated datasets. As it was shown, patterns discovery times have been significantly reduced. Additionally, proposed approach allows to eliminate redundant and noise patterns from the dataset.

Further research problems in the described subject may focus on proposing methods for estimating or simulating appropriate value of the threshold of a pattern significance. Additionally, it may be shown that the above introduced definitions of density and space $V_{N(e)}$ are inappropriate for some datasets. In particular, for scenarios where occurrences of instances of one type attract with significant delay occurrences of other events definition of density is ineffective. Other possible research may propose alternative methods for discovering sequential patterns in spatiotemporal data adopting algorithms developed for mining association rules or considering the problem of efficient mining the K most important patterns.

References

1. Agrawal, R., Srikant, R.: Fast algorithms for mining association rules in large databases. In: Proceedings of the 20th International Conference on Very Large Data Bases, VLDB '94, pp. 487–499. Morgan Kaufmann Publishers Inc., San Francisco (1994). http://dl.acm.org/citation.cfm?id=645920.672836
2. Arge, L., Procopiuc, O., Ramaswamy, S., Suel, T., Vitter, J.S.: Scalable sweeping-based spatial join. In: Proceedings of the 24rd International Conference on Very Large Data Bases, VLDB '98, pp. 570–581. Morgan Kaufmann Publishers Inc., San Francisco (1998). http://dl.acm.org/citation.cfm?id=645924.671340
3. Benkert, M., Gudmundsson, J., Hübner, F., Wolle, T.: Reporting flock patterns. Comput. Geom. **41**(3), 111–125 (2008). http://www.sciencedirect.com/science/article/pii/S092577210700106X
4. Boinski, P., Zakrzewicz, M.: Algorithms for spatial collocation pattern mining in a limited memory environment: a summary of results. J. Intell. Inf. Syst. **43**(1), 147–182 (2014). https://doi.org/10.1007/s10844-014-0311-x
5. Han, J., Pei, J., Yin, Y.: Mining frequent patterns without candidate generation. In: Proceedings of the 2000 ACM SIGMOD International Conference on Management of Data, SIGMOD '00, pp. 1–12. ACM, New York (2000). https://doi.org/10.1145/342009.335372
6. Huang, Y., Zhang, L., Zhang, P.: A framework for mining sequential patterns from spatio-temporal event data sets. IEEE Trans. Knowl. Data Eng. **20**(4), 433–448 (2008)
7. Li, Z., Ding, B., Han, J., Kays, R.: Swarm: mining relaxed temporal moving object clusters. Proc. VLDB Endow. **3**(1–2), 723–734 (2010). https://doi.org/10.14778/1920841.1920934
8. Li, Z., Ding, B., Han, J., Kays, R., Nye, P.: Mining periodic behaviors for moving objects. In: Proceedings of the 16th ACM SIGKDD International Conference on Knowledge Discovery and Data Mining, KDD '10, pp. 1099–1108. ACM, New York (2010). https://doi.org/10.1145/1835804.1835942
9. Mamoulis, N., Cao, H., Kollios, G., Hadjieleftheriou, M., Tao, Y., Cheung, D.W.: Mining, indexing, and querying historical spatiotemporal data. In: Proceedings of the Tenth ACM SIGKDD International Conference on Knowledge Discovery and Data Mining, KDD '04, pp. 236–245. ACM, New York (2004). https://doi.org/10.1145/1014052.1014080
10. Mohan, P., Shekhar, S., Shine, J.A., Rogers, J.P.: Cascading spatio-temporal pattern discovery. IEEE Trans. Knowl. Data Eng. **24**(11), 1977–1992 (2012)
11. Shekhar, S., Evans, M.R., Kang, J.M., Mohan, P.: Identifying patterns in spatial information: a survey of methods. Wiley Interdiscip. Rev. Data Min. Knowl. Discov. **1**(3), 193–214 (2011). https://doi.org/10.1002/widm.25
12. Srikant, R., Agrawal, R.: Mining sequential patterns: generalizations and performance improvements, pp. 1–17. Springer, Berlin (1996). https://doi.org/10.1007/BFb0014140
13. Thabtah, F.: A review of associative classification mining. Knowl. Eng. Rev. **22**(1), 37–65 (2007). https://doi.org/10.1017/S0269888907001026
14. Zaki, M.J., Parthasarathy, S., Ogihara, M., Li, W.: New algorithms for fast discovery of association rules. In: Proceedings of the Third International Conference on Knowledge Discovery and Data Mining, KDD'97, pp. 283–286. AAAI Press (1997). http://dl.acm.org/citation.cfm?id=3001392.3001454
15. Zaki, M.J.: Spade: an efficient algorithm for mining frequent sequences. Mach. Learn. **42**(1), 31–60 (2001). https://doi.org/10.1023/A:1007652502315
16. Zhang, T., Ramakrishnan, R., Livny, M.: Birch: an efficient data clustering method for very large databases. In: Proceedings of the 1996 ACM SIGMOD International Conference on Management of Data, SIGMOD '96, pp. 103–114. ACM, New York (1996). https://doi.org/10.1145/233269.233324

Part IV
Medical Applications and Bioinformatics

Unsupervised Machine Learning in Classification of Neurobiological Data

Konrad A. Ciecierski and Tomasz Mandat

Abstract In many cases of neurophysiological data analysis, the best results can be obtained using supervised machine learning approaches. Such very good results were obtained in detection of neurophysiological recordings recorded within Subthalamic Nucleus (*STN*) during deep brain stimulation (DBS) surgery for Parkinson disease. Supervised machine learning methods relay however on external knowledge provided by an expert. This becomes increasingly difficult if the subject's domain is highly specialized as is the case in neurosurgery. The proper computation of features that are to be used for classification without good domain knowledge can be difficult and their proper construction heavily influences quality of the final classification. In such case one might wonder whether, how much and to what extent the unsupervised methods might become useful. Good result of unsupervised approach would indicate presence of a natural grouping within recordings and would also be a further confirmation that features selected for classification and clustering provide good basis for discrimination of recordings recorded within Subthalamic Nucleus (*STN*). For this test, the set of over 12 thousand of brain neurophysiological recordings with precalculated attributes were used. This paper shows comparison of results obtained from supervised - random forest based - method with those obtained from unsupervised approaches, namely K-Means and Hierarchical clustering approaches. It is also shown, how inclusion of certain types of attributes influences the clustering based results.

K. A. Ciecierski (✉)
Institute of Computer Science, Warsaw University of Technology,
00-655 Warsaw, Poland
e-mail: K.Ciecierski@ii.pw.edu.pl

T. Mandat
Department of Neurosurgery, M. Sklodowska-Curie Memorial Oncology Center,
Warsaw, Poland
e-mail: tomaszmandat@yahoo.com

T. Mandat
Department of Neurosurgery, Institute of Psychiatry and Neurology,
Warsaw, Poland

R. Bembenik et al. (eds.), *Intelligent Methods and Big Data in Industrial Applications*, Studies in Big Data 40, https://doi.org/10.1007/978-3-319-77604-0_15

Keywords STN · DBS · DWT (Discrete Wavelet Transform) decomposition Signal power · Unsupervised learning · K-means clustering · Hierarchical clustering

Introduction

Data used in this paper come from neurosurgical procedures performed during deep brain stimulation treatment of the Parkinson disease. In this surgery the final goal is the placement of stimulating electrodes in deep within the brain placed structure called Subthalamic Nucleus (*STN*) [1–3]. As it is difficult to precisely locate the STN using only the medical imaging techniques (CT and MRI) [1], it is often located neurophysiologically during the first stage of the surgery. During this part of the surgery (described in detail in [4–8]) set of few parallel electrodes is advanced through the brain, towards and through the STN.

As the electrodes are advanced, they record the neurophysiological activity of the surrounding brain tissue. This recording is done in steps, it typically starts about 10 mm above the expected location of the STN and continues for at least 15 mm. As the activity of the STN is distinct from the one seen in tissue adjacent to it, it is possible to create the morphological mapping of this brain area.

This mapping is then used during subsequent part of the surgery for placement of the brain stimulating electrodes.

The special activity of the STN can be distinguished by a neurosurgeon and as it has been shown in [1, 9, 10] it can also be with good certainty discriminated using machine learning techniques – classifiers. Currently the random forest classifier run on such data yields sensitivity over 0.88 and specificity of over 0.96 [4–8].

This is possible due to definition of discriminating attributes that when calculated, differ for recording from the STN [1, 9]. The construction of these attributes provided some challenges as they had to reflect the differences between STN and surrounding tissues that are of neurophysiological nature. Aside from the clinical trials, the quality of those attributes can be in some measure also confirmed using unsupervised machine learning. Highly discriminating attributes should introduce natural partitioning of the recordings in the attribute space. Such partitioning in turn should be noticeable in unsupervised approach, namely in the clustering results.

Attributes mentioned above and also other (temporal) attributes were used in this paper as an input to the clustering attempt. In preliminary tests, the addition of temporal attributes (not used in [4–8]) yielded additional improvement in the quality of classification output. Temporal attributes rely not only on activity at given location within the brain but also on activities registered at locations previously (dorsally) traversed by the recording electrode. Those attributes help to distinguish data recorded before (dorsal) and after (ventral) the location of the STN.

This paper aims to show that with properly calculated and selected set of attributes, data can also be in meaningful way discriminated using only unsupervised machine learning algorithms. This is important as the neurophysiological data is difficult to analyze (recordings contain much noise and other non neurophysiological artifacts from the process of recording) and finding proper attributes might be extremely difficult.

Data used throughout this paper after calculation of the attributes were discriminated by classifier and medical expert into two groups. First group labeled STN were recorded within the STN, others labeled MISS were recorded in other parts of the brain.

1 Input Data and Attributes

For calculations shown in this paper attributes calculated for 12494 brain recordings were taken. Recordings were made during 115 surgical procedures made in Institute of Psychiatry and Neurology in Warsaw. Recordings have been discriminated into two disjoint classes, STN - 2790 (22.3%) recordings and MISS - 9704 (77.7%) recordings.

Two sets of attributes were used in calculation.

First set of 14 attributes

PRC80	80th percentile of amplitude's module
PRCDLT	difference between 95th and 80th percentile of amplitude's module in proportion to 95th percentile of amplitude's module
RMS	Root Mean Square of the signal
LFB	power of the signal's background in range 0–500 Hz
HFB	power of the signal's background in range 500–3000 Hz
MPRC80	five element wide moving average of PRC80
MRMS	five element wide moving average of RMS
MLFB	five element wide moving average of LFB
MHFB	five element wide moving average of HFB
AvgSpkRate	is the average number of spikes detected per second
AvgSpkRateScMax	maximal $AvgSpkRate$ observed for single cell
BurstRatio	percentage of intervals between spikes that are smaller than 33 ms
BurstRatioScMax	maximal $BurstRatio$ observed for single cell
MPWR	power of the derived meta signal.

The LFR and HFB attributes were calculated from wavelet decomposition coefficients [11, 12].

Second set contained all attributes from the first set plus additional 16 temporal attributes holding information about changes of RMS, PRC80, LFB and HFB attributes.

For electrode E at depth D and attribute X, the temporal attribute

$\mathbf{X}_{1U}$ holds maximal increase of attribute X over distance of 1 mm detected by an electrode E at depths dorsal to D.

$\mathbf{X_{2U}}$ holds maximal increase of attribute X over distance of 2 mm detected by an electrode E at depths dorsal to D.
$\mathbf{X_{1D}}$ holds maximal decrease of attribute X over distance of 1 mm detected by an electrode E at depths dorsal to D.
$\mathbf{X_{2D}}$ holds maximal decrease of attribute X over distance of 2 mm detected by an electrode E at depths dorsal to D.

To minimize influence of random fluctuations the change affecting any of above temporal attributes must be above certain threshold. In this way, those attributes in scope of all recordings provided by a single electrode, do change at the points of its entry and leave of the STN.

While the temporal attributes by helping with detection of the STN borders [1, 9] do increase the sensitivity and specificity of the classifier based approach they introduce non locality i.e. attribute at given anatomical location depends not only on its sole properties but also on properties of locations already traversed by a given electrode. It can be therefore expected that temporal attributes might negatively influence the clustering quality. Because of that calculations are made for two sets of attributes, first (14 attributes) without temporal attributes and second (30 attributes) with addition of temporal attributes. This difference caused by non locality of attribute values can be readily seen already when observations are plotted using first two PCA components (Fig. 1). The variance of 2nd PCA component is much greater with 30 attributes.

Still, in case when only non temporal attributes are selected, the plot of the first two PCA components shows significant cluster overlapping (Fig. 2).

While in the PCA component space the observations overlap heavily, the clustering might be able to generate results of good quality when made in full attribute space, especially when one of the classes (MISS) shows much less variance in the value of attributes than the other one (STN), see Fig. 2. This difference comes from the fact that values of attributes is the MISS class are uniformly low regardless of the patient while the values in the STN class depend on patient and severity of the

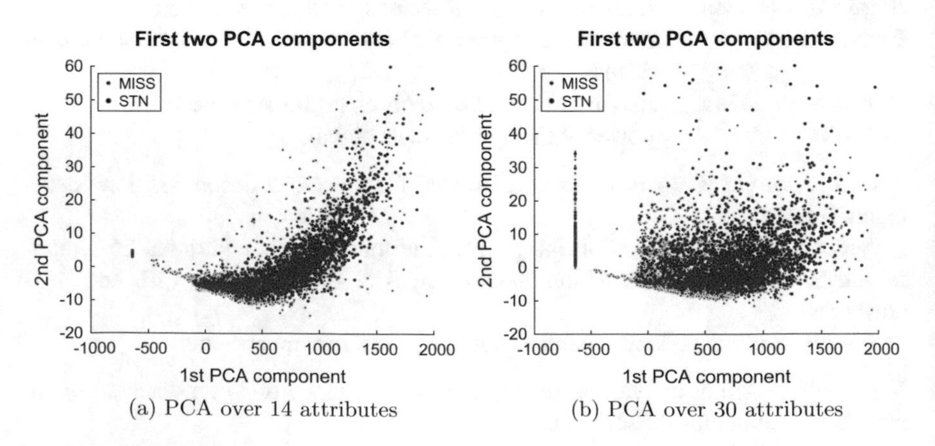

(a) PCA over 14 attributes

(b) PCA over 30 attributes

Fig. 1 PCA plot for 14 and 30 attributes

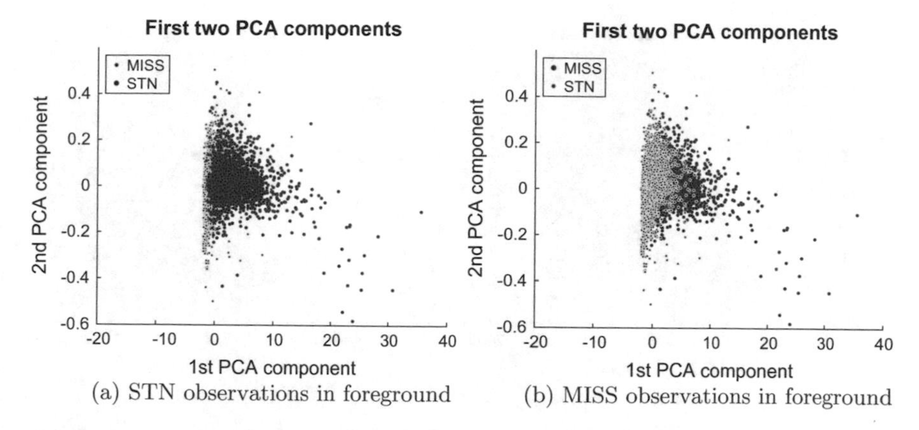

(a) STN observations in foreground (b) MISS observations in foreground

Fig. 2 PCA plot of first two components calculated from RMS and PRC80

disease. On the other hand, large variance within the STN class might be troublesome when clustering is done using Euclidean distance in attribute space of high dimensionality ($\mathbb{R}^{14}$ or $\mathbb{R}^{30}$).

2 Methods of Evaluation, Clustering of Observations

Clustering has been done firstly using the standard K-means algorithm (albeit using not only its standard settings). As a measure of distance between observations not only the Euclidean norm [13] was used, also the applicability of cosine norm [13] has been tested. In the cosine norm the observations are treated as vectors in $\mathbb{R}^{14}$ or $\mathbb{R}^{30}$ spaces. Then the distance between vectors $\overrightarrow{a}$ and $\overrightarrow{b}$ is calculated as $1 - cos(\angle(\overrightarrow{a}, \overrightarrow{b}))$. This measure is focused on directions of vectors in attribute space, not on their length. It might be here especially useful as the STN is differently active in different patients and this in turn influences values of the attributes (even thou they are normalized during calculation).

Assuming that cluster 1 corresponds to MISS class and cluster 2 corresponds to the STN class, the values of sensitivity and specificity are as follows:

What can be immediately seen is that inclusion of temporal attributes causes both sensitivity and specificity to be unacceptable. Taking into account only the clearly location based remaining 14 attributes it is evident that results are highly sensitive but not very specific. In other words there is a high probability that observation originally classified as STN will be placed in cluster 2. Still, low specificity (large number of false positives) does not allow to infer any good quality prediction from an observation being put in cluster 2. One can however with good certainty assume that observation placed is cluster 1 was originally labeled as MISS.

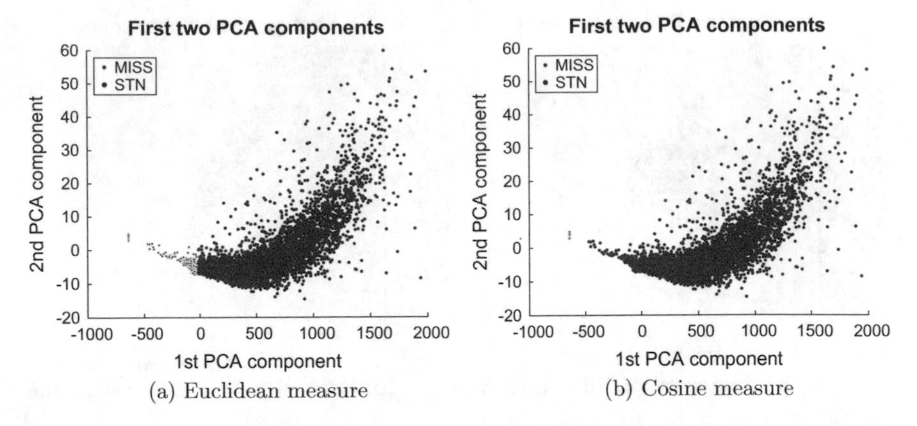

(a) Euclidean measure (b) Cosine measure

Fig. 3 PCA plot of K-means clusters depending on distance measure

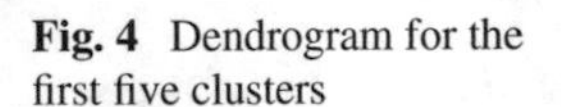

Fig. 4 Dendrogram for the first five clusters

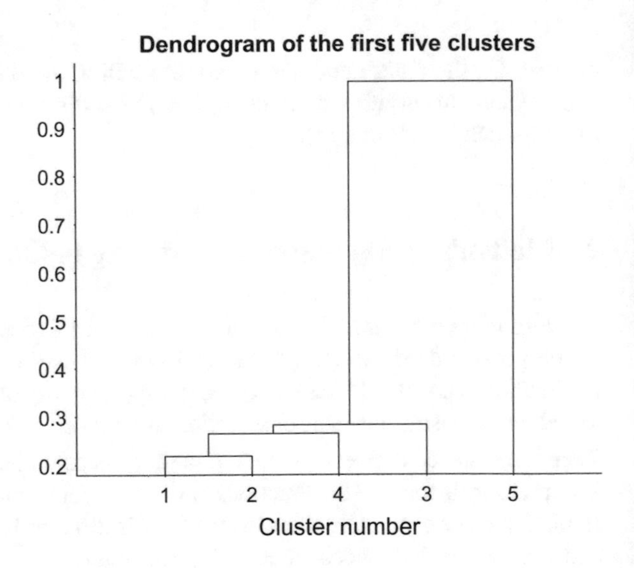

Regarding difference between Euclidean and cosine measure, it can be readily seen on Fig. 3a that some of the STN observations being close to the MISS cluster were wrongly assigned to it when using Euclidean measure.

Results obtained using hierarchical clustering with cosine based distance measure have not improved output generated by the K-means clustering. The quality of the obtained clustering was good as the Cophenet measure was as high as 0.998. In hierarchical clustering one can however easily see if the data can be further subdivided into smaller clusters. In the Fig. 4 one can immediately see that distance between cluster 1 and 5 is much greater than between any of clusters numbered 1–4.

This finds also confirmation in number of elements in each of the clusters. Clearly the clusters with numbers 2–4 are forced by selected number of clusters and should be parts of cluster 1. In this way it is shown that observations clearly do cluster into two clusters and that there are no further evident subdivisions.

2.1 Interpretation of the Poor Specificity

Looking at the Table 2(b) on can see that while most of the STN observations have fallen into the cluster 2 the MISS class was poorly discriminated. It might be interesting to look at the observations from the MISS class that nonetheless have fallen into cluster 2 (STN). There are 4861 of such observations (Tables 1, 3 and 4).

First thing to check is how distant from the STN during surgery were those observations. As it was described in Introduction, in each surgery a set of parallel electrodes is used where perpendicular distance between them is close to 1.5 mm.

Table 1 Statistics for attributes

	MISS		STN	
Attribute	μ	σ^2	μ	σ^2
PRC80	1.092	0.054	2.169	0.303
PRCDLT	0.005	0.005	0.057	0.007
RMS	1.095	0.073	2.277	0.383
LFB	1.254	0.800	5.134	16.529
HFB	1.264	0.478	5.287	10.943
MPRC80	1.135	0.056	1.952	0.193
MRMS	1.143	0.069	2.033	0.229
MLFB	1.413	0.949	4.302	7.325
MHFB	1.432	0.726	4.461	5.926
AvgSpkRate	6.080	109.435	22.375	221.912
AvgSpkRateScMax	3.870	44.0318	13.684	88.420
BurstRatio	0.209	0.072	0.571	0.048
BurstRatioScMax	0.176	0.056	0.487	0.045
MPWR	479.624	315723.472	1313.641	218764.850

Table 2 K-means clustering results

		Observation group	
		MISS	*STN*
Cluster	1	5936	143
	2	4193	2222

(a) Euclidean norm, 14 attributes

		Observation group	
		MISS	*STN*
Cluster	1	5268	107
	2	4861	2258

(b) Cosine norm, 14 attributes

		Observation group	
		MISS	*STN*
Cluster	1	5005	114
	2	5124	1251

(c) Euclidean norm, 30 attributes

		Observation group	
		MISS	*STN*
Cluster	1	4388	987
	2	5741	1378

(d) Cosine norm, 30 attributes

Table 3 Sensitivity and specificity

Clustering	Sensitivity	Specificity
Euclidean norm, 14 attributes	0.940	0.590
Cosine norm, 14 attributes	0.955	0.520
Euclidean norm, 30 attributes	0.530	0.490
Cosine norm, 30 attributes	0.583	0.433

Table 4 Count of elements in hierarchical clusters and K-means correspondence

Cluster	Element count	Comment
1	5351	Corresponds to cluster 1 from K-means
2	12	Corresponds to cluster 1 from K-means
3	9	Corresponds to cluster 1 from K-means
4	3	Corresponds to cluster 1 from K-means
5	7119	Corresponds to cluster 2 from K-means

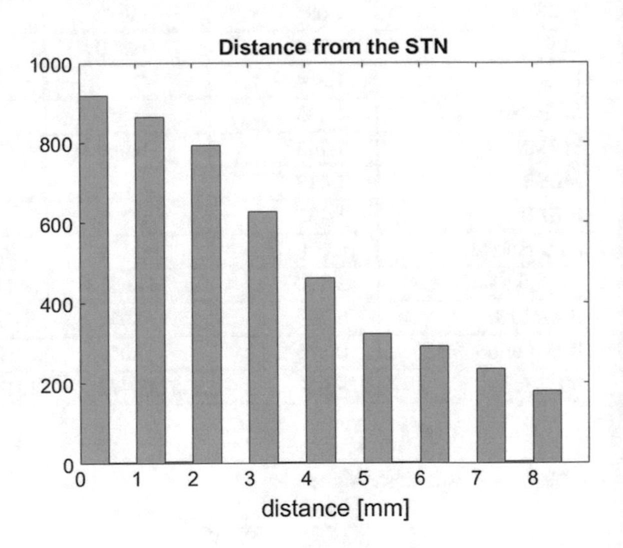

Fig. 5 Distance from the STN histogram

From this set, not all of the electrodes always do pass through the STN. In such occasions one or even two electrodes misses the STN but are still in its neighborhood. Values of attributes in such occasions might not be high enough for recording to be automatically classified as STN, but still would be higher. For purpose of this statistic, as the extent of the STN for given set of electrodes is taken its maximal extend detected on all electrodes from that set.

Knowing that recordings were made every 1 mm this will give one some information about their placement and distribution. The mean distance between such observation and STN is 2.8 mm. As many as 3212 out of 4861 (66%) observations

Table 5 μ and σ^2 for attributes for observations from MISS class that have been placed in cluster 2

Attribute	Mean (μ)	Variance (σ^2)
AvgSpkRate	12.669	144.561
AvgSpkRateScMax	8.065	52.927
BurstRatio	0.436	0.052
BurstRatioScMax	0.3675	0.047
MPWR	999.406	138372

are no farther away from the STN than 3 mm. The histogram of the distance is shown on Fig. 5. For clarity the distances larger than 8 mm have not been shown.

From this one can infer that most of the observations from MISS class that were clustered together with observations from the STN class come from the 3 mm wide vicinity of the STN.

Additionally when one is to look at mean and variance of spike based attributes for observations from MISS class clustered into cluster 2 (shown in Table 5), one can notice that those values (especially μ) are much closer to those of STN class than to those of the MISS class as seen in Table 1.

3 Results

In conclusion it has been shown that with properly calculated and selected set of attributes neurobiological data from STN DBS surgeries can be in meaningful way discriminated using only unsupervised machine learning algorithms.

This has also confirmed that attributes defined in [4–8] provide natural partitioning of recordings in attribute space and do provide a good basis for discrimination of recordings into MISS and STN classes.

It has been also shown that selection of attributes and measure used to tell distance between observations are of utmost importance for the quality of the clustering.

Inclusion of temporal attributes – especially those of values inherited from previous depths – disrupts heavily the quality of the clustering.

Also in observation where one group (STN) have greater values of attributes but also with bigger variance, the use of Euclidean distance measure might lead of inclusion of some STN observations into cluster housing mainly MISS observations. Use of cosine measure which is focused not so much on magnitude of attributes but rather on their direction in attribute space provided much better results.

STN observation were clustered together in way very much similar to the original labeling of the input data, namely sensitivity 0.965. MISS observations were distributed almost equally among both clusters giving poor specificity of 0.520. Still further investigation showed that those of observations from the MISS class that were assigned to the cluster housing STN data were in fact largely recorded in the vicinity of the STN and thus have had values of some attributes similar to those of the STN class.

Clustering results might be with good certainty used to asses that given observation is not from the STN class. High quality discrimination of the observation from the MISS class could not have be here achieved as recordings made near STN were often mislabeled. To achieve high both sensitivity and specificity the supervised machine learning approach should be used.

References

1. Israel, Z., Burchiel, K.J.: Microelectrode Recording in Movement Disorder Surgery. Thieme Medical Publishers, New York (2004)
2. Nieuwenhuys, R., Huijzen, C., Voogd, J.: The Human Central Nervous System. Springer, Berlin (2008)
3. Nolte, J.: The Human Brain: An Introduction to Its Functional Anatomy. Mosby (2002)
4. Ciecierski, K., Raś, Z.W., Przybyszewski, A.W.: Foundations of recommender system for STN localization during DBS surgery in Parkinson's patients. Foundations of Intelligent Systems, ISMIS 2012 Symposium, LNAI, vol. 7661, pp. 234–243. Springer (2012)
5. Ciecierski, K., Raś, Z.W., Przybyszewski, A.W.: Discrimination of the micro electrode recordings for STN localization during DBS surgery in Parkinson's patients. Flexible Query Answering Systems, FQAS 2013 Symposium, LNAI, vol. 8132, pp. 328–339. Springer (2013)
6. Ciecierski, K., Raś, Z.W., Przybyszewski, A.W.: Foundations of automatic system for intrasurgical localization of subthalamic nucleus in Parkinson patients. Web Intelligence and Agent Systems, 2014/1, pp. , 63–82. IOS Press (2014)
7. Ciecierski, K.: Decision Support System for surgical treatment of Parkinsons disease, Ph.D. thesis, Warsaw University of technology Press (2013)
8. Ciecierski, K., Mandat, T., Rola, R., Raś, Z.W., Przybyszewski, A.W.: Computer aided subthalamic nucleus (STN) localization during deep brain stimulation (DBS) surgery in Parkinson's patients. Annales Academiae Medicae Silesiensis, vol. 68, 5, pp. 275–283 (2014)
9. Mandat, T., Tykocki, T., Koziara, H., et al.: Subthalamic deep brain stimulation for the treatment of Parkinson disease. Neurologia i neurochirurgia polska **45**(1), 32–36 (2011)
10. Novak, P., Przybyszewski, A.W., Barborica, A., Ravin, P., Margolin, L., Pilitsis, J.G.: Localization of the subthalamic nucleus in Parkinson disease using multiunit activity. J. Neurol. Sci. **310**(1), 44–49 (2011)
11. Jensen, A.: A Ia Cour-Harbo. Ripples in Mathematics. Springer, Berlin (2001)
12. Smith, S.W.: Digital Signal Processing. Elsevier (2003)
13. Cha, S.-H.: Comprehensive survey on distance/similarity measures between probability density functions. City **1**(2), 1 (2007)

Incorporating Fuzzy Logic in Object-Relational Mapping Layer for Flexible Medical Screenings

Bożena Małysiak-Mrozek, Hanna Mazurkiewicz and Dariusz Mrozek

Abstract Introduction of fuzzy techniques in database querying allows for flexible retrieval of information and inclusion of imprecise expert knowledge into the retrieval process. This is especially beneficial while analyzing collections of patients' biomedical data, in which similar results of laboratory tests may lead to the same conclusions, diagnoses, and treatment scenarios. Fuzzy techniques for data retrieval can be implemented in various layers of database client-server architecture. However, since in the last decade, the development of real-life database applications is frequently based on additional object-relational mapping (ORM) layers, inclusion of fuzzy logic in data analysis remains a challenge. In this paper, we show our extensions to the Doctrine ORM framework that supply application developers with the possibility of fuzzy querying against collections of crisp data stored in relational databases. Performance tests prove that these extensions do not introduce a significant slowdown while querying data and can be successfully used in development of applications that benefit from fuzzy information retrieval.

Keywords Databases · Fuzzy sets · Fuzzy logic · Querying · Information retrieval · Biomedical data analysis · Object-relational mapping · ORM

1 Introduction

Laboratory tests are basic methods that support healthcare tasks such as monitoring of patient's general health condition, disease detection, diagnosis, treatment, and prevention of disease. Laboratory tests analyze a sample of patient's blood, urine, or body tissues with respect to selected biomedical markers to check if they fall within the normal range. Ranges are used to describe normal values of particular

B. Małysiak-Mrozek · H. Mazurkiewicz · D. Mrozek (✉)
Institute of Informatics, Silesian University of Technology, ul. Akademicka 16,
44-100 Gliwice, Poland
e-mail: dariusz.mrozek@polsl.pl

B. Małysiak-Mrozek
e-mail: bozena.malysiak@polsl.pl

R. Bembenik et al. (eds.), *Intelligent Methods and Big Data in Industrial Applications*, Studies in Big Data 40, https://doi.org/10.1007/978-3-319-77604-0_16

markers because what is normal depends on many factors and differs from person to person. Doctors usually compare results of laboratory tests of a patient to results from previous tests, or they may compare results of many patients in order to draw general conclusions on the applied treatment. They may also recommend measuring some markers, like blood sugar, or other health indicators, like resting heart rate, at home and collecting measurements for later analysis in special health journals or measurement logs. Laboratory tests, such as: morphology, erythrocyte sedimentation rate (ESR), blood sugar, lipid profile, are often part of a routine inspection to look for changes in patients' health. They also help doctors diagnose medical conditions, plan or evaluate treatments, and monitor diseases. Almost 70% of health care decisions are guided by laboratory test results.

Biomedical laboratories must keep information about the examined people – their personal data and the results of laboratory tests together with ranges for particular test types. For example, normal levels for *Blood sugar* are as follows:

- Between 4.0 and 6.0 mmol/L (72–108 mg/dL) when fasting
- Up to 7.8 mmol/L (140 mg/dL) 2 h after eating.

But for people with diabetes, blood sugar level targets are as follows:

- Before meals: 4–7 mmol/L for people with type 1 or type 2 diabetes
- After meals: under 9 mmol/L for people with type 1 diabetes and under 8.5mmol/L for people with type 2 diabetes.

Results of laboratory tests more often have numeric character and they are usually stored in relational databases. Their volume can be large, especially, when the data are collected remotely by an external system, like a telemedicine system that monitors patients and provides clinical health care from a distance. While developing software applications that operate on such data, it seems reasonable to include methods of approximate information retrieval in one of the layers of the client–server database architecture. This may help in finding similar cases of patients, whose health indicators or biomedical markers, e.g., blood pressure, BMI, cholesterol, age, have similar, but not the same values. Results of various laboratory tests fall into certain ranges or may exceed them. People, including medical doctors, usually use common terms, like *normal*, *above* or *below* to describe levels of particular biochemical markers. This provides a good motivation to apply such a logic in computer processing of the data, which would be able to appropriately assign particular values to a certain range, or mark them as going beyond the range: above or below. These conditions can be met by using fuzzy logic [31, 32]. However, since in the last decade, the development of real-life database applications is frequently based on additional object-relational mapping (ORM) layers, inclusion of fuzzy logic in data analysis remains a challenge and requires additional extensions to the ORM application production layer. Object-relational mapping (ORM) brought evolution in software development by delivering programming technique that allows for automatic conversions of data between relational databases and object-oriented programming languages. ORM was introduced to programming practice in recent decade in response to the incompatibility between relational model of database systems and object-oriented model of client applications.

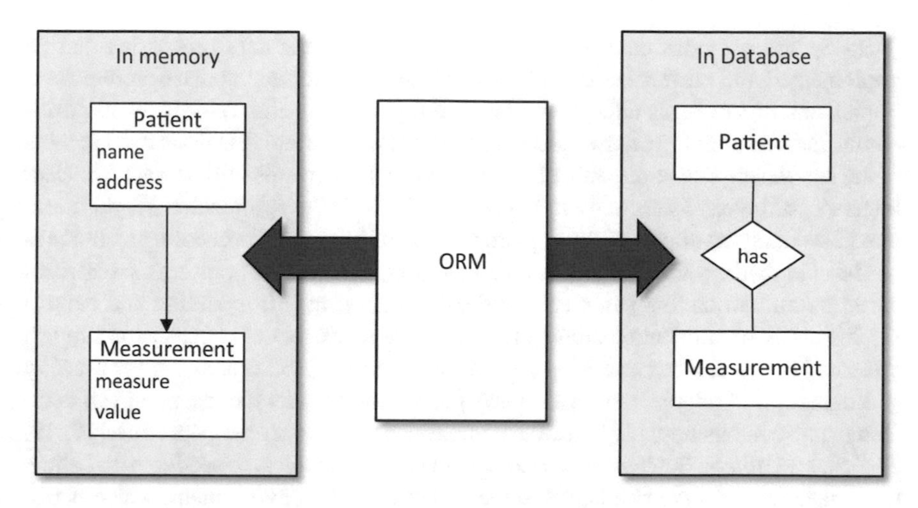

Fig. 1 Role and place of object-relational mapping layer while developing database applications

This incompatibility, which is often referred to as the object-relational impedance mismatch, covers various difficulties while mapping objects or class definitions in the developed software application to database tables defined by relational schema. Object-relational mapping tools mitigate the problem of OR impedance mismatch and simplify building software applications that access data in relational database management systems (RDBMSs). Figure 1 shows the role and place of the ORM tools while developing database applications. ORM tools, however, do not provide solutions for people involved in the development of applications that make use of fuzzy data processing techniques.

In the paper, we show extensions to the Doctrine object-relational mapping tool that allow fuzzy processing of crisp data stored in relational databases. Doctrine ORM framework is one of several PHP libraries developed as a part of the Doctrine Project, which is primarily focused on database storage and object mapping. Doctrine has greatly benefited from concepts of the Hibernate ORM and has adapted these concepts to fit the PHP language. One of the Doctrine's key features is the option to write database queries in Doctrine Query Language (DQL), an object-oriented dialect of SQL, which we extended with the capability of fuzzy data processing.

2 Related Works

A typical architecture of database application contains at least two parts - a client and a server. The client part (front-end) runs the database application that retrieves database information and interacts with a user usually through the Graphical User Interface (GUI). The server (back-end) hosts the Database Management System

(DBMS) and provides data for client applications. Fuzzy data processing can be implemented and performed in both parts of the client-server database architecture. Implementation of fuzzy data processing techniques on the client side involves development and embedding of processing logic as procedures and functions that are parts of the software application, either bound to particular controls of the Graphical User Interface, or invoked internally from other modules of the application. World literature gives many examples of the implementation of fuzzy data processing techniques on the client side while processing data in various domains, e.g. in: risk assessment based on human factors [2], database flexible querying [3], modeling and control [9], historical monuments searching [10], damage assessment [11], decision support systems [16], searching candidates for a date, human profile analysis for missing and unidentified people, automatic news generation for stock exchange [21], decision making in business [29], flexible comparison of protein energy profiles [18, 19, 23–25], and others. Such an approach has several advantages. Procedures that allow fuzzy data processing are adapted to the object-oriented environment, which is frequently used in the development of such applications. They are also adjusted to the specificity of the application and to the specificity of processed data. The procedures and functions are usually tailored to data being processed. The huge disadvantage of such a solution is a tight coupling to a particular application, and consequently, a negligible re-usability of these procedures in other applications and in the analysis of other type of data.

Alternatively, the same fuzzy processing functionality can be implementation as additional procedures and functions on the server side, extending the processing capabilities of the database management system and querying capabilities of the SQL language. This involves implementation of the procedures and functions in the programming language native for the particular DBMS. Examples of such implementations are: SQLf [5, 30], FQUERY [14], Soft-SQL [4], FuzzyQ [21], fuzzy SQL extensions for relational databases [17, 20, 28], possibilistic databases [26], and for data warehouses [1, 22]. Proposed fuzzy extensions to the DBMSs allow to perform various operations on relational data, including in particular fuzzy selection [12, 14, 17, 21, 22, 27, 28], division of fuzzy relations [7], several non-commutative connectives [6], fuzzy grouping [8, 20], filtering groups with fuzzy linguistic quantifiers [13, 15], and others. These solutions, which extend existing relational DBMSs and the syntax of the SQL language, provide universal routines that can be utilized in fuzzy processing of various data coming from different domains. This versatility is a great asset. On the other hand, such an implementation binds users and software developers to particular database management system and its native language, which also has some limitations. One of them is low portability between various DBMSs as it requires re-implementation of the fuzzy procedures in the native language of a particular DBMS.

In both mentioned approaches, the prevalent problem is mapping between classes of the client software application and database tables, which is necessary for applications that manipulate and persist data. In the past, each application created its own mapping layer in which the client application's specific classes were mapped to specific tables in the relational database using dedicated SQL statements. Object-relational mapping tools mitigate this problem by providing additional layer in the data access stack (Fig. 3) that maps relational data to class objects. However, flexible operating on objects residing in memory with the use of fuzzy techniques requires extensions to the ORM layer. In this paper, we propose such an approach that delivers selected functionalities of fuzzy data processing, in particular fuzzy selection and fuzzy data filtering, in a form of additional module extending the Doctrine ORM framework.

3 Fuzzy Selection over Collections of Objects

Fuzzy selection allows to filter out data on the basis of fuzzy search conditions (fuzzy predicates). In our extension to the Doctrine ORM framework we provide various fuzzy selection modes, which are formally defined in this section. Since ORM layer maps database tables to object classes, we will operate on classes and objects while presenting the implemented selection modes.

Given a class C with n attributes:

$$C = \{A_1 A_2 A_3 \dots A_n\}, \tag{1}$$

which corresponds with a relation R of the same schema in a relational database, and given a collection of objects of the class C:

$$O^C = \{o_i^C | i = 1 \dots m, m \in \mathbb{N}_+\}, \tag{2}$$

where m is the number of objects in the collection corresponding with tuples of the relation R in the database, a *fuzzy selection* $\tilde{\sigma}$ is a unary operation that denotes a subset of a collection O^C on the basis of fuzzy search condition:

$$\tilde{\sigma}_{A_i \overset{\lambda}{\approx} v}(O^C) = \{o : o \in O^C, o(A_i) \overset{\lambda}{\approx} v\}, \tag{3}$$

where: $A_i \overset{\lambda}{\approx} v$ is a fuzzy search condition, A_i is one of attributes of the class C for $i = 1 \dots n$, n is the number of attributes of the class C, v is a fuzzy set (e.g., *young* person, *tall* man, *normal* blood pressure, *age near* 30), $\approx$ is a comparison operator used to compare crisp value of attribute A_i for each object o from the collection O^C with fuzzy set v, λ is a minimum membership degree for which the search condition is satisfied.

The selection $\tilde{\sigma}_{A_i \overset{\lambda}{\approx} v}(O^C)$ denotes all objects in O^C for which $\approx$ holds between the attribute A_i and the fuzzy set v with the membership degree greater or equal to λ. Therefore,

$$\tilde{\sigma}_{A_i \overset{\lambda}{\approx} v}(O^C) = \{o : o \in O^C, \mu_v(o(A_i)) \geq \lambda\}, \tag{4}$$

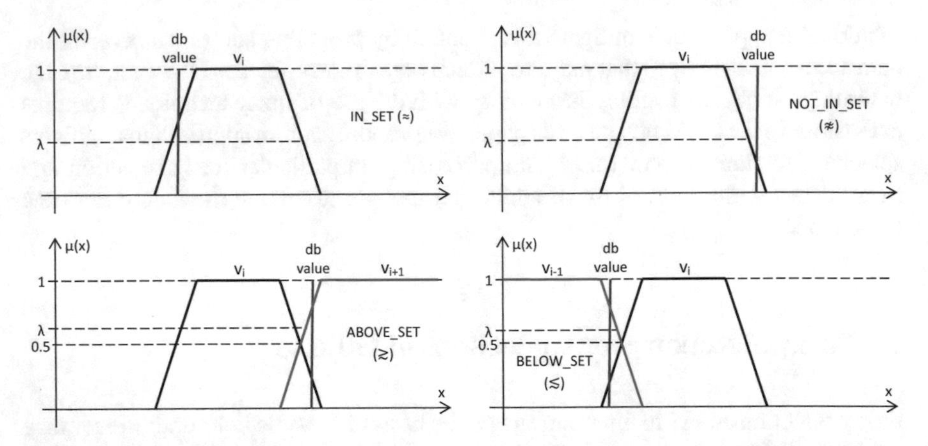

Fig. 2 Various selection modes showing a database value satisfying the selection criteria for the given λ cut-off similarity degree: (top left) *in set* selection ($value \approx v_i$, e.g., normal diastolic blood pressure), (top right) *not in set* selection ($value \not\approx v_i$, e.g., not normal diastolic blood pressure), (bottom left) *above set* selection ($value \gtrsim v_i$, e.g., diastolic blood pressure above the normal level), (bottom right) *below set* selection ($value \lesssim v_i$, e.g., diastolic blood pressure below the normal level)

where μ_v is a membership function of a fuzzy set v.

The fuzzy set v can be defined by various types of membership functions, including triangular, trapezoidal, and Gaussian. Fuzzy selection can be performed on the basis of multiple fuzzy search conditions (or mixed with crisp search conditions), e.g.:

$$\tilde{\sigma}_{A_i \overset{\lambda_i}{\approx} v_i \ \Theta \ldots \Theta \ A_j \overset{\lambda_j}{\approx} v_j}(O^C) = \{o : o \in O^C, o(A_i) \overset{\lambda_i}{\approx} v_i \ \ \Theta \ldots \Theta \ \ o(A_j) \overset{\lambda_j}{\approx} v_j\}, \quad (5)$$

where: A_i, A_j are attributes of class C, $i, j = 1 \ldots n, i \neq j$, v_i, v_j are fuzzy sets, λ_i, λ_j are minimum membership degrees for particular fuzzy search conditions, and Θ can be any of logical operators of conjunction or disjunction $\Theta \in \{\wedge, \vee\}$. Therefore:

$$\begin{aligned} \tilde{\sigma}_{A_i \overset{\lambda_i}{\approx} v_i \ \Theta \ldots \Theta \ A_j \overset{\lambda_j}{\approx} v_j}(O^C) = \{o : o \in O^C, \mu_{v_i}(o(A_i)) \geq \lambda_i \\ \Theta \ldots \Theta \ \ \mu_{v_j}(o(A_j)) \geq \lambda_j\}, \end{aligned} \quad (6)$$

where: μ_{v_i}, μ_{v_j} are membership functions of fuzzy sets v_i, v_j.

Standard comparison operator $\approx$ used in fuzzy selection allows to check whether a database value (mapped to attribute of a class) is in certain range (Fig. 2 top left), e.g., blood pressure is normal, body mass index is around 28, a person is young, or outside of the range (Fig. 2 top right). We also give the possibility to perform fuzzy selections that allow to check whether some database values are above or below a certain range, represented by a fuzzy set. To this purpose we use different comparison operators $\gtrsim$ and $\lesssim$.

Selection with the *above set* fuzzy search condition $\tilde{\sigma}_{A_i \gtrsim v}^{\lambda}(O^C)$ denotes all objects in O^C for which $\gtrsim$ holds between the attribute A_i and the fuzzy set v_i with the membership degree greater or equal to λ (Fig. 2 bottom left). The selection is defined as follows:

$$\tilde{\sigma}_{A_i \gtrsim v_i}^{\lambda}(O^C) = \{o : o \in O^C, \mu_{v_{i+1}}(o(A_i)) \geq \lambda\}, \tag{7}$$

where μ_{v_i} is a membership function of a fuzzy set v_i that represents a certain range (e.g., normal diastolic blood pressure), $\mu_{v_{i+1}}$ is a membership function of a fuzzy set v_{i+1} that represents a range of values above the certain range v_i (e.g., diastolic blood pressure above the normal). Therefore:

$$\tilde{\sigma}_{A_i \gtrsim v_i}^{\lambda}(O^C) = \tilde{\sigma}_{A_i \approx v_{i+1}}^{\lambda}(O^C). \tag{8}$$

Similarly, selection with the *below set* fuzzy search condition $\tilde{\sigma}_{A_i \lesssim v}^{\lambda}(O^C)$ denotes all objects in O^C for which $\lesssim$ holds between the attribute A_i and the fuzzy set v_i with the membership degree greater or equal to λ (Fig. 2 bottom right). The selection is defined as follows:

$$\tilde{\sigma}_{A_i \lesssim v_i}^{\lambda}(O^C) = \{o : o \in O^C, \mu_{v_{i-1}}(o(A_i)) \geq \lambda\}, \tag{9}$$

where μ_{v_i} is a membership function of a fuzzy set v_i that represents a certain range (e.g., normal diastolic blood pressure), $\mu_{v_{i-1}}$ is a membership function of a fuzzy set v_{i-1} that represents a range of values below the certain range v_i (e.g., diastolic blood pressure below the normal). Therefore:

$$\tilde{\sigma}_{A_i \lesssim v_i}^{\lambda}(O^C) = \tilde{\sigma}_{A_i \approx v_{i-1}}^{\lambda}(O^C). \tag{10}$$

Fuzzy sets v_{i-1} and v_{i+1} are parts of the *complement* $\hat{v_i}$ of the fuzzy set v_i:

$$\hat{v_i} = v_{i-1} \cup v_{i+1}, \tag{11}$$

represented by a membership function:

$$\mu_{\hat{v_i}}(o(A_i)) = 1 - \mu_{v_i}(o(A_i)). \tag{12}$$

The complement operation is used to implement logical negation (NOT, $\not\approx$).

4 Extensions to Doctrine ORM

Presented fuzzy selection modes were implemented in the *Fuzzy* module extending standard capabilities of the Doctrine ORM library. The library with the *Fuzzy* extension enables fuzzy processing of crisp data retrieved from relational database and mapped to class objects. Figure 3 shows location of Doctrine ORM and the *Fuzzy* module in the data access stack. PHP applications can access data through Doctrine ORM that maps relational tables stored persistently in Database to objects kept in memory. The PHP Data Objects (PDO) provides a data-access abstraction layer and defines a consistent interface for accessing databases in PHP. The Doctrine database abstraction and access layer (Doctrine DBAL) provides a lightweight and thin runtime layer over the PDO and a lot of additional, horizontal features like database schema introspection and manipulation through an object-oriented application programming interface. In this section, we describe the most important classes extending standard functionality of the Doctrine ORM library with the possibility of fuzzy data processing. Additionally, we present a sample usage of the library classes in a real application, while performing a simple fuzzy analysis of ambulatory data.

4.1 Class Model of the **Fuzzy** *Module*

The *Fuzzy* module is a dedicated programming module that extends the Doctrine object-relational mapping tool and provides the capability of fuzzy processing of data stored in a relational database. The module is available at https://gitlab.com/auroree/fuzzy. The *Fuzzy* module is universal, i.e., independent of the domain of

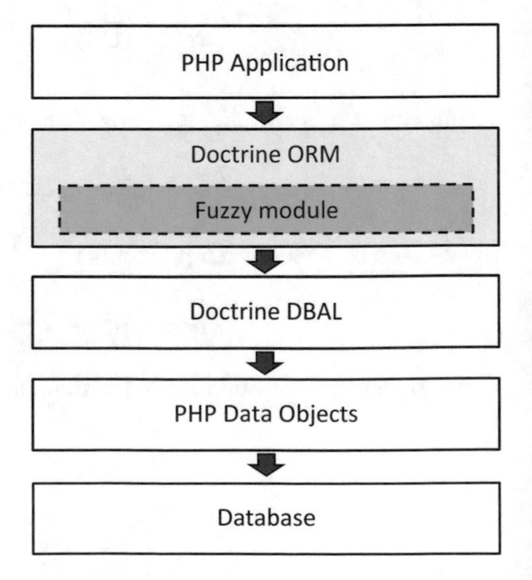

Fig. 3 Doctrine object-relational mapping layer and location of *Fuzzy* module proposed in the paper within the data access stack

developed application and data stored in a database. Software developers can use the implemented functionality for fuzzy processing of any values stored in the database, e.g., atmospheric pressure, body temperature, person's age, or the number of hours spent watching television. It is necessary to select the type of membership function defining a fuzzy set and provide the relevant parameters.

The *Fuzzy* module consists of classes that provide implementations of functions executed while performing fuzzy data processing, including those used for defining fuzzy sets used in fuzzy selection, those used to control how to perform the selection, i.e., how to compare data, and those extending the Doctrine Query Language (DQL) – the query language of the Doctrine ORM library. Classes delivered by the *Fuzzy* module are presented in Fig. 4. They are marked in green, in contrast to native PHP classes and classes of the Doctrine ORM/DBAL library that are marked in white. In order to enable fuzzy data processing in the ORM layer we had to implement a set of classes and methods that lead to proper generation of SQL queries for particular functions of fuzzy data processing. To this purpose, we have extended the *Doctrine\ORM\Query\AST\Functions/FunctionNode* class provided by the Doctrine ORM library (Fig. 4). Classes that inherit from the *FunctionNode* class are divided into two groups placed in separate namespaces: membership functions (e.g., *InRange*, *Near*, *RangeUp*) and general-purpose functions (e.g., *Floor*, *Date*). They all implement two important methods: *parse* and *getSql*. The *parse* method detects function parameters in the DQL expression and stores them for future use, then the *getSql* method generates an expression representing a particular mathematical function in the native SQL for a database.

The *InRange* class represents classical (*LR*) trapezoidal membership function and is suitable to describe values of a domain, e.g., a fuzzy set of *normal* blood pressure, but also *near optimal* LDL Cholesterol (which are in certain ranges of values). The *RangeUp* and *RangeDown* classes represent special cases of trapezoidal membership functions - L-functions (with parameters $n = p = +\infty$) and R-functions (with parameters $l = m = -\infty$), respectively. They are both defined automatically with respect to the fuzzy set of chosen value of a domain (as a complement of the fuzzy set, Eqs. 11 and 12). They are suitable to formulate selection conditions, such as *HDL below the norm* or *slow heart rate* with the use of R-functions (according to Eqs. 9 and 10) and *LDL above the norm* or *high blood pressure* by using L-functions (according to Eqs. 7 and 8). Classes used to represent particular cases of trapezoidal membership functions and names of DQL functions delivered by them are presented in Fig. 5.

The *Near* class (Fig. 4) represents triangular membership functions and is suitable, e.g., in formulating fuzzy search conditions, like *age about 35*. The *NearGaussian* class represents Gaussian membership function and has similar purpose to triangular membership function.

The *Fuzzy* module also provides a function factory (*FuzzyFunctionFactory* class, Fig. 4), which creates instances of classes for the selected membership functions, based on the specified type of the function (one of the values of *FuzzyFunctionTypes*: IN_RANGE, NEAR, NEAR_GAUSSIAN), e.g., instance of the *NearFunction* class for the *Near* characteristic function. The function factory class generates appropri-

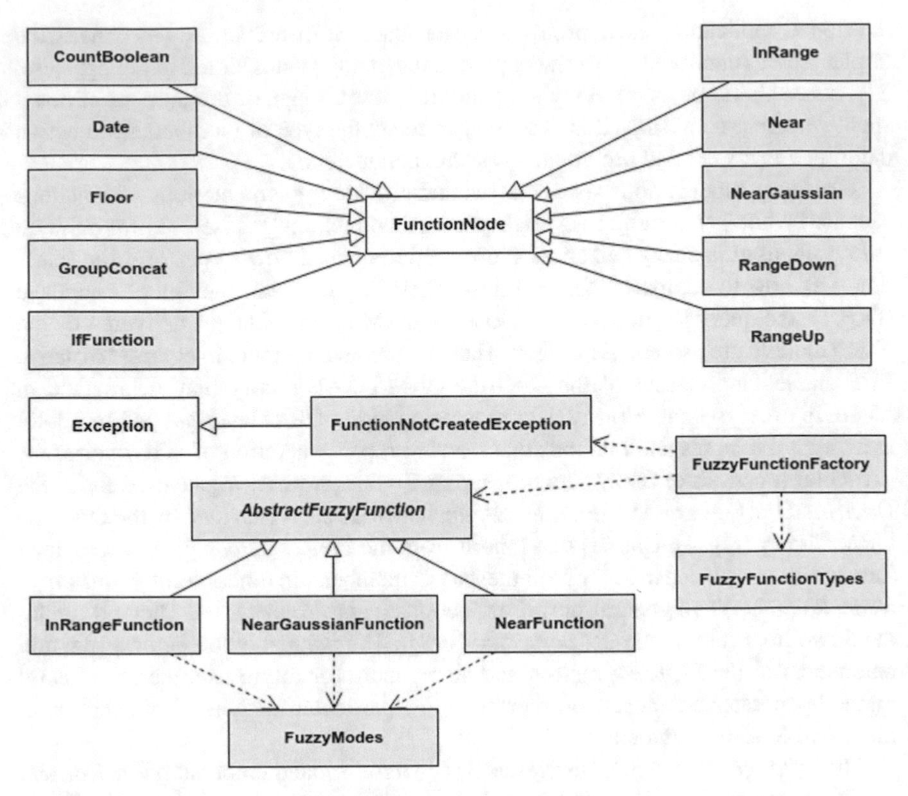

Fig. 4 Main classes provided by the *Fuzzy* module extending the Doctrine ORM library

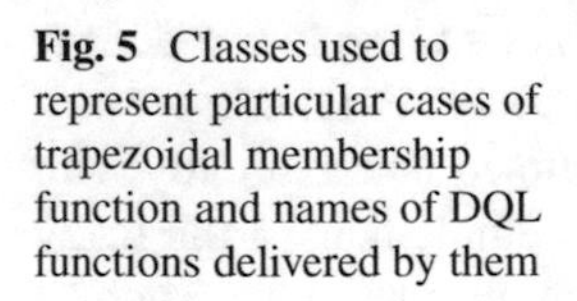

Fig. 5 Classes used to represent particular cases of trapezoidal membership function and names of DQL functions delivered by them

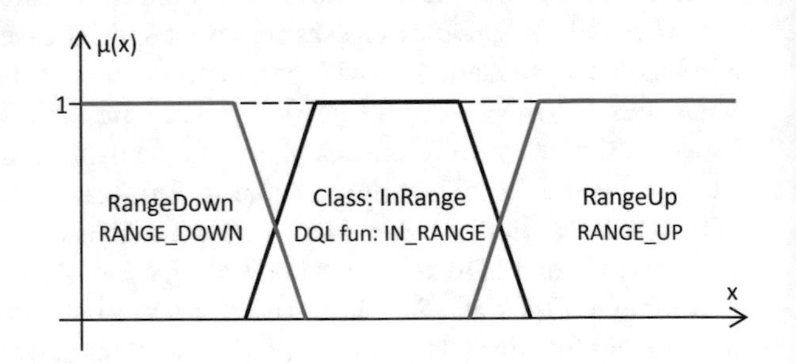

ate DQL expression together with fuzzy selection condition (as formally defined in Sect. 3) on the basis of declared selection mode/query type. Accepted query types are constants of the class *FuzzyModes*. The selection mode/query type (one of the *FuzzyModes*: NONE, IN_SET, NOT_IN_SET, ABOVE_SET, BELOW_SET) decides how to compare a database value with the defined fuzzy set (Eqs. 3, 4, 7 and 9). Selection modes are graphically presented in Fig. 2 in Sect. 3.

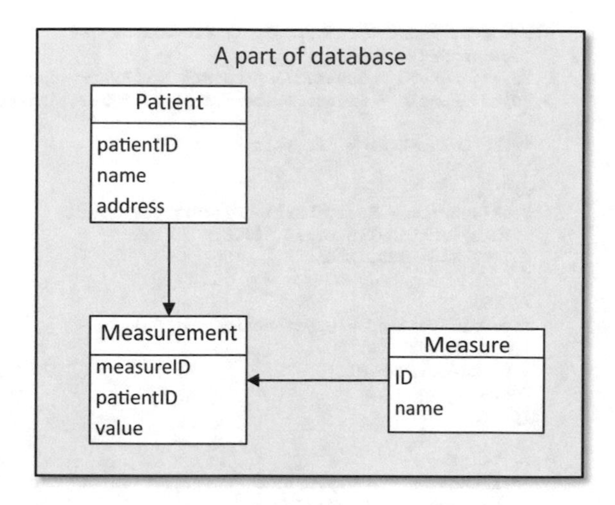

Fig. 6 A part of the database used in sample codes

4.2 Sample Usage of the Doctrine ORM Library with Fuzzy Extensions

This section shows sample usage of the Doctrine ORM library with the developed *Fuzzy* module in the analysis of ambulatory data. We show how the fuzzy extensions are utilized in the PHP code of our software application that allows reporting on measurements stored in MySQL relational data repository by calling appropriate DQL query through the Doctrine ORM library. Finally, we present the form of the generated SQL query executed in the relational database that corresponds to the DQL query.

Presented sample of the code refers to a part of the database presented in Fig. 6. The database consists of relational table *measurement* containing ambulatory measurements in the *value* attribute for particular measures identified by *measure_id* attribute and particular patients identified by *patient_id*.

In the ORM layer, the *measurement* table is mapped to a class called *Measurement*, which attributes correspond to fields (columns) of the *measurement* table. In the sample code presented in Fig. 7 we impose the following fuzzy search conditions:

$$\tilde{\sigma}_{\substack{(value \overset{0.5}{\approx} normal \ AND \ measureID=2) \\ OR \ (value \overset{0.5}{\gtrsim} normal \ AND \ measureID=3)}} (measurement), \tag{13}$$

where: $measureID = 2$ corresponds to systolic blood pressure, $measureID = 3$ corresponds to diastolic blood pressure, the *normal* fuzzy set is represented by trapezoidal membership function with proper parameters specific for both measures (Fig. 7, sections (3) and (4)).

```
/** @var MeasurementRepository $repository */
$repository = ...;
$sysMeasureId = $measureRepository->getByName(systolic)->getId();  // (1)
$diaMeasureId = $measureRepository->getByName(diastolic)->getId(); // (1)

$valueColumnName = 'mm.value'; // (2)

// (3)
$sysFunction = FuzzyFunctionFactory::create(
  FuzzyFunctionTypes::IN_RANGE,
  [90, 110, 130, 135]
);
// (8)
$sysDqlCondition = $sysFunction->getDql(
  FuzzyModes::IN_SET,
  $valueColumnName,
  0.5
);

// (4)
$diaFunction = FuzzyFunctionFactory::create(
  FuzzyFunctionTypes::IN_RANGE,
  [50, 65, 80, 90]
);
// (8)
$diaDqlCondition = $diaFunction->getDql(
  FuzzyModes::ABOVE_SET,
  $valueColumnName,
  0.5
);

// (9)
$query = $repository->createQueryBuilder('mm')
  ->select('p.id, m.name, mm.value')
  ->join('mm.patient', 'p') // (5)
  ->join('mm.measure', 'm') // (6)
  ->where("(mm.measure = {$sysMeasureId} AND {$sysDqlCondition})
    OR (mm.measure = {$diaMeasureId} AND {$diaDqlCondition})") // (7)
  ->getQuery();

$result = $query->getResult();
```

Fig. 7 Sample usage of the *Fuzzy* module in PHP code

Fuzzy search conditions, created by means of appropriate classes of the *Fuzzy* module, are imposed on the *value* attribute - Fig. 7, section (2) - for systolic blood pressure and diastolic blood pressure measure selected in section (1). In the presented example we assume that domains of both measures are divided into three fuzzy sets: *normal*, *low*, and *high* blood pressure, according to applicable standards for systolic and diastolic blood pressure. The starting point in this case is to define fuzzy sets for *normal* systolic and diastolic blood pressure with respect to which we define *low* and *high* fuzzy sets for both measures. To represent *normal* blood pressure we use trapezoidal membership functions (specified by *IN_RANGE* function type in the *Fuzzy* module) with 90, 110, 130, 135 parameters for systolic blood pressure (3) and with 50, 65, 80, 90 parameters for diastolic blood pressure (4). We define both membership functions by invocation of the *create* function of the *FuzzyFunctionFactory*

```
SELECT p.id, m.name, mm.value
FROM Mazurkiewicz\TrackerBundle\Entity\Measurement mm
   INNER JOIN mm.patient p
   INNER JOIN mm.measure m
WHERE (mm.measure = 2 AND IN_RANGE(mm.value, 90, 110, 130, 135) >= 0.5)
    OR (mm.measure = 3 AND RANGE_UP(mm.value, 80, 90) >= 0.5)
```

Fig. 8 DQL query with two fuzzy search conditions for PHP code shown in Fig. 7

class. We are interested in selecting patients, whose systolic blood pressure is normal and diastolic blood pressure is elevated (*high*), i.e., above the *normal* value, with the minimum membership degree equal to 0.5 (8). Therefore, we have to define fuzzy search conditions by using *getDql* method of the *InRangeFunction* class instance returned by the *FuzzyFunctionFactory*. Then, we have to use the *IN_SET* and the *ABOVE_SET* selection modes in the *getDql* method in order to get *normal* values of systolic blood pressure and elevated values of diastolic blood pressure (above the *normal*). In such a way, we obtain two fuzzy search conditions (constructed according to Eqs. 4 and 7) for DQL query that will be used in the *WHERE* clause. To build the whole analytical report we formulate the query by using Doctrine's QueryBuilder that allows to construct queries through a fluent interface (9). We join the *Patient* (5) and the *Measure* (6) classes to add data about patients and measure types (laboratory tests). Finally, we add fuzzy search conditions in (7). The query will return only those rows for which values of systolic blood pressure belongs to the *normal* fuzzy set and values of diastolic blood pressure belongs to the *high* fuzzy set (i.e., *above the normal*).

The PHP code presented in Fig. 7 produces the DQL query shown in Fig. 8, which will be translated to SQL query for relational database (Fig. 9) by the Doctrine

```
SELECT p0_.id AS id_0, m1_.name AS name_1, m2_.value AS value_2
FROM measurement m2_
   INNER JOIN patient p0_ ON m2_.patient_id = p0_.id
   INNER JOIN measure m1_ ON m2_.measure_id = m1_.id
WHERE
      (m2_.measure_id = 2 AND CASE
        WHEN m2_.value <= 90 THEN 0
        WHEN m2_.value <= 110 THEN (m2_.value-90)/(110-90)
        WHEN m2_.value <= 130 THEN 1
        WHEN m2_.value <= 135 THEN (135-m2_.value)/(135-130)
        ELSE 0 END >= 0.5
     )
     OR (m2_.measure_id = 3 AND CASE
         WHEN m2_.value <= 80 THEN 0
         WHEN m2_.value <= 90 THEN (m2_.value-80)/(90-80)
         ELSE 1 END >= 0.5
     )
```

Fig. 9 SQL query translated by fuzzy extension of the Doctrine library from DQL query presented in Fig. 8

library. Translation of the query built up with the PHP Query Builder (section (9) in Fig. 7) to DQL query (Fig. 8) produces WHERE clause containing invocations of the IN_RANGE and RANGE_UP functions (cf. Fig. 5) with appropriate parameters of LR-type and L-type membership functions representing *normal* and *high* fuzzy sets for particular measures.

These invocations are then translated to the CASE ... WHEN ... THEN statements in the WHERE clause of the SQL query command (Fig. 9).

5 Experimental Results

We tested performance of the fuzzy extension for the Doctrine ORM library in several series of tests. We were primarily interested in verification of how the necessity of calculation of the value of a membership function influences the execution time of particular fuzzy queries with respect to classical queries that operate on given ranges of values (appropriately chosen intervals). For this purpose, we used a database containing 2,500,000 records in the *measurement* table coming from laboratory tests.

Results of performance tests are presented in Fig. 12 for three fuzzy queries (Q1–Q3) retrieving measurement data for patients having:

- Q1 - *normal* systolic blood pressure (Fig. 10a),
- Q2 - systolic blood pressure *above the normal* (Fig. 10a),
- Q3 - *normal* platelet count (PLT) (Fig. 10b),

with a minimum membership degree $\lambda = 0.5$. Queries Q1–Q3 contain fuzzy search conditions. Definitions of fuzzy sets used in these search conditions are presented in Fig. 10.

Additionally, for queries Q1–Q3 we created their classical counterparts with precise search criteria based on intervals, where left and right boundaries of the intervals were calculated for the membership degree $\lambda = 0.5$ in fuzzy queries Q1–Q3. Particular fuzzy queries and their precise counterparts (Fig. 11) returned the same sets of

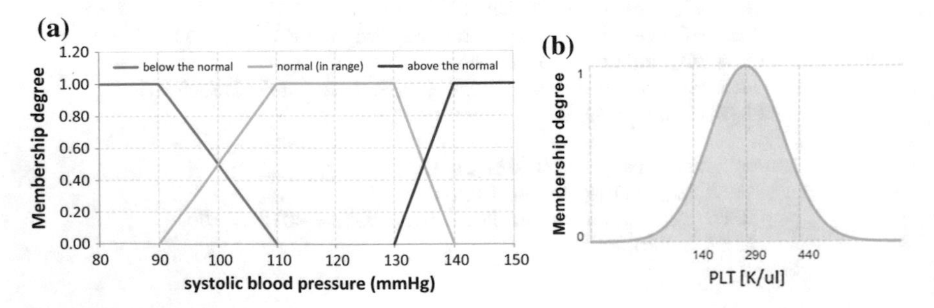

Fig. 10 Membership functions for *systolic blood pressure* (**a**). Membership function for the fuzzy set *normal PLT* (**b**)

```
-- Q1. normal systolic blood pressure

-- Classic SQL query (defines an interval)
SELECT m0_.id AS id_0, m0_.observation_date AS observation_date_1, m0_.value AS value_2,
   m0_.measure_id AS measure_id_8, m0_.patient_id AS patient_id_9
FROM measurement m0_
WHERE m0_.measure_id = 2 AND m0_.patient_id = 2
    AND m0_.value <= 135 AND m0_.value >= 100;

-- SQL query translated from DQL
SELECT m0_.id AS id_0, m0_.observation_date AS observation_date_1, m0_.value AS value_2,
   m0_.measure_id AS measure_id_8, m0_.patient_id AS patient_id_9
FROM measurement m0_
WHERE m0_.measure_id = 2 AND m0_.patient_id = 2
    AND CASE
        WHEN m0_.value <= 90 THEN 0
        WHEN m0_.value <= 110 THEN (m0_.value-90)/(110-90)
        WHEN m0_.value <= 130 THEN 1
        WHEN m0_.value <= 140 THEN (140-m0_.value)/(140-130)
        ELSE 0 END >= 0.5;

-- Q2. systolic blood pressure above the normal

-- Classic SQL query (defines an interval)
SELECT m0_.id AS id_0, m0_.observation_date AS observation_date_1, m0_.value AS value_2,
   m0_.measure_id AS measure_id_8, m0_.patient_id AS patient_id_9
FROM measurement m0_
WHERE m0_.measure_id = 2 AND m0_.patient_id = 2
    AND m0_.value >= 135;

-- SQL query translated from DQL
SELECT m0_.id AS id_0, m0_.observation_date AS observation_date_1, m0_.value AS value_2,
   m0_.measure_id AS measure_id_8, m0_.patient_id AS patient_id_9
FROM measurement m0_
WHERE m0_.measure_id = 2 AND m0_.patient_id = 2
    AND CASE
        WHEN m0_.value <= 130 THEN 0
        WHEN m0_.value <= 140 THEN (m0_.value-130)/(140-130)
        ELSE 1 END >= 0.5;

-- Q3. normal platelet count (PLT)

-- Classic SQL query (defines an interval)
SELECT m0_.id AS id_0, m0_.observation_date AS observation_date_1, m0_.value AS value_2,
   m0_.measure_id AS measure_id_8, m0_.patient_id AS patient_id_9
FROM measurement m0_
WHERE m0_.measure_id = 10 AND m0_.patient_id = 2
    AND m0_.value >= 165.117 AND m0_.value <= 414.883;

-- SQL query translated from DQL
SELECT m0_.id AS id_0, m0_.observation_date AS observation_date_1, m0_.value AS value_2,
   m0_.measure_id AS measure_id_8, m0_.patient_id AS patient_id_9
FROM measurement m0_
WHERE m0_.measure_id = 10 AND m0_.patient_id = 2
    AND EXP(-(m0_.value-290)*(m0_.value-290)/(150*150)) >= 0.5;
```

Fig. 11 Sample queries used in performance tests

results, but were parametrized in a different way - precise queries need exact values of left and right boundaries of intervals, while fuzzy queries need only the minimum membership degree λ.

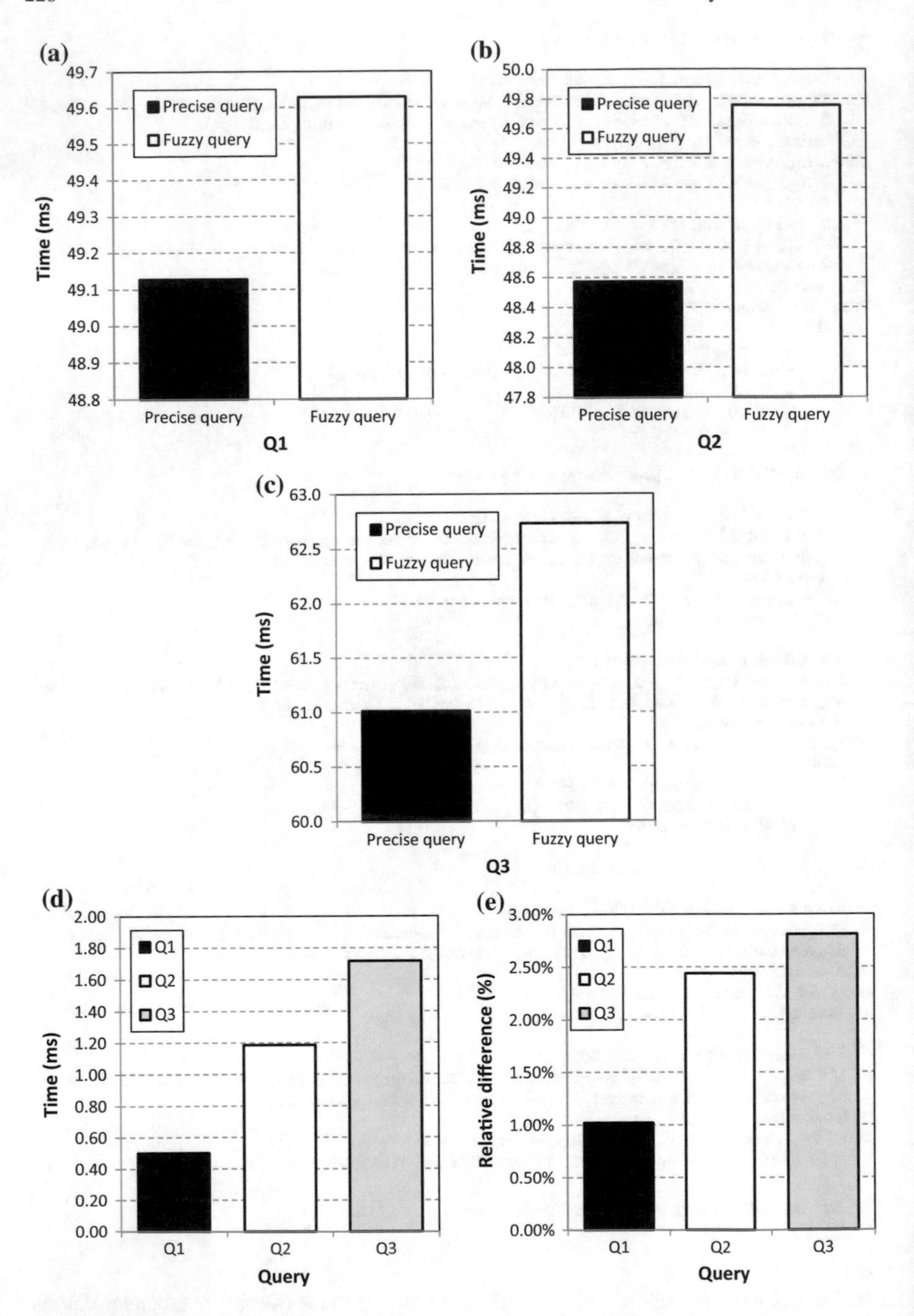

Fig. 12 Results of performance tests for fuzzy queries Q1–Q3 and their precise counterparts for the minimum membership degree $\lambda = 0.5$: **a**–**c** execution times for fuzzy queries Q1–Q3, **d** execution time difference, **e** relative difference

(a)

Observation date	Value	
2017-05-12	110	Edit Delete
2017-05-11	141	Edit Delete
2017-05-10	146	Edit Delete
2017-05-09	136	Edit Delete
2017-05-08	105	Edit Delete
2017-05-07	111	Edit Delete
2017-05-06	107	Edit Delete
2017-05-05	125	Edit Delete
2017-05-04	104	Edit Delete
2017-05-03	114	Edit Delete

Fig. 13 Sample application of *Fuzzy* module in *Measurement log for monitoring biomedical indicators*: **a** a form showing sample daily measurements of *systolic blood pressure*, **b** sample personal measurement report with fuzzy filtering capability on the basis of fuzzy selection

Results of performance tests presented in Fig. 12 proved that execution times of fuzzy queries were only slightly worse than execution times of precise queries that returned the same sets of results. Fuzzy queries were executed relatively 1–3% longer than their precise counterparts. This means that for users of the ORM library with fuzzy extensions the difference in execution time is almost imperceptible.

Presented *Fuzzy* extensions to the Doctrine ORM library were utilized and tested while developing several systems for collecting measurements, including measurement log for monitoring biomedical indicators, nutrition diary, and training activity log. Sample forms from *Measurement log for monitoring biomedical indicators* showing sample daily measurements of systolic blood pressure and fuzzy filtering are presented in Fig. 13.

(b)

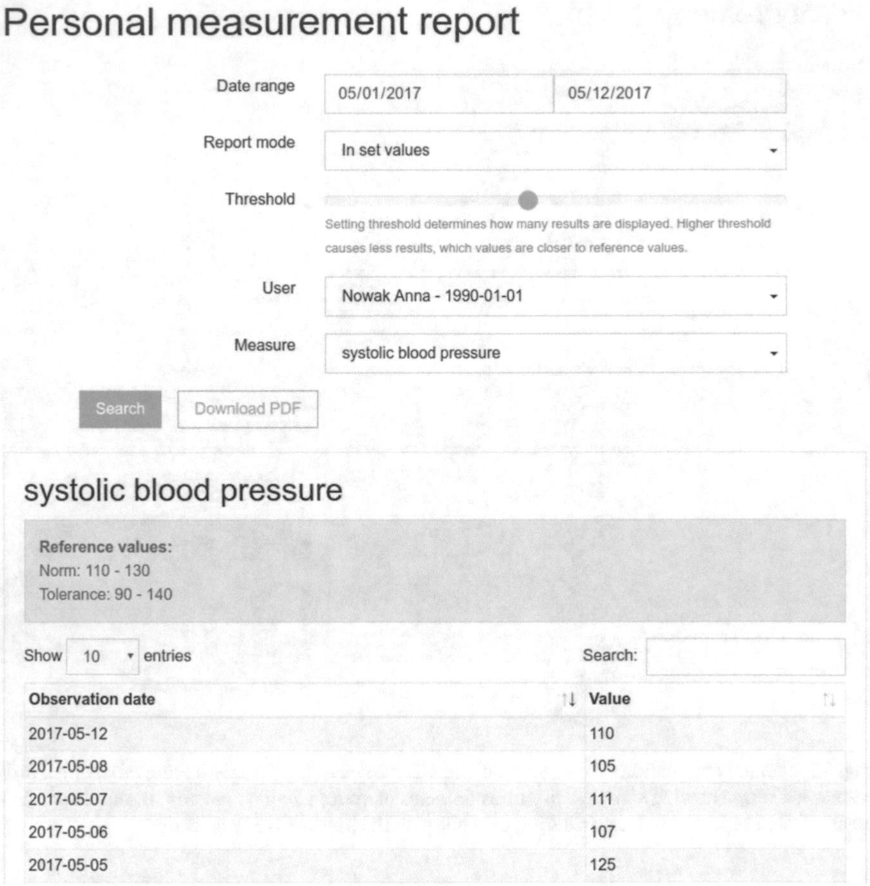

Personal measurement report

Date range 05/01/2017 05/12/2017

Report mode In set values

Threshold

Setting threshold determines how many results are displayed. Higher threshold causes less results, which values are closer to reference values.

User Nowak Anna - 1990-01-01

Measure systolic blood pressure

Search Download PDF

systolic blood pressure

Reference values:
Norm: 110 - 130
Tolerance: 90 - 140

Show 10 entries Search:

Observation date	Value
2017-05-12	110
2017-05-08	105
2017-05-07	111
2017-05-06	107
2017-05-05	125

Fig. 13 (continued)

6 Discussion and Concluding Remarks

Extensions of the Doctrine object-relational mapping framework toward fuzzy data processing allow to enhance standard database querying with new capabilities of imprecise, proximity-based or similarity-based searching. This makes the analysis of crisp, numerical data stored in databases more flexible, allows to broaden the set of results, and incorporate similar cases. Such a flexibility is important when processing large volumes of biomedical or ambulatory data, which usually have a numerical nature. Our works moved database searching with the use of fuzzy search conditions from front-end layer and database server-side layer to the ORM layer, which is important for software developers. As proved by our experiments, performance cost caused by incorporation of fuzzy data processing techniques in data retrieval is negligible compared to the additional analytic possibilities that are obtained by developers of database applications.

Our fuzzy extensions to the Doctrine library mitigate the problem of object-relational impedance mismatch for those software developers that want to perform fuzzy searches while working in the object-oriented model, regardless of the data domain being analyzed. It is limited to the PHP technology of building client software tools, but universal in terms of the type of analyzed data and built application. The *Fuzzy* module for the Doctrine ORM framework presented in the paper enables re-usability of procedures for fuzzy data processing for any client application that is developed and any data that is analyzed, which was a limitation of client-based solutions mentioned in Sect. 2. On the other hand, software developers are not bound to a particular database management system and its native query language, which was a weakness of server-side solutions presented in Sect. 2. This ensures broader portability of our fuzzy extension. In such a way, our solution complements the collection of existing solutions and, to the best of our knowledge, is the first attempt to incorporate fuzzy data processing in ORM frameworks.

Acknowledgements This work was supported by Statutory Research funds of Institute of Informatics, Silesian University of Technology, Gliwice, Poland (grant No BK-230/RAu2/2017).

References

1. Appelgren Lara, G., Delgado, M., Marín, N.: Fuzzy multidimensional modelling for flexible querying of learning object repositories. In: Larsen, H.L., Martin-Bautista, M.J., Vila, M.A., Andreasen, T., Christiansen, H. (eds.) Flexible Query Answering Systems. FQAS 2013, LNCS, vol. 8132, pp. 112–123. Springer, Berlin (2013). https://doi.org/10.1007/978-3-642-40769-7_10
2. Aras, F., Karakas, E., Bicen, Y.: Fuzzy logic-based user interface design for risk assessment considering human factor: a case study for high-voltage cell. Saf. Sci. **70**, 387–396 (2014)
3. Ben Hassine, M.A., Ounelli, H.: IDFQ: an interface for database flexible querying. In: Atzeni, P., Caplinskas, A., Jaakkola, H. (eds.) Advances in Databases and Information Systems. ADBIS 2008, LNCS, vol. 5207, pp. 112–126. Springer, Berlin (2008)
4. Bordogna, G., Psaila, G.: Customizable flexible querying in classical relational databases. In: Galindo, J. (ed.) Handbook of Research on Fuzzy Information Processing in Databases, pp. 191–217. IGI Global (2008)
5. Bosc, P., Pivert, O.: SQLf query functionality on top of a regular relational database management system. In: Pons, O., Vila, M.A., Kacprzyk, J. (eds.) Knowledge Management in Fuzzy Databases, Studies in Fuzziness and Soft Computing, vol. 39, pp. 171–190. Physica HD, Heidelberg (2000)
6. Bosc, P., Pivert, O.: On four noncommutative fuzzy connectives and their axiomatization. Fuzzy Sets Syst. **202**, 42–60 (2012)
7. Bosc, P., Pivert, O., Rocacher, D.: About quotient and division of crisp and fuzzy relations. J. Intell. Inf. Syst. **29**(2), 185–210 (2007)
8. Bosc, P., Pivert, O., Smits, G.: On a fuzzy group-by and its use for fuzzy association rule mining. In: Catania, B., Ivanović, M., Thalheim, B. (eds.) Advances in Databases and Information Systems. ADBIS 2010, LNCS, vol. 6295, pp. 88–102. Springer, Berlin (2010)
9. Cheng, S., Dong, R., Pedrycz, W.: A framework of fuzzy hybrid systems for modelling and control. Int. J. Gen. Syst. **39**(2), 165–176 (2010). https://doi.org/10.1080/03081070903427358

10. Czajkowski, K., Olczyk, P.: Fuzzy interface for historical monuments databases. In: Kozielski, S., Mrozek, D., Kasprowski, P., Małysiak-Mrozek, B., Kostrzewa, D. (eds.) Beyond Databases, Architectures, and Structures. BDAS 2014, CCIS, vol. 424, pp. 271–279. Springer International Publishing, Cham (2014)
11. Furuta, H., Shiraishi, N.: Fuzzy Data Processing in Damage Assessment, pp. 381–392. Vieweg+Teubner Verlag, Wiesbaden (1988)
12. Hudec, M.: An approach to fuzzy database querying, analysis and realisation. Comput. Sci. Inf. Syst. **12**, 127–140 (2009)
13. Kacprzyk, J., Ziólkowski, A.: Database queries with fuzzy linguistic quantifiers. IEEE Trans. Syst. Man Cybern. **16**(3), 474–479 (1986)
14. Kacprzyk, J., Zadrożny, S.: Data mining via fuzzy querying over the internet. In: Pons, O., Vila, M.A., Kacprzyk, J. (eds.) Knowledge Management in Fuzzy Databases, Studies in Fuzziness and Soft Computing, vol. 39, pp. 211–233. Physica HD, Heidelberg (2000)
15. Kacprzyk, J., Zadrożny, S.: Queries with fuzzy linguistic quantifiers for data of variable quality using some extended OWA operators. In: Andreasen, T., et al. (eds.) Flexible Query Answering Systems 2015, Advances in Intelligent Systems and Computing, vol. 400, pp. 295–305. Springer, Cham (2016)
16. Macwan, N., Sajja, P.S.: Fuzzy logic: an effective user interface tool for decision support system. Int. J. Eng. Sci. Innov. Technol. **3**(3), 278–283 (2014)
17. Małysiak, B., Mrozek, D., Kozielski, S.: Processing fuzzy SQL queries with flat, context-dependent and multidimensional membership functions. In: Hamza, M.H. (ed.) IASTED International Conference on Computational Intelligence, Calgary, Alberta, Canada, July 4–6, 2005. pp. 36–41. IASTED/ACTA Press (2005)
18. Małysiak, B., Momot, A., Kozielski, S., Mrozek, D.: On using energy signatures in protein structure similarity searching. In: Rutkowski, L., Tadeusiewicz, R., Zadeh, L.A., Zurada, J.M. (eds.) Artificial Intelligence and Soft Computing. ICAISC 2008, LNCS, vol. 5097, pp. 939–950. Springer, Berlin (2008)
19. Małysiak-Mrozek, B., Mrozek, D.: An improved method for protein similarity searching by alignment of fuzzy energy signatures. Int. J. Comput. Intell. Syst. **4**(1), 75–88 (2011). https://doi.org/10.1080/18756891.2011.9727765
20. Małysiak-Mrozek, B., Mrozek, D., Kozielski, S.: Data grouping process in extended SQL language containing fuzzy elements. In: Cyran, K.A., Kozielski, S., Peters, J.F., Stańczyk, U., Wakulicz-Deja, A. (eds.) Man-Machine Interactions, AISC, vol. 59, pp. 247–256. Springer, Berlin (2009). https://doi.org/10.1007/978-3-642-00563-3_25
21. Małysiak-Mrozek, B., Kozielski, S., Mrozek, D.: Modern software tools for researching and teaching fuzzy logic incorporated into database systems. In: Proceedings of the iNEER International Conference on Engineering Education, Gliwice, Poland. pp. 1–8. iNEER (2010). http://www.ineer.org/Events/ICEE2010/papers/T11D/Paper_954_1141.pdf
22. Małysiak-Mrozek, B., Mrozek, D., Kozielski, S.: Processing of crisp and fuzzy measures in the fuzzy data warehouse for global natural resources. In: García-Pedrajas, N., Herrera, F., Fyfe, C., Benítez, J.M., Ali, M. (eds.) Trends in Applied Intelligent Systems. IEA/AIE 2010, LNCS, vol. 6098, pp. 616–625. Springer, Berlin (2010). https://doi.org/10.1007/978-3-642-13033-5_63
23. Mrozek, D., Małysiak, B., Kozielski, S.: EAST: energy alignment search tool. In: Wang, L., Jiao, L., Shi, G., Li, X., Liu, J. (eds.) Fuzzy Systems and Knowledge Discovery. FSKD 2006, LNCS, vol. 4223, pp. 696–705. Springer, Berlin (2006). https://doi.org/10.1007/11881599_85
24. Mrozek, D., Małysiak, B., Kozielski, S.: An optimal alignment of proteins energy characteristics with crisp and fuzzy similarity awards. In: 2007 IEEE International Fuzzy Systems Conference, pp. 1513–1518 (2007)
25. Mrozek, D., Malysiak-Mrozek, B., Kozielski, S., Swierniak, A.: The Energy Distribution Data Bank: collecting energy features of protein molecular structures. In: 2009 Ninth IEEE International Conference on Bioinformatics and BioEngineering, pp. 301–306 (2009)
26. Myszkorowski, K.: Inference rules for fuzzy functional dependencies in possibilistic databases. In: Kozielski, S., Mrozek, D., Kasprowski, P., Małysiak-Mrozek, B., Kostrzewa, D. (eds.)

Beyond Databases, Architectures and Structures. Advanced Technologies for Data Mining and Knowledge Discovery. BDAS 2016, CCIS, vol. 613, pp. 181–191. Springer International Publishing, Cham (2016)
27. Pivert, O., Bosc, P.: Fuzzy Preference Queries to Relational Databases. Imperial College Press, London (2012)
28. Portinale, L., Montani, S.: A fuzzy logic approach to case matching and retrieval suitable to SQL implementation. In: Proceedings of the 2008 20th IEEE International Conference on Tools with Artificial Intelligence, vol. 02, pp. 241–245. ICTAI '08, IEEE Computer Society, Washington, DC, USA (2008)
29. Ribeiro, R.A., Moreira, A.M.: Fuzzy query interface for a business database. Int. J. Hum. Comput. Stud. **58**(4), 363–391 (2003)
30. Smits, G., Pivert, O., Girault, T.: Towards reconciling expressivity, efficiency and user-friendliness in database flexible querying. In: 2013 IEEE International Conference on Fuzzy Systems (FUZZ-IEEE), pp. 1–8 (2013)
31. Zadeh, L.: Fuzzy sets. Inf. Control **8**, 338–353 (1965)
32. Zadeh, L.: Fuzzy logic. Computer **21**(4), 83–93 (1988)

Multimodal Learning Determines Rules of Disease Development in Longitudinal Course with Parkinson's Patients

Andrzej W. Przybyszewski, Stanislaw Szlufik, Piotr Habela and Dariusz M. Koziorowski

Abstract Parkinson's disease (PD) is neurodegenerative disease (ND) related to the lost of dopaminergic neurons that elevates first by motor and later also by non-motor (dementia, depression) disabilities. Actually, there is no cure for ND as we are not able to revive death cells. Our purpose was to find, with help of data mining and machine learning (ML), rules that describe and predict disease progression in two groups of PD patients: 23 BMT patients that are taking only medication; 24 DBS patients that are on medication and on DBS (deep brain stimulation) therapies. In the longitudinal course of PD there were three visits approximately every 6 months with the first visit for DBS patients before electrode implantation. We have estimated disease progression as UPDRS (unified Parkinson's disease rating scale) changes on the basis of patient's disease duration, saccadic eye movement parameters, and neuropsychological tests: PDQ39, and Epworth tests. By means of ML and rough set theory we found rules on the basis of the first visit of BMT patients and used them to predict UPDRS changes in next two visits (global accuracy was 70% for both visits). The same rules were used to predict UPDRS in the first visit of DBS patients (global accuracy 71%) and the second (78%) and third (74%) visit of DBS patients during stimulation-ON. These rules could not predict UPDRS in DBS patients during stimulation-OFF visits. In

A. W. Przybyszewski (✉) · P. Habela
Polish-Japanese Academy of Information Technology, 02-008 Warsaw, Poland
e-mail: przy@pja.edu.pl

P. Habela
e-mail: piotr.habela@pja.edu.pl

S. Szlufik · D. M. Koziorowski
Faculty of Health Science, Department of Neurology,
Medical University Warsaw, Warsaw, Poland
e-mail: stanislaw.szlufik@gmail.com

D. M. Koziorowski
e-mail: dkoziorowski@esculap.pl

A. W. Przybyszewski
Department of Neurology, University of Massachusetts Medical School,
Worcester, MA 01655, USA

R. Bembenik et al. (eds.), *Intelligent Methods and Big Data in Industrial Applications*, Studies in Big Data 40, https://doi.org/10.1007/978-3-319-77604-0_17

summary, relationships between condition and decision attributes were changed as result of the surgery but restored by electric brain stimulation.

Keywords Neurodegenerative disease · Rough set · Decision rules · Granularity

1 Introduction

We have very limited knowledge about brain's plastic properties and compensatory mechanisms related to continuous death of neuron in the Central Nervous System. Till now, we did not achieve to construct an artificial NN with a similar to the brain compensatory mechanisms. Late diagnoses of the neurodegenerative diseases (ND) such as Alzheimer (AD) or Parkinson's (PD) is side effect of the brain plasticity as patients for a decade or two do not notice that cells in their brains are dying several time faster than in other people. As a consequence, the first symptoms are diagnosed when large parts of their brain are dead and we do not know how to recover dead cells. We can only make precise diagnoses of symptoms and in the PD case use medication to supplement lack of the neurotransmitter—dopamine.

Specialized in PD neurologists after many tests and by using their experience can implement individually adjusted therapy. However, in many cases therapy should be corrected and adjust with the disease development, but doctors have very limited time for each patient. Also their tests and approaches to patients may differ and changing doctor may lead to confusions and changes in the therapy. We propose to improve doctors' diagnoses by additional more automatic eye movement measurements. In addition, data mining and machine learning (ML) procedures based on rough set theory may improve prediction of disease development and optimize medications.

We have developed intelligent methods of symptom classification [1] that are similar to that found in the visual system for the complex objects recognition [2]. A fast and precise object classification in the visual system is possible as certain patterns are in-born and others are changing by continuous learning processes (brain plastic changes) [2]. Our algorithms follow visual system intelligent approach [2].

It is important to estimate the disease stage because it determines different sets of therapies. The neurological standards are based on Hoehn and Yahr and the UPDRS (Unified Parkinson's Disease Rating) scales. The last one is more precise and it will be used in this study. We would like to estimate disease progression in different groups of patients that were tested during three visits every half-year. Our method may lead to introduce more precise follow up and introduction of the possible internet-treatment.

2 Methods

Our data are from 47 Parkinson Disease (PD) patients divided into two groups: (1) 23 BMT (best medical treatment) patients that were only on medications; (2) 24 DBS patients on medication and DBS therapies. These went for the Deep Brain

Stimulation (DBS) implantation to the Institute of Neurology and Psychiatry WUM. The main indication for DBS are "wearing off' medication. All patients were tested in the following sessions: MedON/MedOFF sessions (sessions with or without medication). In addition DBS patients were also tested in StimON/StimOFF session were DBS stimulation was switched ON or OFF. All combinations gave four sessions: (1) MedOFFStimOFF; (2) MedOFFStimON; (3) MedONStimOFF; (4) MedONStimON. Details of these procedures were described earlier [2]. Tests of different motor and non-motor tasks (UPDRS = $\Sigma_{\mathrm{I}}^{\mathrm{IV}}$ UPDRS$_{\mathrm{i}}$) and neuropsychological tests were performed by neurologists from Warsaw Medical University. Fast eye movements (EM)—reflexive saccades (RS) were recorded as described in details before [1, 3]. Each patient sat watching a computer screen and has to follow randomly in delay and direction, horizontally moving to the right or the left dot after fixating on the starting point in the middle of the screen [3]. These EM tests were repeated ten times in each described above session. The following parameters of RS were measured: the delay (latency) related to time difference between the beginning of the light spot movements and the beginning of the eye movement; saccade's amplitude in comparison to the light spot amplitude; max velocity of the eye movement; duration of saccade defined as the time from the beginning to the end of the saccade.

Detailed procedures and orders of sessions were also described before [1].

Institutional Ethic Committee at the Warsaw Medical University approved all procedures.

2.1 Theoretical Basis

Our data mining analysis follows rough set theory after Zdzislaw Pawlak [4]). Our data are represented as a decision table where rows represented different measurements (may be obtained from the same or different patients) and columns were related to different attributes. An information system [4] is as a pair $S = (U, A)$, where U, A are finite sets: U is the universe of objects; and A is the set of attributes. The value $a(u)$ is a unique element of V (where V is a value set) for $a \in A$ *and* $u \in U$.

We define as in [4] the *indiscernibility relation* of any subset B of A or *IND*(B) as: $(x, y) \in IND(B)$ or $xI(B)y$ iff $a(x) = a(y)$ for every a $\in B$ where the value of $a(x) \in V$. It is an equivalence relation $[u]_B$ that we understand as a *B-elementary granule*. The family of $[u]_B$ gives the partition U/B containing u will be denoted by $B(u)$. The set $B \subset A$ of information system S is a reduct $IND(B) = IND(A)$ and no proper subset of B has this property [5]. In most cases, we are only interested in such reducts that are leading to expected rules (classifications). On the basis of the reduct we have generated rules using four different ML methods (RSES 2.2): exhaustive algorithm, genetic algorithm [6], covering algorithm, or LEM2 algorithm [7].

A **lower approximation** of set $X \subseteq U$ in relation to an attribute B is defined as $\underline{B}X = \{u \in U : [u]_B \subseteq X\}$. The **upper approximation** of X *is* defined as $\overline{B}X = \{u \in U : [u]_B \cap X \neq \phi\}$. The difference of $\overline{B}X$ *and* $\underline{B}X$ is the boundary region of

X that we denote as $BN_B(X)$. If $BN_B(X)$ is empty then set than *X* is *exact* with respect to *B*; otherwise if $BN_B(X)$ is not empty and *X* is not *rough* with respect to *B*.

A decision table (training sample in ML) for *S* is the triplet: $S = (U, C, D)$ where: *C, D* are condition and decision attributes [8]. Each row of the information table gives a particular rule that connects condition and decision attributes for a single measurements of a particular patient. As there are many rows related to different patients and sessions, they gave many particular rules. Rough set approach allows generalizing these rules into universal hypotheses that may determine optimal treatment options for an individual PD patient. The decision attribute *D* is giving a particular object (patient's state) classification by an expert (neurologist). Therefore a decision table classifies data by *supervised learning (ML)* where teaching is related to decisions made by neurologist(s). Each raw is an example of the teacher's decision.

It is well known that neurodegenerative processes start to accelerate a decade or two before the first PD symptoms and these processes are not exactly same in different patients. As a consequence, different patients need different treatments. The most neurologists use their intuition based on general medical rules and experiences to adapt an individual treatment plan. We would like to find rules more precisely dependent on an individual patient symptoms but also enough universal to describe symptoms of many different patients. Different rules' granularities (abstraction) are similar to complex objects recognition [2] and may simulate association processes of an ideal neurologist.

In our previous works [1, 3] we have divided experimental tests into several subsets, learned rules, tested each subset in several terms and averaged precisions of our predictions in order to get global precision (classical n-fold approach). In present study, we have used data from different treatments and group of patients for training and testing. The purpose was to find what are limits of rules that may predict symptoms development of patients with different treatments in different disease stages.

We have used the RSES 2.2 (Rough System Exploration Program) [9] with implementation of RS rules to process our data.

3 Results

All 47 PD patients have mean age of 56 ± 11.7 (SD) years with mean disease duration of 8.3 ± 3.7 years and BMT patients (only medication) and DBS patients (medication and with implanted electrodes in the basal ganglia, in our case in the subthalamic nucleus [3]).

In the BMT group of 23 patients with mean age of 57.8 ± 13 (SD) years; disease duration was 7.1 ± 3.5 years. The second DBS group of 24 patients with mean age of 53.7 ± 9.3 (SD) years; disease duration was 9.5 ± 3.5 years (statically significant longer disease duration than BMT-group: $p < 0.025$). These statistical data are related to the data obtained during the first visit for each group: so-called BMT W1 (visit one) and DBS W1 (visit one).

3.1 Rules for Longitudinal Study in Same Population of BMT Patients

Only on medication—BMT patients were tested in two sessions (session 1: without, and session 3: with medication) three times every half-year. In Table 2 are data from three patients for the first visit.

The full table has 23 (subjects) × 2 (sessions) = 46 objects (measurements). In the Table 1 are values of nine attributes for three subjects where: P# is the patient number; t_dur is the duration of the disease; UPDRS—unified PD rating scale, which is the best indicator of the disease stage; PDQ39—PD quality of life test; Epworth—sleep disturbances test; there are four parameters describing saccades: SccDur is the mean duration of 10 saccades; SccLat is the mean latency of saccades, SccAmp is the mean amplitude of 10 saccades, SccVel is the mean velocity of saccades; and S#—session number (Table 1).

In the next step, using RSES, we have completed reduction and discretization of all attributes except of the patient number (see reduct in [1, 3] and Method section). In the table below (Table 2) for the same data as Table 1 we have performed discretization as range of attributes' values and reductions marked by '*'. Notice that only latency of saccades was significant as other parameters of saccades: duration, amplitude and velocity were reduced.

In the first column patient's number (P#) is symbolic attribute as well as S# (session number in the third column) and they are not discretized; in the second column is patient's disease duration divided in two groups longer and shorter than 9.7 year; in the fourth column is PDQ39—39 questions related to PD quality life divided into two groups; the next Epworth sleeping test—two groups; and in the next four columns are parameters of saccades, but only the latency was important and divided into three ranges; in the last column is UPDRS (decision attribute) divided into four ranges. Each row gives a particular rule, e.g. the first one:

$$('P\#' = 4) \& ('t\_dur' =" (-Inf, 9.7)") \& ('S\#' = 1) \& ('PDQ39' =" (-Inf, 55.0)") \\ \& ('Epworth' =" (-Inf, 3.0)")) \& ('SccLat' =" (-Inf, 181.5)") \Rightarrow ('UPDRS' =" (36.0, 45.0)") \quad (1)$$

We can read Eq. 1 that for patient #4 *and* with disease duration below 9.7 years *and* in session #1 *and* with PDQ39 below 55 *and* with Epworth below *and* with saccade latency shorten than 181.5 ms *then* patient's UPDRS is between 36.0 and 45.0. We found with the RSRS help 70 rules with UPDRS decision values in 4 ranges e.g.:

$$(S\# = 3) \& (PDQ39 =" (-Inf, 50.5)") \Rightarrow (UPDRS =" (-Inf, 33.5)" [12]) \quad (2)$$

$$(t\_dur =" (-Inf, 5.65)") \& (S\# = 3) \& (Epworth =" (-Inf, 14.0)") => (UPDRS =" (-Inf, 33.5)" [7]) \quad (3)$$

Both rules (Eq. 2—12 cases, Eq. 3—7 cases) have one condition attribute session number (S#) and the same decision attribute limits. A simplified interpretation is that if a patient is on appropriate dose of medication (S# = 3) and fulfills some additional conditions then his/her UPDRS will be below 33.5.

Table 1 Part of the information table for BMT patients

P#	t_dur	UPDRST	PDQ39	Epworth	SccLat	SccDur	SccAmp	SccVel	S#
4	5.2	44	7	2	152.3	53.7	10	340.4	1
4	5.2	14	7	2	162.1	57.3	9.1	319.8	3
5	10.2	40	46	6	253	51.1	12.1	454.3	1
5	10.2	31	46	6	221.7	50.4	10.6	467.5	3
7	6.4	44	70	3	233	62.4	10	302.4	1
7	6.4	22	70	3	247	56	9.7	371	3

Table 2 Discretized-table extract for BMT patients

P#	t_dur	S#	PDQ39	Epworth	SccLat	SccDur	SccAmp	SccVel	UPDRS
4	"(−Inf,9.7)"	1	"(−Inf,55.0)"	"(−Inf,3.0)"	"(−Inf,181.5)"	*	*	*	"(36.0,45.0)"
4	"(−Inf,9.7)"	3	"(−Inf,55.0)"	"(−Inf,3.0)"	"(−Inf,181.5)"	*	*	*	"(−Inf,24.0)"
5	"(9.7,Inf)"	1	"(−Inf,55.0)"	"(3.0,Inf)"	"(181.5,395.0)"	*	*	*	"(36.0,45.0)"
5	"(9.7,Inf)"	3	"(−Inf,55.0)"	"(3.0,Inf)"	"(181.5,395.0)"	*	*	*	"(24.0,36.0)"
7	"(−Inf,9.7)"	1	"(55.0,Inf)"	"(3.0,Inf)"	"(181.5,395.0)"	*	*	*	"(36.0,45.0)"
7	"(−Inf,9.7)"	3	"(55.0,Inf)"	"(3.0,Inf)"	"(181.5,395.0)"	*	*	*	"(−Inf,24.0)"

Table 3 Confusion matrix for UPDRS of **BMT W2** by rules obtained from **BMT W1**

		Predicted				
		"(36.0, 45.0)"	"(−Inf, 24.0)"	"(24.0, 36.0)"	"(45.0, Inf)"	ACC
Actual	"(36.0, 45.0)"	0.0	1.0	1.0	4.0	0.0
	"(−Inf, 24.0)"	0.0	11.0	0.0	0.0	0.65
	"(24.0, 36.0)"	0.0	5.0	3.0	1.0	0.5
	"(45.0, Inf)"	0.0	0.0	2.0	18.0	0.78
	TPR	0.0	1.0	0.33	0.9	

Table 4 Confusion matrix for UPDRS of **BMT W3** by rules obtained from **BMT W1**

		Predicted				
		"(36.0, 45.0)"	"(−Inf, 24.0)"	"(24.0, 36.0)"	"(45.0, Inf)"	ACC
Actual	"(36.0, 45.0)"	0.0	2.0	3.0	3.0	0.0
	"(−Inf, 24.0)"	0.0	11.0	0.0	0.0	1.0
	"(24.0, 36.0)"	0.0	5.0	1.0	0.0	0.17
	"(45.0, Inf)"	0.0	0.0	1.0	20.0	0.95
	TPR	0.0	0.61	0.2	0.87	

We have used machine learning and rough set theory [6] in order to predict (confusion matrix) precision of rules obtained from the first visit W1 to data from the second (half-year later W2—Table 3) and the third (one year later W3—Table 4) visits.

TPR: True positive rates for decision classes; ACC: Accuracy for decision classes: the global coverage was 1.0 and the **global accuracy was 0.7**.

Cross validation (sixfold) of the first visit BMT W1 data gave the global accuracy 0.896 and global coverage 0.35. TPR for above digitalization were 0, 0.38. 0.56, 0.44, accuracy: 0, 0.312, 0.625, and 0.5. In this *train-and-test* procedure we have used ML classifier based on the decomposition tree.

TPR: True positive rates for decision classes; ACC: Accuracy for decision classes: the global coverage was 1.0 and the **global accuracy was 0.7**.

In BMT (medication only) patients UPDRS increases with time, as medications cannot cure the disease. Our predictions are consistent, even with continues disease development rules (mechanisms) are the same.

Table 5 Confusion matrix for UPDRS of **DBS W1** by rules obtained from **BMT W1**

		Predicted				
		"(43.0, 63.0)"	"(−Inf, 33.5)"	"(33.5, 43.0)"	"(63.0, Inf)"	ACC
Actual	"(43.0, 63.0)"	6.0	1.0	4.0	4.0	0.429
	"(−Inf, 33.5)"	0.0	16.0	1.0	0.0	0.941
	"(33.5, 43.0)"	2.0	1.0	2.0	0.0	0.4
	"(63.0, Inf)"	2.0	0.0	0.0	10.0	0.833
	TPR	0.6	0.94	0.29	0.71	

3.2 *Rules for Longitudinal Study Between Different Patients Populations*

As mentioned above, the second group of PD patients (DBS patients) was under medication and brain stimulation treatments. In the most cases DBS procedure is performed in the later stage of PD development, which means that patients have larger UPDRS. Therefore, we have obtained rules from the same data as above but with higher UPDRS values. The first visit test results from DBS patients (DBSW1) were before the electrodes implantation surgery—patients were tested with or without medications (Table 5).

TPR: True positive rates for decision classes; ACC: Accuracy for decision classes: the global coverage was 1 and the **global accuracy was 0.708**, coverage for decision classes: 1, 1, 1, 1. Classification was based on rules generated from the reduct calculated with genetic algorithm method.

Notice that there are problems with predicting UPDRS values higher than 63 as from tables above (Tables 3 and 4) these values do not appear in the BMT patients.

TPR: True positive rates for decision classes; ACC: Accuracy for decision classes: the global coverage was 0.2 and the **global accuracy was 0.78**, coverage for decision classes: 0.8, 0.11, 0.125, 0.0.

As in Table 5 BMT group results could not predict UPDRS values above 63 as such values are spare in the BMT group.

TPR: True positive rates for decision classes; ACC: Accuracy for decision classes: the global coverage was 0.562 and the **global accuracy was 0.741**, coverage for decision classes: 0.75, 0.528, 0.625, 0.0.

For DBS W3 group predictions for both UPDSR ranges: high above 63 and also (33.5, 43.0) are bad (0). As predictions for the second range were very good (1) in DBS W1 and W2 groups, it looks that longer period of stimulation might change relationships between different attributes.

Table 6 Confusion matrix for UPDRS of **DBS W2** by rules obtained from **BMT W1**

		Predicted				
		"(43.0, 63.0)"	"(−Inf, 33.5)"	"(33.5, 43.0)"	"(63.0, Inf)"	ACC
Actual	"(43.0, 63.0)"	4.0	0.0	0.0	0.0	1.0
	"(−Inf, 33.5)"	2.0	2.0	0.0	0.0	0.5
	"(33.5, 43.0)"	0.0	0.0	1.0	0.0	1.0
	"(63.0, Inf)"	3.0	0.0	0.0	0.0	0.0
	TPR	0.67	1.0	1.0	0.0	

Table 7 Confusion matrix for UPDRS of **DBS W3** by rules obtained from **BMT W1**

		Predicted				
		"(43.0, 63.0)"	"(−Inf, 33.5)"	"(33.5, 43.0)"	"(63.0, Inf)"	ACC
Actual	"(43.0, 63.0)"	2.0	1.0	0.0	0.0	0.667
	"(−Inf, 33.5)"	1.0	18.0	0.0	0.0	0.947
	"(33.5, 43.0)"	3.0	2.0	0.0	0.0	0.0
	"(63.0, Inf)"	0.0	0.0	0.0	0.0	0.0
	TPR	0.33	1.0	0.0	0.0	

In Tables 6 and 7 we have compared predictions based on only medication patients (BMT) with DBS patient population when their brain stimulation was ON. We were unable to predict UPDRS development when these patients were without stimulation. As many rules have patient numbers, global coverage in Table 6 was 20% and in Table 7 only 21%, but obtained accuracies were sufficient. It is not the case for the data from visits 2 and 3 obtained without electric brain stimulation.

4 Discussion

There are standard neurological procedures that are changing every several years (as e.g. UPDRS procedures) based on statistical approach for many PD cases and effectiveness of their treatments. However, different clinical centers may have different rules that are also not in the same way interpreted by different neurologists. New procedures, new technologies and data constantly improve PD patients' treatments, but one may still doubt (as some patients do) if the actual procedures are optimal for this individual case (for me). We propose to use the data mining and machine learning in order to compare different neurological protocols and their effectiveness. But still there are several problems related to their precision and objectivities. Probably, the best future approach will be to perform all tests automatically, process them with intelligent algorithms and to submit results to the doctor for his/her decision. Another, more advanced approach that we were testing in this work, would be to create a new 'Golden Standard' for each new case on the basis of already successfully treated patients. As each individual has different set of symptoms, it is probably more optimal to gather data from many different patients with some similarities in symptoms to that actually treated subject. Alternative problem is how symptoms are developing in time that is another additional challenge. We have demonstrated, in the present work, that we can estimate symptoms and their time development (longitudinal course) in one population treated in a similar way (e.g. the most popular in PD is only medication treatment). This result may give the basic (locally optimal) follow-ups. If patient is doing significantly worse then others (rules), his/her treatment is not optimal, and should be changed. In the next step, we may use rules obtained from different clinics to make them even more universal and optimal. It was the first part of our approach. In the second part, we have tested different patient population with different treatments. Can we in this case find optimal way of different treatments? The second group of patients were in more advanced stage of disease so it was not possible to get 100% coverage like in the first case. The second group with longitudinal study had a new treatment (brain stimulation) that started from the second visit. We have tested if the same treatment in different populations gives similar results. We have covered 50% cases and got 64% accuracy (Table 6). In next two visits, patients got two treatments: medication (medication ON and OFF) and electric brain stimulation (ON and OFF). We have analyzed these treatments as two different sets: (1) StimOFF: medication ON and OFF; (2) StimON: medication ON and OFF. As a result, it was not possible to get sufficient accuracy in set 1, but we got good accuracy for the set 2—with the brain stimulation. Coverage was about 20% (different patients) but accuracy for the second visit DBSW2 was 78 and 60% for the second visit. In the future, we may look for additional condition attributes in order to improve global accuracy. The reason that our rules did not apply to symptoms of patients without brain stimulation was probably related to the surgery. Inserting electrodes in basal ganglia (into or near STN) may destroy connections between different structures. Functions of these connections are expressed by our rules, disturbing them destroyed functionality of our rules. It is interesting that electric stimulation can revoke our rules again.

5 Conclusions

Presented work is a continuation of our previous findings [1, 3], comparing classical approach used by most neurologists and based on their partly subjective experience and intuitions with the intelligent data processing (machine learning, data mining) classifications. We have analyzed two longitudinal studies that have patients in different disease stages and with different treatments in order to predict UPDRS changes in time. BMT patients have only medication treatment and more advanced DBS group has medication and brain stimulation treatments. We began by finding rules for the first visit of BMT group of patients and we have applied successfully these rules to match symptoms and treatments of all other visits of BMT patients. We have also applied these rules to individual patients belonging to the DBS group. We were only successful with prediction of symptoms for patients before surgery, or after surgery only when stimulation was ON. Without stimulations, after surgery, rules were changed that was probably related to the side effects of the procedure. We have demonstrated that the parameters of eye movements and neuropsychological data are sufficient to predict longitudinal symptom developments (UPDRS) in different groups of PD patients.

Acknowledgements This work was partly supported by projects Dec-2011/03/B/ST6/03816, from the Polish National Science Centre.

References

1. Przybyszewski, A.W., Kon, M., Szlufik, S., Szymanski, A., Koziorowski, D.M.: Multimodal learning and intelligent prediction of symptom development in individual Parkinson's patients. Sensors **16**(9), 1498 (2016). https://doi.org/10.3390/s16091498
2. Przybyszewski, A.W.: Logical rules of visual brain: From anatomy through neurophysiology to cognition. Cogn. Syst. Res. **11**, 53–66 (2010)
3. Przybyszewski, A.W., Kon, M., Szlufik, et al.: Data mining and machine learning on the basis from reflexive eye movements can predict symptom development in individual Parkinson's patients. In: Gelbukh et al. (eds.) Nature-Inspired Computation and Machine Learning, pp. 499–509. Springer (2014)
4. Pawlak, Z.: Rough Sets: Theoretical Aspects of Reasoning About Data. Kluwer, Dordrecht (1991); Springer, pp. 499–509 (2014)
5. Bazan, J., Nguyen, H.Son, Nguyen, Trung T., Skowron, A., Stepaniuk, J.: Desion rules synthesis for object classification. In: Orłowska, E. (ed.) Incomplete Information: Rough Set Analysis, pp. 23–57. Physica-Verlag, Heidelberg (1998)
6. Bazan, J., Nguyen, H.S., Nguyen, S.H., Synak, P., Wróblewski, J.: Rough set algorithms in classification problem. In: Polkowski, L., Tsumoto, S., Lin, T. (eds.) Rough Set Methods and Applications, pp. 49–88. Physica-Verlag, Heidelberg, New York (2000)
7. Grzymała-Busse, J.: A new version of the rule induction system LERS. Fundamenta Informaticae **31**(1), 27–39 (1997)
8. Bazan, J., Szczuka, M.: The rough set exploration system. In: Peters, J.F., Skowron, A. (eds.) Transactions on Rough Sets III. LNCS, vol. 3400, pp. 37–56 (2005)
9. Bazan, J., Szczuka, M.: RSES and RSESlib—a collection of tools for rough set computations. In: Ziarko, W., Yao, Y. (eds.) RSCTC 2000, LNAI 2005, pp. 106−113 (2001)

Comparison of Methods for Real and Imaginary Motion Classification from EEG Signals

Piotr Szczuko, Michał Lech and Andrzej Czyżewski

Abstract A method for feature extraction and results of classification of EEG signals obtained from performed and imagined motion are presented. A set of 615 features was obtained to serve for the recognition of type and laterality of motion using 8 different classifications approaches. A comparison of achieved classifiers accuracy is presented in the paper, and then conclusions and discussion are provided. Among applied algorithms the highest accuracy was achieved with: Rough Set, SVM and ANN methods.

Keywords EEG · Electroencephalography · Imaginary motion · Classification

1 Introduction

The classification of EEG signals is mandatory for the brain-computer interface (BCI) application. It is expected to be highly accurate and capable to be performed with a low latency, thus such methods need to maintain an efficient interaction between a disabled person and a computer application [1, 2]. Applying a dedicated method of signal processing to EEG recordings allows for determining emotional states, mental conditions, and motion intents. Numerous experiments of imagery motion recognition deal with unilateral, i.e. of left or right, hand motion. Such a classification is useful for locked-in-state or paralyzed subjects, and is successfully applied to controlling computer applications [3–11] or a wheelchair [12, 13].

P. Szczuko (✉) · M. Lech · A. Czyżewski
Faculty of Electronics, Telecommunications and Informatics, Gdańsk University of Technology, Gdańsk, Poland
e-mail: szczuko@sound.eti.pg.gda.pl

M. Lech
e-mail: mlech@sound.eti.pg.gda.pl

A. Czyżewski
e-mail: andcz@sound.eti.pg.gda.pl

R. Bembenik et al. (eds.), *Intelligent Methods and Big Data in Industrial Applications*, Studies in Big Data 40, https://doi.org/10.1007/978-3-319-77604-0_18

The aims of this work is to evaluate accuracy of selected classifiers on the task of motion intent detection. The used database described in Sect. 2 contains recordings of untrained and inexperienced 106 persons of undefined origin, and in unknown conditions, requiring individual approach to each case during the classification. Results are provided in Sect. 3, and Sect. 4 concludes the observation that part of the test group is able to interact with a computer by imagining motion intents, but other persons are not able to produce recognizable brain wave patterns suitable to be classified with any of proposed methods, as observed previously by Vidaurre and Blankertz [43]. Authors' contribution is verification of numerous classifiers on the same database in several test scenarios. Thus the comparison of accuracies is presented, and applicability of classifier is judged. The Rough Set method previously not applied on attributes of EEG recordings is proven to be the most accurate in all considered cases.

The motion intent classification is performed in a synchronous or asynchronous mode. The former uses a flashing visual cue (an icon) on the screen in timed intervals, and it verifies user's focus and reaction by means of the P300 potential [14–17]. The latter approach is suited for self-paced interaction, but first it requires classification between a rest and action, and then determining the type of the action [18–20]. The asynchronous approach is evaluated in our work, the classification of left and right, and up and down motion intents are assessed and evaluated by various decision algorithms applied.

Hajibabazadeh et al. [21] employs Support Vector Machine (SVM) in the process of classifying EEG signals representing left and right hand imaginary motion. The low-pass filtered 6-channel signals in their solution are decomposed using wavelet transform into frequency sub-bands and then treated as features. The authors claim to obtain 75% accuracy with training set containing 360 vectors and testing set containing 90 vectors.

SVM has been also used by Sun et al. [22] in on-line EEG signals classification for Brain-Computer interface. Also, RBF kernel was employed with $C = 0.8$ and $\gamma = 0.1$. Features representing imaginary left and right hand movements were extracted using Common Spatial Pattern (CSP). For high-resolution 62 channel system the performance examination conducted in a shielded room enabled the authors to achieve accuracy between 86.3 and 92% for two classes.

Sonkin et al. [23] compared the performance of SVM and Artificial Neural Network (ANN) classifiers in the task of real and imaginary thumb and index finger movements of one hand. The subjects in their experiments pressed a button with a thumb and next with an index finger of the right hand, in both cases. The same activity was imagined (with eyes open) in the imagery tasks. Two approaches to feature extraction were applied, i.e. accumulated trial variant in which task performing trials of the same type were accumulated, and single-trial variant. ANNs with the backward propagation of errors, 1 input layer, 2 hidden layers and 1 output layer were employed. The sigmoid (hyperbolic tangent) function was used as an activation function for the neurons in the hidden layers, and a linear function was used for neurons in the output layer. For the SVM, RBF kernel was used. The ANN classifier provided the best decoding accuracy with single-trial variant (average value: 38%). The average decoding accuracy of SVM classification grown linearly with the number of summarized trials (average value of 45% with 20 trials; the best rate equaled to 62%).

Kayikcioglu and Aydemir [24] compared performance of k-NN, Multiple Layer Perceptron, which is a type of Artificial Neural Network tested herein, and SVM with RBF kernel. Training datasets were created based on one-channel EEG signal. The authors claim that the best accuracy was obtained for k-NN classifier but the presentation of the results is vague and seems manipulative.

Schwarz et al. [25] aimed at developing BCI system that generates control signals for users with severe motor impairments, based on EEG signals processed using filter-bank common spatial patterns (fbCSP) and then classified with Random Forest which is a type of a random tree classifier, tested in experiments presented in their paper. In their experiments users were asked to perform right hand and feet motor imagery for 5 s according to the cue on the screen. For motor imagery of the right hand, each user was instructed to imagine sustained squeezing of a training ball. For motor imagery of both feet, the user was instructed to imagine repeated plantar flexion of both feet. The median accuracy of 81.4% over the feedback period (presenting information to the user about the motion intention) was achieved.

Siuly et al. [26] employed a conjunction of an optimal allocation system and two-class Naïve Bayes classifier in the process of recognizing hand and foot movements. Data was partitioned in such a way that right hand movements were analyzed along with the right foot (first set) movements and left hand movements were analyzed also with right foot movements (second set). Left foot movements were not performed in the experiment. The global average accuracy over tenfolds, for the first and the second set, equaled to 96.36% and to 91.97%, respectively. The authors claimed to obtain the higher accuracy for the two-class Naïve Bayes classifier than for the LS-SVM, both CC and CT based, examined in their earlier works [27, 28].

It was repeatedly observed that real and imagery motion are reflected with similar neural activity [8], particularly decrease of alpha wave power in a motor cortex on the hemisphere opposite to the movement side [7, 8, 10, 29–31]. This is justified by event-related de-synchronization phenomena (ERD) [32–34].

Beside observing ERD occurrences, the motion intent classification is performed by: linear discriminant analysis (LDA) [9, 10, 34, 35], k-means clustering and Principal Component Analysis (PCA) [36], or Regularized Fisher's Discriminant (RFD) [37].

2 Classification Method

The EEG signals are parameterized based on signal characteristics in defined frequency bands, experimentally associated with mental and physical conditions [38, 39]. Several frequency ranges are used: delta (2–4 Hz, consciousness and attention), theta (4–7 Hz) and alpha (8–15 Hz, thinking, focus, and attention). Electrodes are positioned over crucial brain regions, and thus can be used for assessing activity of motor cortex, and visual cortex, facilitating motor imagery classification [40].

Recordings of EEG are polluted with various artifacts, originating from eye blinks, movement, and heartbeat. Dedicated methods were developed for detecting artifacts, filtering and improving signal quality. A Signal-Space Projection (SSP) [34, 41, 42], involving spatial decomposition of the EEG signals is used for determining contaminated samples. Artifact repeatedly originates from a given location, e.g. from eye muscles and is being recorded with the same characteristics, amplitudes, and phases. Such a pattern is detected and filtered. Signal improvements are also achieved by Independent Component Analysis (ICA) [32, 41–43].

Meanwhile, the research approach presented in this paper assumes a simple parametrization of original signals, and a classification based on training simple classifiers with sample dataset.

For the experiment reported in this paper a large EEG database was used: EEG Motor Movement/Imagery Dataset [44], collected with BCI2000 system [45, 46] and published by PhysioNet [44]. This database includes 106 persons and exceeds the amount of data collected by Authors' themselves up to date, thus is better suitable for training and examining classification methods over a large population.

The dataset contains recordings of 4 tasks: real movement of left-right hand; real movement of upper-lower limbs; imagery left-right hand; and imagery upper-lower limbs. 64 electrodes were used located following a 10–20 standard, sampling rate 160 Sa/s, with timestamps denoting start and end of particular movement and one of 3 classes: rest, left/up, right/down. Among the available channels, only 21 were used, obtained from motor cortex: $FC_{Z,1,2,3,4,5,6}$, $C_{Z,1,2,3,4,5,6}$, $CP_{Z,1,2,3,4,5,6}$ (Fig. 1).

All 21 signals were processed in a similar manner, decomposed into the time-frequency domain (TF): delta (2–4 Hz), theta (4–7 Hz), alpha (8–15 Hz), beta (15–29 Hz), and gamma (30–59 Hz). Next, each subband's envelope was obtained by Hilbert transform [47], reflecting activity in the given frequency band. 615 various features of envelopes were collected:

1. The sum of squared samples of the signal envelope (1), mean (2), variance (3), minimum (4), and maximum of signal envelope values (5).
2. For symmetrically positioned electrodes k_L and k_R (e.g. $k_L = C_1$, and $k_R = C_2$) the sum of envelopes differences (6), showing asymmetry in hemispheres activities:

$$\mathrm{SqSum}_{j,k} = \sum_{i=1}^{N} \left(e_{j,k}[i]\right)^2 \tag{1}$$

$$\mathrm{Mean}_{\mathrm{j,k}} = \frac{1}{N} \sum_{i=1}^{N} \left(e_{j,k}[i]\right) \tag{2}$$

$$\mathrm{Var}_{\mathrm{j,k}} = \frac{1}{N} \sum_{i=1}^{N} \left(e_{j,k}[i] - Mean_{j,k}\right)^2 \tag{3}$$

$$\mathrm{Min}_{j,k} = \min\left(e_{j,k}[i]\right) \tag{4}$$

$$\mathrm{Max}_{j,k} = \max\left(e_{j,k}[i]\right) \tag{5}$$

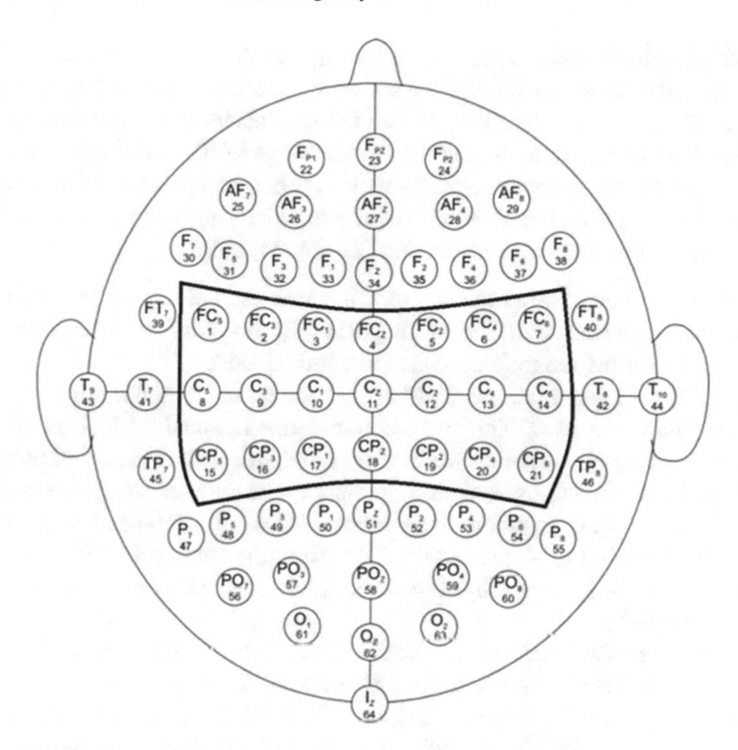

Fig. 1 Electrodes in 10–20 setup, motor cortex channels are marked (*Source* [44])

$$\mathrm{SumDiff}_{\mathrm{j,kL,kR}} = \sum\nolimits_{i=1}^{N} \left(e_{j,kL}\,[i] - e_{j,kR}\,[i] \right) \tag{6}$$

where, $e_{\mathrm{j,k}}$ [i] is an envelope of the signal from particular subband *j* and electrode *k*.

EEG classification is hampered by personal biological and neurological differences, or other characteristics influencing EEG signal quality and features. It was verified that set of attributes with high information gain determined for one person is not informative enough to be applied in classification of other's person recordings. Therefore each individual is treated as separate classification problem, and thus customized classifiers are created.

2.1 Data Classification Procedure

Data classification was performed in WEKA software package offering various data mining techniques [48], and in R programming environment [49] with RoughSets package [50, 51]. From available methods 8 were chosen for testing and described below. Each classifier was tested in the wide range of values for all of the parameters using grid search method (2 to the power of *i*, where *i* was integer taking both negative and positive values). Such a method of testing was possible thanks to developing JAVA application in which WEKA classes were included and used. The process run automatically for all the classifiers and all the parameters sets. Additionally, it was checked for SVM that normalizing data to the range of [0, 1] did not result in any improvement. Below, the parameters for which the best performance of the classifiers was obtained are given. The chosen classifiers have been reported to perform well in the process of brain signals classification [21–23, 25, 26].

- **Naïve Bayes**. Naïve Bayes method uses numeric estimator with precision values chosen based on analysis of the training data [52]. A supervised discretization was applied, converting numeric attributes to nominal ones.
- **Support Vector Machine**. An SVM classifier with sequential minimal optimization algorithm was used. This method transforms nominal attributes into binary ones, normalizes all attributes. Multi-class problems, such as recognition between a rest, left motion, and right motion, in this case, are solved using pair-wise classification [53, 54]. Complexity parameter was C = 1.0, tolerance L = 0.001.
- **Classifier trees—J48**. A pruned C4.5 decision tree was applied [55], with confidence factor used for pruning C = 0.25. A minimum number of instances for a leaf was set to M = 15.
- **RandomTree**. This method constructs a tree considering K = 100 randomly selected attributes at each node. Pruning is not performed.
- **AdaBoost M1**. A method for boosting a nominal class classifier using the Adaboost M1 method was applied [56], with a number of iterations I = 10, and weight threshold for weight pruning P = 100. The used classifier was a mean-squared error regression.
- **Nearest-Neighbors NNge**. An algorithm of Nearest-neighbors using non-nested generalized exemplars (hyperrectangles, reflecting if-then rules) was used [57, 58]. A parameter of a number of attempts for generalization was set to G = 3, and a number of folder for mutual information was set to I = 2.
- **Artificial Neural Network**. A neural network trained by a backpropagation method. The neuron activation function was a sigmoid, numbers of neurons in the input/hidden/output layers were set to 615/12/3, accordingly. The learning rate for weights updating was set to L = 0.3, the momentum for weights values was M = 0.2, number of epochs to train was N = 100.
- **Rough Set classifier**. A method applying Pawlak's Rough Set theory [51, 59] was employed to classification. It applies maximum discernibility method for data discretization and it selects a minimal set of attributes (a reduct) maintaining discernibility between different classes, by applying greedy heuristic algorithm [60–62]. A reduct is finally used to generate decision rules describing objects of the testing set, and applying these to the testing set.

All methods were applied in a 10 cross-validation runs, with a training and testing sets selected randomly in a 65/35 ratio split. These sets contain 1228 and 662 signals for a single person performing a particular task of 3 different action classes (rest, up/left motion, and down/right motion). The process is repeated for every 106 persons, while achieved average classification accuracy records are collected.

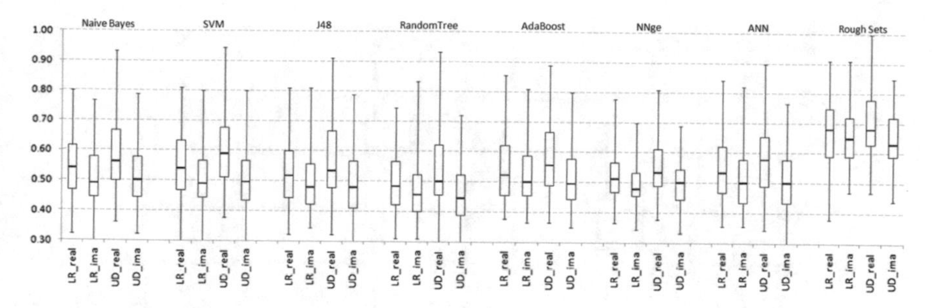

Fig. 2 Classification accuracy of 4 performed tasks of real and imagery motion for 8 classification methods (boxes mark 1st and 3rd quartile, thick line median, whiskers are 1.5 inter-quartile)

3 Classification Results

Obtained classification accuracies from 8 classifiers for all 106 persons were grouped into quartiles and plotted as box-whiskers Tukey plots [63] (Fig. 2).

It can be observed that Rough Sets are significantly more accurate in classification than other methods. SVM and ANN are next in line, both performing on a similar level. Decision trees are usually the worst in accuracy. There are a few cases of very high accuracy > 0.9, but also few persons' action were impossible to classify (for 3 classes accuracy < 0.33 reflects inability to classify, i.e. results are random). In each case the imagery motion classification is not as accurate as classification of real motion. This can be justified by inability to perform a task restricted to only mental activity in a repeated way, and by subjects fatigue as well. Classification of real upper/lower limbs movement is the easiest one for every method. The Rough Set method was further employed in a modified classification scenarios: the distinction between rest and motion, and recognition of motion direction were evaluated by a proper combining of decision classes and repeating the training procedure (Fig. 3).

A large increase of accuracy can be noticed in scenarios involving only two classes. Distinction between movement directions is the easiest task, but still, classification of real motion is more accurate than imagery one. This result justifies a common approach of staged classification: the user action can be first assigned to a class of rest or motion, and in the latter case a second classifier is employed to determine the motion direction.

4 Conclusion

A joint method of EEG signal pre-processing, parametrization, and classification with selected classifiers was presented. Among applied algorithms the following ones: Rough Set, SVM and ANN achieved the highest accuracy. The Rough Set – based algorithm was proven to be the best choice in terms of accuracy, and convenience

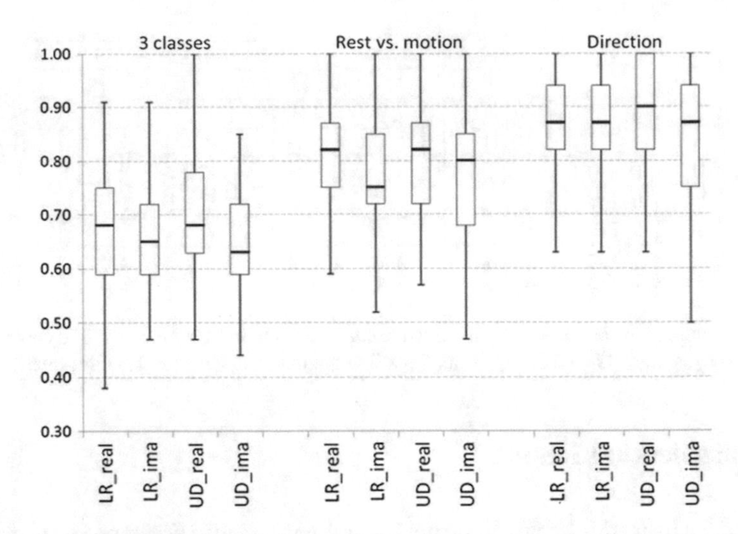

Fig. 3 Accuracy of Rough Set method in classification of 3 classes (rest, motion left/up, motion right/down), presence of motion (rest vs. motion), and direction (left/up vs. right/down)

related to application of rules generated in this method, employing up to 7 parameters instead of all 615 acquired.

The proposed parameterisation method based on signal time-frequency wavelet decomposition and straightforward statistics was proven to yield accurate results for part of the test group, so called BCI illiterates [43].

The presented method can be employed in a simple, yet practical system for motion classification by EEG signals analysis. It opens a possibility to develop computer applications to be interacted by performing actions of rest, left, right, up, and down motion intent. These five binary input controls are sufficient to perform complex actions as: selecting, confirming, and cancelling options in a graphical user interface. As it was stated in Sect. 2, each person in this experiment was treated as a separate classification case. Therefore for practical application of this method in user interface or medical systems, one must start with a classifier training for particular person. In the described research random parts of the EEG signals summing up to 78 s were used for training, therefore the actual training session could be of the same length.

Future work will focus on verifying similar approaches on other EEG sensor setups, especially employing a low number of sparsely positioned electrodes, instead of 10–20 standard setup described in this paper.

Acknowledgements The research is funded by the National Science Centre of Poland on the basis of the decision DEC-2014/15/B/ST7/04724.

References

1. Choi, K.: Electroencephalography (EEG)-based neurofeedback training for brain–computer interface (BCI). Exp. Brain Res. **231**, 351–365 (2013). https://doi.org/10.1007/s00221-013-3699-6
2. He, B., Gao, S., Yuan, H., Wolpaw, J.R.: Brain–computer interfaces. In: He, B. (ed.) Neural Engineering, pp. 87–151 (2012). https://doi.org/10.1007/978-1-4614-5227-0_2
3. Bek, J., Poliakoff, E., Marshall, H., Trueman, S., Gowen, E.: Enhancing voluntary imitation through attention and motor imagery. Exp. Brain Res. **234**, 1819–1828 (2016). https://doi.org/10.1007/s00221-016-4570-3
4. Doud, A.J., Lucas, J.P., Pisansky, M.T., He, B.: Continuous three-dimensional control of a virtual helicopter using a motor imagery based brain-computer interface. PLoS ONE **6**(10), e26322 (2011). https://doi.org/10.1371/journal.pone.0026322
5. Kumar, S.U., Inbarani, H.: PSO-based feature selection and neighborhood rough set-based classification for BCI multiclass motor imagery task. Neural Comput. Appl. 1–20 (2016). https://doi.org/10.1007/s00521-016-2236-5
6. LaFleur, K., Cassady, K., Doud, A.J., Shades, K., Rogin, E., He, B.: Quadcopter control in three-dimensional space using a noninvasive motor imagery based brain-computer interface. J. Neural Eng. **10** (2013). https://doi.org/10.1088/1741-2560/10/4/046003
7. Leeb, R., Pfurtscheller, G.: Walking through a virtual city by thought. In: Proceedings of the 26th Annual International Conference of the IEEE EMBS (2004)
8. Leeb, R., Scherer, R., Lee, F., Bischof, H., Pfurtscheller, G.: Navigation in virtual environments through motor imagery. In: Proceedings of the 9th Computer Vision Winter Workshop, pp. 99–108 (2004)
9. Pfurtscheller, G., Brunner, C., Schlogl, A., Lopes, F.H.: Mu rhythm (de)synchronization and EEG single-trial classification of different motor imagery tasks. NeuroImage **31**, 153–159 (2006)
10. Pfurtscheller, G., Neuper, C.: Motor imagery and direct brain–computer communication. Proc. IEEE **89**, 1123–1134 (2001). https://doi.org/10.1109/5.939829
11. Velasco-Alvarez, F., Ron-Angevin, R., Lopez-Gordo, M.A.: BCI-based navigation in virtual and real environments. IWANN. LNCS **7903**, 404–412 (2013)
12. Corralejo, R., Nicolas-Alonso, L.F., Alvarez, D., Hornero, R.: A P300-based brain–computer interface aimed at operating electronic devices at home for severely disabled people. Med. Biol. Eng. Comput. **52**, 861–872 (2014). https://doi.org/10.1007/s11517-014-1191-5
13. Faller, J., Scherer, R., Friedrich, E., Costa, U., Opisso, E., Medina, J., Müller-Putz, G.R.: Non-motor tasks improve adaptive brain-computer interface performance in users with severe motor impairment. Front. Neurosci. **8** (2014). https://doi.org/10.3389/fnins.2014.00320
14. Bhattacharyya, S., Konar, A., Tibarewala, D.N.: Motor imagery, P300 and error-related EEG-based robot arm movement control for rehabilitation purpose. Med. Biol. Eng. Comput. **52**, 1007 (2014). https://doi.org/10.1007/s11517-014-1204-4
15. Chen, S., Lai, Y.: A signal-processing-based technique for P300 evoked potential detection with the applications into automated character recognition, EURASIP. J. Adv. Signal Process. **152** (2014). https://doi.org/10.1186/1687-6180-2014-152
16. Iscan, Z.: Detection of P300 wave from EEG data for brain-computer interface applications. Pattern Recognit. Image Anal. **21**, 481 (2011)
17. Postelnicu, C., Talaba, D.: P300-based brain-neuronal computer interaction for spelling applications. IEEE Trans. Biomed. Eng. **60**, 534–543 (2013). https://doi.org/10.1109/TBME.2012.2228645
18. Diez, P.F., Mut, V.A., Avila Perona, E.M.: Asynchronous BCI control using high-frequency SSVEP. J. NeuroEng. Rehabil. **8**, 39 (2011). https://doi.org/10.1186/1743-0003-8-39
19. Silva, J., Torres-Solis, J., Chau, T.: A novel asynchronous access method with binary interfaces. J. NeuroEng. Rehabil. **5**, 24 (2008). https://doi.org/10.1186/1743-0003-5-24
20. Xia, B., Li, X., Xie, H.: Asynchronous brain–computer interface based on steady-state visual-evoked potential. Cogn. Comput. **5**, 243 (2013). https://doi.org/10.1007/s12559-013-9202-7

21. Hajibabazadeh, M., Azimirad, V.: Brain-robot interface: distinguishing left and right hand EEG signals through SVM. In: Proceedings of the 2nd RSI/ISM International Conference on Robotics and Mechatronics, Tehran, Iran, 15–17 October 2014
22. Sun, H., Xiang, Y., Sun, Y., Zhu, H., Zeng, J.: On-line EEG classification for brain-computer interface based on CSP and SVM. In: 3rd International Congress on Image and Signal Processing (2010)
23. Sonkin, K., Stankevich, L., Khomenko, J., Nagornova, Z., Shemyakina, N.: Development of electroencephalographic pattern classifiers for real and imaginary thumb and index finger movements of one hand. Artif. Intell. Med. **63**, 107–117 (2015)
24. Kayikcioglu, T., Aydemir, O.: A polynomial fitting and k-NN based approach for improving classification of motor imagery BCI data. Pattern Recognit. Lett. **31**(11), 1207–1215 (2010)
25. Schwarz, A., Scherer, R., Steyrl, D., Faller, J., Müller-Putz, G.: Co-adaptive sensory motor rhythms brain-computer interface based on common spatial patterns and random forest. In: 37th Annual International Conference of the Engineering in Medicine and Biology Society (EMBC) (2015)
26. Siuly, S., Wang, H., Zhang, Y.: Detection of motor imagery EEG signals employing Naïve Bayes based learning process. J. Meas **86**, 148–158 (2016)
27. Siuly, S., Li, Y.: Improving the separability of motor imagery EEG signals using a cross correlation-based least square support vector machine for brain computer interface. IEEE Trans. Neural Syst. Rehabil. Eng. **20**(4), 526–538 (2012)
28. Zhang, R., Xu, P., Guo, L., Zhang, Y., Li, P., Yao, D.: Z-score linear discriminant analysis for EEG based brain–computer interfaces. PLoS ONE, **8**:**9**, e74433 (2013)
29. Shan, H., Xu, H., Zhu, S., He, B.: A novel channel selection method for optimal classification in different motor imagery BCI paradigms. BioMed. Eng. OnLine **14** (2015). https://doi.org/10.1186/s12938-015-0087-4
30. Suh, D., Sang Cho, H., Goo, J., Park, K.S., Hahn, M.: Virtual navigation system for the disabled by motor imagery. Adv. Comput. Inf. Syst. Sci. Eng. 143–148 (2006). https://doi.org/10.1007/1-4020-5261-8_24
31. Yang, J., Singh, H., Hines, E., Schlaghecken, F., Iliescu, D.: Channel selection and classification of electroencephalogram signals: an artificial neural network and genetic algorithm-based approach. Artif. Intell. Med. **55**, 117–126 (2012). https://doi.org/10.1016/j.artmed.2012.02.001
32. Kasahara, T., Terasaki, K., Ogawa, Y.: The correlation between motor impairments and event-related desynchronization during motor imagery in ALS patients. BMC Neurosci. **13**, 66 (2012). https://doi.org/10.1186/1471-2202-13-66
33. Nakayashiki, K., Saeki, M., Takata, Y.: Modulation of event-related desynchronization during kinematic and kinetic hand movements. J. NeuroEng. Rehabil. **11**, 90 (2014). https://doi.org/10.1186/1743-0003-11-90
34. Yuan, H., He, B.: Brain-computer interfaces using sensorimotor rhythms: current state and future perspectives. IEEE Trans. Biomed. Eng. **61**, 1425–1435 (2014). https://doi.org/10.1109/tbme.2014.2312397
35. Krepki, R., Blankertz, B., Curio, G., Muller, K.R.: The Berlin brain-computer interface (BBCI)—towards a new communication channel for online control in gaming applications. Multimed. Tools Appl. **33**, 73–90 (2007). https://doi.org/10.1007/s11042-006-0094-3
36. Tesche, C.D., Uusitalo, M.A., Ilmoniemi, R.J., Huotilainen, M., Kajola, M., Salonen, O.: Signal-space projections of MEG data characterize both distributed and well-localized neuronal sources. Electroencephalogr. Clin. Neurophysiol. **95**, 189–200 (1995)
37. Uusitalo, M.A., Ilmoniemi, R.J.: Signal-space projection method for separating MEG or EEG into components. Med. Biol. Eng. Comput. **35**, 135–140 (1997)
38. Solana, A., Martinez, K., Hernandez-Tamames, J.A., San Antonio-Arce, V., Toledano, R.: Altered brain rhythms and functional network disruptions involved in patients with generalized fixation-off epilepsy. Brain Imaging Behav. **10**, 373–386 (2016). https://doi.org/10.1007/s11682-015-9404-6

39. Wu, C.C., Hamm, J.P., Lim, V.K., Kirk, I.J.: Mu rhythm suppression demonstrates action representation in pianists during passive listening of piano melodies. Exp. Brain Res. **234**, 2133–2139 (2016). https://doi.org/10.1007/s00221-016-4615-7
40. Alotaiby, T., El-Samie, F.E., Alshebeili, S.A.: A review of channel selection algorithms for EEG signal processing. EURASIP. J. Adv. Signal Process. **66** (2015). https://doi.org/10.1186/s13634-015-0251-9
41. Jung, T.P., Makeig, S., Humphries, C., Lee, T.W., McKeown, M.J., Iragui, V., Sejnowski, T.J.: Removing electroencephalographic artifacts by blind source separation. Psychophysiology **37**, 163–178 (2000)
42. Ungureanu, M., Bigan, C., Strungaru, R., Lazarescu, V.: Independent component analysis applied in biomedical signal processing. Meas. Sci. Rev. **4**, 1–8 (2004)
43. Vidaurre, C., Blankertz, B.: Towards a cure for BCI illiteracy. Brain Topogr. **23**, 194–198 (2010). https://doi.org/10.1007/s10548-009-0121-6
44. Goldberger, A.L., Amaral, L.A., Glass, L., Hausdorff, J.M., Ivanov, P.C., Mark, R.G., Mietus, J.E., Moody, G.B., Peng, C.K., Stanley, H.E.: PhysioBank, PhysioToolkit, and PhysioNet: components of a new research resource for complex physiologic signals. Circulation **101**, 215–220 (2000). [ocirc.ahajournals.org/cgi/content/full/101/23/e215c]; physionet.org/pn4/eegmmidb. Accessed 2 Feb 2017
45. BCI2000 instrumentation system project. www.bci2000.org. Accessed 2 Feb 2017
46. Schalk, G., McFarland, D.J., Hinterberger, T., Birbaumer, N., Wolpaw, J.R.: BCI2000: a general-purpose brain-computer interface (BCI) system. IEEE Trans. Biomed. Eng. **51**, 1034–1043 (2004)
47. Marple, S.L.: Computing the discrete-time analytic signal via FFT. IEEE Trans. Signal Proc. **47**, 2600–2603 (1999)
48. Witten, I.H., Frank, E., Hall, M.A.: Data Mining: Practical Machine Learning Tools and Techniques. Morgan Kaufmann Series in Data Management Systems (2011). www.cs.waikato.ac.nz/ml/weka/. Accessed 2 Feb 2017
49. Gardener, M.: Beginning R: The Statistical Programming Language (2012). https://cran.r-project.org/manuals.html. Accessed 2 Feb 2017
50. Riza, S.L., Janusz, A., Ślęzak, D., Cornelis, C., Herrera, F., Benitez, J.M., Bergmeir, C., Stawicki, S.: RoughSets: data analysis using rough set and fuzzy rough set theories (2015). https://github.com/janusza/RoughSets. Accessed 2 Feb 2017
51. https://cran.r-project.org/web/packages/RoughSets/index.html. Accessed 2 Feb 2016
52. John, G.H., Langley, P.: Estimating continuous distributions in Bayesian classifiers. In: 11th Conference on Uncertainty in Artificial Intelligence, San Mateo, pp. 338–345 (1995)
53. Platt, J.: Fast training of support vector machines using sequential minimal optimization. In: Schoelkopf, B., et al. (eds.) Advances in Kernel Methods - Support Vector Learning (1998)
54. Keerthi, S.S., Shevade, S.K., Bhattacharyya, C., Murthy, K.R.K.: Improvements to Platt's SMO algorithm for SVM classifier design. Neural Comput. **13**(3), 637–649 (2001)
55. Quinlan, R.: C4.5: Programs for Machine Learning. Morgan Kaufmann (1993)
56. Freund, Y., Schapire, R.E.: Experiments with a new boosting algorithm. In: Thirteenth International Conference on Machine Learning, San Francisco, pp. 148–156 (1996)
57. Martin, B.: Instance-Based Learning: Nearest Neighbour with Generalization. Hamilton, New Zealand (1995)
58. Roy, S.: Nearest Neighbor with Generalization. Christchurch, New Zealand (2002)
59. Pawlak, Z.: Rough sets. Int. J. Comput. Inf. Sci. **11**, 341–356 (1982)
60. Janusz, A., Stawicki, S.: Applications of approximate reducts to the feature selection problem. Proc. Int. Conf. Rough Sets Knowl. Technol. (RSKT) **6954**, 45–50 (2011)
61. Szczuko, P.: Rough set-based classification of EEG signals related to real and imagery motion. In: Proceedings of the Signal Processing Algorithms, Architectures, Arrangements, and Applications, Poznań (2016)
62. Szczuko, P.: Real and Imagery Motion Classification Based on Rough Set Analysis of EEG Signals for Multimedia Applications. Multimedia Tools and Applications (2017)
63. Tukey, J.W.: Exploratory Data Analysis. Addison-Wesley (1977)

Part V
Multimedia Processing

Procedural Generation of Multilevel Dungeons for Application in Computer Games using Schematic Maps and L-system

Izabella Antoniuk and Przemysław Rokita

Abstract This paper presents a method for procedural generation of multilevel dungeons, by processing set of schematic input maps and using L-system for shape generation. Existing solutions usually focus on generation of 2D systems or only consider creation of cave-like structures. If any 3D underground systems are considered, they tend to require large amount of computation, usually not allowing user any considerable level of control over generation process. Because of that, most of existing solutions are not suitable for applications such as computer games. We propose our solution to that problem, allowing generation of multilevel dungeon systems, with complex layouts, based on simplified maps. User can define all key properties of generated dungeon, including its layout, while results are represented as easily editable 3D meshes. Final objects generated by our algorithm can be used in computer games or similar applications.

Keywords Computer games · L-systems · Procedural content generation
Procedural dungeon generation

1 Introduction

Computer games provide player with different terrains and various challenges, specific for each map type. Among such areas especially dungeon-like structures offer content interesting from gameplay point of view: mazes that player needs to navigate, passages, spaces and treasures that are hidden inside created system, traps

I. Antoniuk (✉) · P. Rokita
Institute of Computer Science, Warsaw University of Technology,
Nowowiejska 15/19, Warsaw, Poland
e-mail: iantoniu@mion.elka.pw.edu.pl

P. Rokita
e-mail: P.Rokita@ii.pw.edu.pl

R. Bembenik et al. (eds.), *Intelligent Methods and Big Data in Industrial Applications*, Studies in Big Data 40, https://doi.org/10.1007/978-3-319-77604-0_19

and obstacles hindering progress, as well as many more challenges characteristic for such environments. In recent years, quality of objects required to provide user with believable and visually appealing structures has grown considerably. Completing even small maps can take huge amount of work, prolonging production time or increasing number of required designers.

Procedural content generation can provide some solution to that problem. Available algorithms cover creation of different objects and areas. At the same time, such procedures can generate huge amounts of content, much faster than human designer. Main disadvantage is that application of such algorithms to actual content generation can pose some difficulties. Main issues concern low level of control over final object shape as well as lack of possibility to supervise generation process or edit final object. Because of those problems, such solutions are often discarded.

When procedural generation of underground systems is considered another problem is representing and maintaining height data. Such areas often contain overlapping elements, or areas placed above one another. Most of existing solutions only considers 2D layouts, omitting 3rd dimension entirely [3, 9, 11, 25]. If 3rd dimension is considered, it often involves larger amount of computation, without guaranteeing acceptable or even easily editable results [5–7, 12, 17]. Such disadvantages are unacceptable in applications such as computer games, where final terrain often needs to meet series of strict constraints.

If we consider layout of different dungeon-like structures, both from computer games and real life, it is possible to notice that they can be divided into few basic elements: walls, obstacles, passages and spaces of different size. Placement and relations of such components are much easier to define, than layout of entire structure. Taking that into account, we propose a method for procedural generation of multilevel dungeon systems.

In our approach, we use schematic maps to define overall layout of final system. Algorithm used to generate shapes in each part of terrain with data obtained from input maps, depends from its type, and can incorporate different procedures, including L-system and other algorithms customized for each purpose. Final, 3D terrain is generated and visualized in Blender application.

The rest of this work is organized as follows. Section 2 contains review and classification of existing solutions. Section 3 outlines initial assumptions and overview of our procedural dungeon generation algorithm. We discuss obtained results and some areas of future work in Sect. 4. Finally, we conclude our work in Sect. 5.

2 Related Work

Procedural content generation, especially in recent years is growing fast due to its vast applications and possibilities. Existing procedures vary greatly, both in their complexity and offered possibilities [8, 10, 23]. Some of approaches can even produce diverse worlds, with sets of properties and parameters describing shape of created content [20]. While in most cases generated content is interesting visually, level of control user has over shape of final object is another issue. This property is especially important when created content will be used in computer games or similar applications. In that case there is also additional matter of playability of generated elements. Even very interesting object, when applied to computer games might prove unusable, to complicated or unwieldy.

In "Introduction" chapter of [18] authors discuss procedural content generation in general, with its application to computer games specifically and various classifications. When it comes to content playability, it can be defined in various ways, depending from game type and its application. In general though it can be said, that playable content should allow player to use it according to its purpose (i.e. climb up procedurally generated staircase), and finish entire level (either procedurally generated as a whole, or containing elements obtained in such way). Some of existing solutions ensure that any generated objects will meet series of strict properties [24]. Another issue is way in which such control can be maintained, like using parameters to describe desired results [20], using simplified elements as base for generation method [14] or incorporating story to guide entire process [13]. There are also some procedures, that use schematic maps, to assign different properties to final terrain [21, 22].

When it comes to procedural underground generation, we have different sets of algorithms, that need to take into account properties specific for such structures, like generation of overlapping elements. This problem can be simplified by considering such terrains in 2D, for example while generating mazes [9]. Different methods were used to achieve interesting shapes, like cellular automata [11], predefined shapes and fitness function checking quality of generated system [25], or checkpoints also paired with appropriate fitness function [3].

Generation of similar systems in 3D contains additional set of constraints and problems. Some of existing solutions focus on cave-like structures and try to produce geologically correct shapes with realistic features [5–7]. There are interesting approaches for generation of buildings [15]. Some approaches also focus on ensuring terrain playability.

In [17] authors ensure, that terrain generated by their procedure is playable and meets series of constraints. User can define such elements as size of generated space, number of existing branches and relationships between structures incorporated into entire system. Unfortunately, due to volumetric terrain representation used by authors, incorporating such elements into most computer games would be difficult. In this solution, there is no way for user do define overall layout of generated terrain and there is no easy way to transfer this algorithm to generation of dungeon like structures.

Another interesting solution also focusses on playability of generated cave system, although it takes different approach [12]. After defining overall layout of generated terrain with L-system and generating structural points, authors then create tunnels by wrapping meta-ball around defined path, generating final shapes. After extracting mesh data from voxel representation, final results are covered with some textures, presenting interesting shapes. Unfortunately, since the same textures tend to cover large areas, entire system can look repeatable. Authors also do not provide user with any tool to define overall layout of generated system.

Procedural content generation contains different procedures and can produce various elements, with various properties. For detailed study of existing methods see [8–10, 18, 20, 23].

3 Procedural Dungeon Generation

In this section we present an algorithm used to generate multilevel dungeons. We obtain information about terrain from schematic maps either presented by user or generated automatically. Our approach is implemented using Python scripting language, while final terrain is visualized using Blender application (since it has in-build python interpreter and its sufficiency was proven in our previous experiments [1, 2]). We use 2.78 version of this 3D modelling environment. For Blender documentation see [4].

Presented procedure consists of three main steps: processing input files and extracting terrain data, using obtained information to generate basic system shapes in tiles and transferring resulting shapes to 3D object.

3.1 Processing Input Files

In presented procedure, similarly as in our previous works (see [1, 2]), we use schematic maps to describe properties of our final dungeon system. Since generation algorithm needs to allow representation of vertical transitions, we use space representation as in our previous work concerning underground system generation in general [1].

Entire system is divided into square levels, containing terrain that is not overlapping. Each level is then divided into square tiles. Single tile contains portion of terrain, that can be assigned to one of defined categories: space (containing rooms of different size), corridor (containing set of passages) or empty (representing a wall in dungeon system). Using this space representation, we define shape of dungeon system with set of schematic maps, where each file stores different information about desired shape of final terrain.

First of our input maps (terrain map), stored in RGB image file, contains definition of terrain type in each tile. Each single pixel in that file represents single region in defined system. Colour of that pixel stores information about terrain category in currently processed file. We define large space as white, small space as green, corridor as red and empty tile as black.

Second input map, stored in grayscale image defines relative placement of tiles inside each level. Similarly as in terrain map, each pixel in height map corresponds to single tile in final terrain, defining its placement in related level.

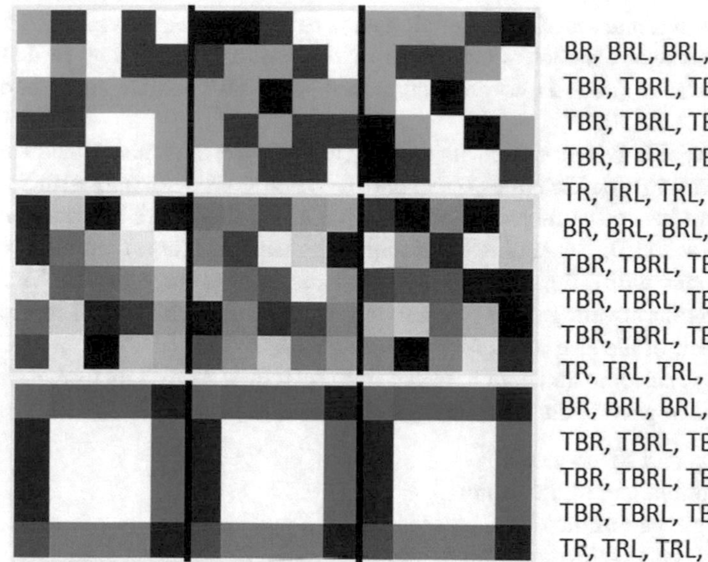

Fig. 1 Example set of input maps for dungeon generation algorithm: Terrain map (top left), height map (middle left), connection map represented as image (bottom left) and the same connection map represented as text (right). Vertical lines on images mark borders between succeeding level data

Final two maps define transitions between terrain regions. Connection maps (separate for connections inside single level, and transitions between levels) can be stored either in text file or in RGB image file. Each tile can be connected to other, neighbouring region (excluding tiles placed diagonally).

Since we decided to store data from all levels in single file for each map type, we also needed to organize file structure, to represent multilevel data. In case of image files, we place highest region on left side with succeeding levels placed to the right. In case of text file, we add lower level data at the end of the file. For example set of input maps see Fig. 1.

3.2 Generating Terrain Tiles

After data concerning terrain type and its parameters is obtained from input files, we proceed to generating dungeon shapes in each tile. For that purpose, we use two separate sets of algorithms: first for generating room shapes and second for obtaining corridor layouts. In our tile visualizations, we use white for spaces and red for passages.

Room Generation.

In our procedure, we obtain room shapes in each tile using L-system. L-system is a type of formal grammar, that consists of set of symbols (alphabet) that can be connected into strings and set of rules defining ways to exchange each symbol with larger set of characters. Characters can be either nonterminal (they allow further expansion) or terminal (there is no production rule expanding that character into larger set).

Transformation of L-system starts with defining initial string of symbols and then expanding it according to defined rules, either until L-system consists of only terminal characters or until desired number of iterations is reached. For more information about L-systems see [16]. Depending from room type (small or large), number of performed iterations will differ, with higher value assigned to large rooms. This results in large rooms covering most of space in processed tile, while small rooms usually occupy half of tile size or slightly above.

With given type of room and data obtained from map files, as well as from user input, we proceed to generation. Our starting key set is organized as follows:

- ET - extend top part of the room,
- EB - extend bottom part of the room,
- ER - extend right part of the room,
- EL - extend left part of the room,
- ETR - extend top right part of the room,
- ETL - extend top left part of the room,
- EBR - extend bottom right part of the room,
- EBL - extend bottom left part of the room.

In our case tile size is an user defined parameter that determines size of tile shape image (in pixels) and in case of 3D object, number of vertices along sides of grid representing portion of terrain contained in each region. With defined tile size, we first generate transformation rules for L-system.

When we consider man-made structures it is possible to notice, that they can be symmetrical along horizontal or vertical axis, both of them and in some cases, neither. Taking that property into account, we produce four versions of generation rules, each enforcing chosen type of symmetry if it applies to currently processed tile, and one for case with no symmetry present. For each key from starting key set, we generate few transformation rules, that can be used interchangeably during L-system expansion. We wanted to ensure, that results generated by given sets of rules are interesting. At the same time, to high number of rules generated for our L-system in smaller tiles would increase computational time, without much improvement in shape quality. We decided to set this parameter according to Eq. (1). Such value ensures that we have at least two rules, and at the same time, their number is adequate to single tile size. For example set of transformation rules for all eight starting keys, as well as results given by applying them to room shape see Fig. 2.

$$NumberOfRules = \left\lfloor \frac{TileSize}{10} \right\rfloor + 2 \tag{1}$$

EBR: [[EB, ER], [ETL, ETL, ER], [ET, EBR, ET], [EL, EBL, EB], [ET, EL, ETR]]
EL: [[ETR, EBL], [ETL, ETR], [EL, ETL, ETL], [ETL, EL, EL], [ER, ETR]]
EBL: [[EL, ETL, EL], [ER, ET, EB], [EB, ETL], [EBR, EL, ET], [ET, ETL, ETL]]
ETR: [[ET, ETL, ER], [ER, EBL, ER], [ER, EBR, EL], [ER, EB, EBL], [ET, ER]]
ET: [[EB, EB, ER], [EL, ER, EBL], [ET, ER, ET], [EBR, EBR, EL], [EB, ETL, ET]]
ER: [[ER, EBR, ETR], [EBR, ET, ET], [EL, EB, ETL], [ET, EBR], [ET, EB]]
EB: [[ETL, EBR], [ETR, ER, EBR], [EBL, ET], [EBR, ET, EB], [ETR, ER, ETL]]
ETL: [[EL, ETR], [ETL, EBL], [EBR, EBR], [EB, EBR], [EBR, ET]]

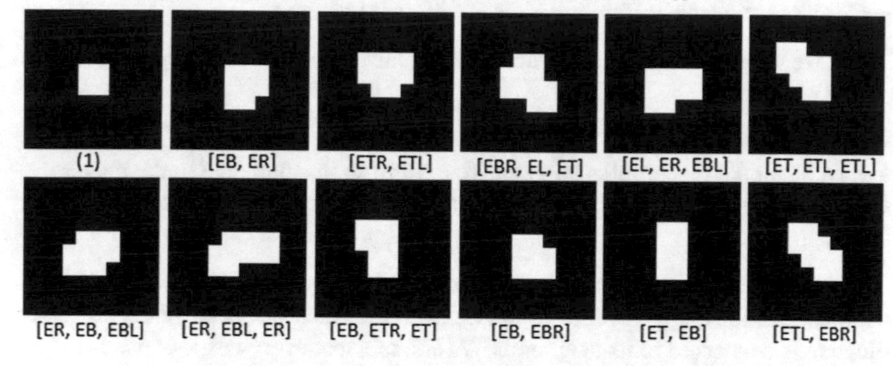

Fig. 2 Example set of transformation rules for starting keys (Tile size is set at 31 and number of alternative rules for single key equals 5). Image (1) shows initial tile shape with basic shape set at middle. Other images show example room expansions according to transformation rules (we only use square shape for expansions in those examples)

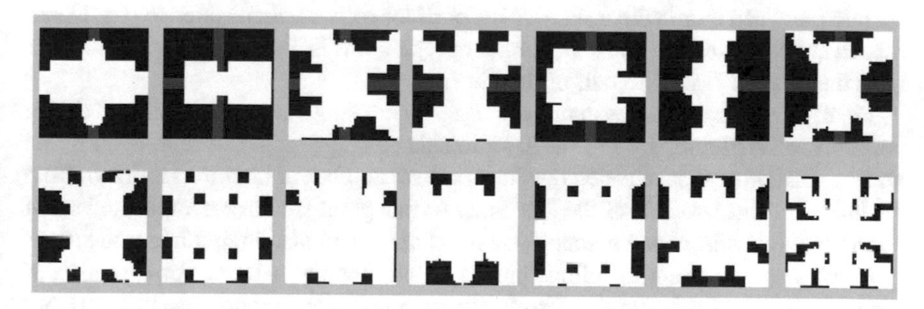

Fig. 3 Example tile shapes for small (top) and large rooms (bottom)

After we obtained all versions of rulesets, we proceed to creating actual shapes of defined spaces. With data obtained from terrain map, we use L-system to extend single room inside each tile according to those rules, by placing simple shapes (rectangle, square or circle) in indicated location. Shape placed at each point is randomly chosen before each expansion. If currently processed tile is defined as Large Space, number of iterations for L-system will be defined as 20% of tile size while for Small Spaces this value will be equal to 10% of tile size. Example room shapes generated by our procedure for different parameters are presented at Fig. 3.

Corridor Generation.

Since with randomly generated rules L-system tends to be unpredictable, and often circles around some local minimum, we decided to use separate procedure for deciding layout of corridors. When it comes to passages in man-made structures, they usually have following properties:

- Corridors join at right angle.
- Corridors lead from starting point to exit.
- Corridors can have different widths at different locations.

When we consider applications such as computer games few more properties also comes to mind:

- There should be places where enemies could lurk, or traps can be hidden.
- Layout should be complicated enough, to form some challenge to the player.
- Corridors should not have any isolated spaces, that cannot be accessed from main system.

While generating our corridor shapes we take those properties into account. First, using connection maps we place an entry/exit point at each side of currently processed tile, that is connected to its neighbour. We assume, that every corridor needs to pass through middle of the tile, so we add this point to our initial set.

After initial points selection we then generate main corridor, with random twists and turns, that leads from selected point at tile edge, to its middle. Those shapes are achieved by choosing two points along current corridor path, and then exchanging corridor between them with u shape. New corridor cells are then added to considered set and entire process is repeated. Number of iterations is defined as random number from the interval from 1 to half of the tile size.

Finally, we expand our system by randomly choosing some points along existing corridors, and extending them further into side passages with different lengths and widths. That process is repeated few times, where number of iterations is drawn from the interval from 1 to half of the Tile Size. At this point we have complicated set of corridors, interesting from gameplay point of view, and also maintaining directions in which we want them to follow, but not in an obvious pattern. For evolution of corridors inside single tile see Fig. 4. Example corridor layouts generated by our procedure are presented at Fig. 5. For example dungeon layout with used input maps see Fig. 6.

3.3 Generating 3D Terrain Model

With tile maps already generated for each part of final system, we proceed, to transiting those shapes into 3D space. At this point we have both basic height transitions in each level (obtained from input height map), as well as layout of processed dungeon system. While tile generation and map processing can be done using basic python

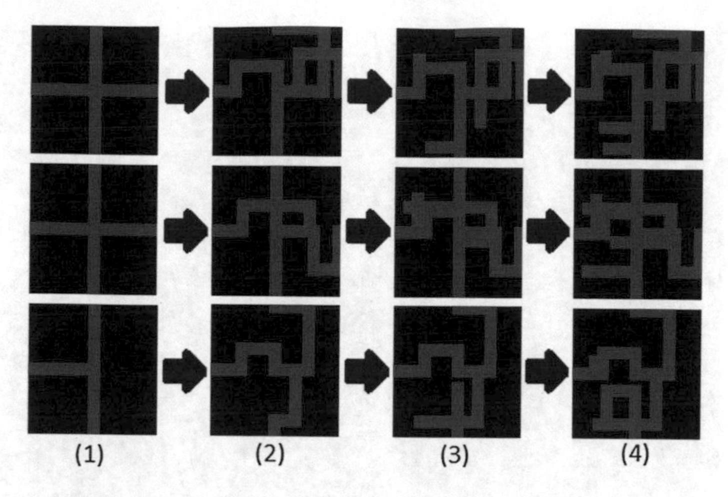

Fig. 4 Example generation of corridor sets inside single terrain tile. (1) Initial main corridors, received from tile connections. (2) Main corridors after alteration of their course. (3) First iteration of additional corridors placement. (4) Final, generated corridor set

scripts, this part of procedure is implemented with build-in Blender functionality concerning basic 3D operations.

First step in this part of our algorithm is placing basic objects (planes or grids) at appropriate heights obtained from height map. Each tile defined as empty will not appear at this point, while rest of them shows general transitions inside designed terrain and its layout. Number of vertices in each grid corresponds directly to tile size (so if tile size is defined as 11, 3D object at this point will be composed of 11^2 vertices).

After basic 3D tiles are placed, we start assigning shapes generated for each of them in previous step. Using tile maps we select and remove all vertices that are not classified as room space or corridor, leaving only those elements that belong to final system (this procedure also simplifies mesh of final object). We use this method, since it can easily translate shapes from generated tile maps to 3D object and does not require additional computation for edge and face placement.

At this point we have set of objects representing tile shapes, that are not yet connected along z axis in any way. We start similarly as in our previous algorithms (see [1, 2]), by aligning edges of neighbouring tiles, that share connection. One of necessary changes, while adapting our previous procedures to dungeon generation concerns overall properties of such structures. Usually, any transitions in heights will be executed in corridors while rooms stay mostly levelled. Especially any kind of sharp transition should not occur in such spaces. Therefore, instead of just aligning terrain starting from edge, we first check type of each shape and only even cells classified as corridors. There are three main cases to consider:

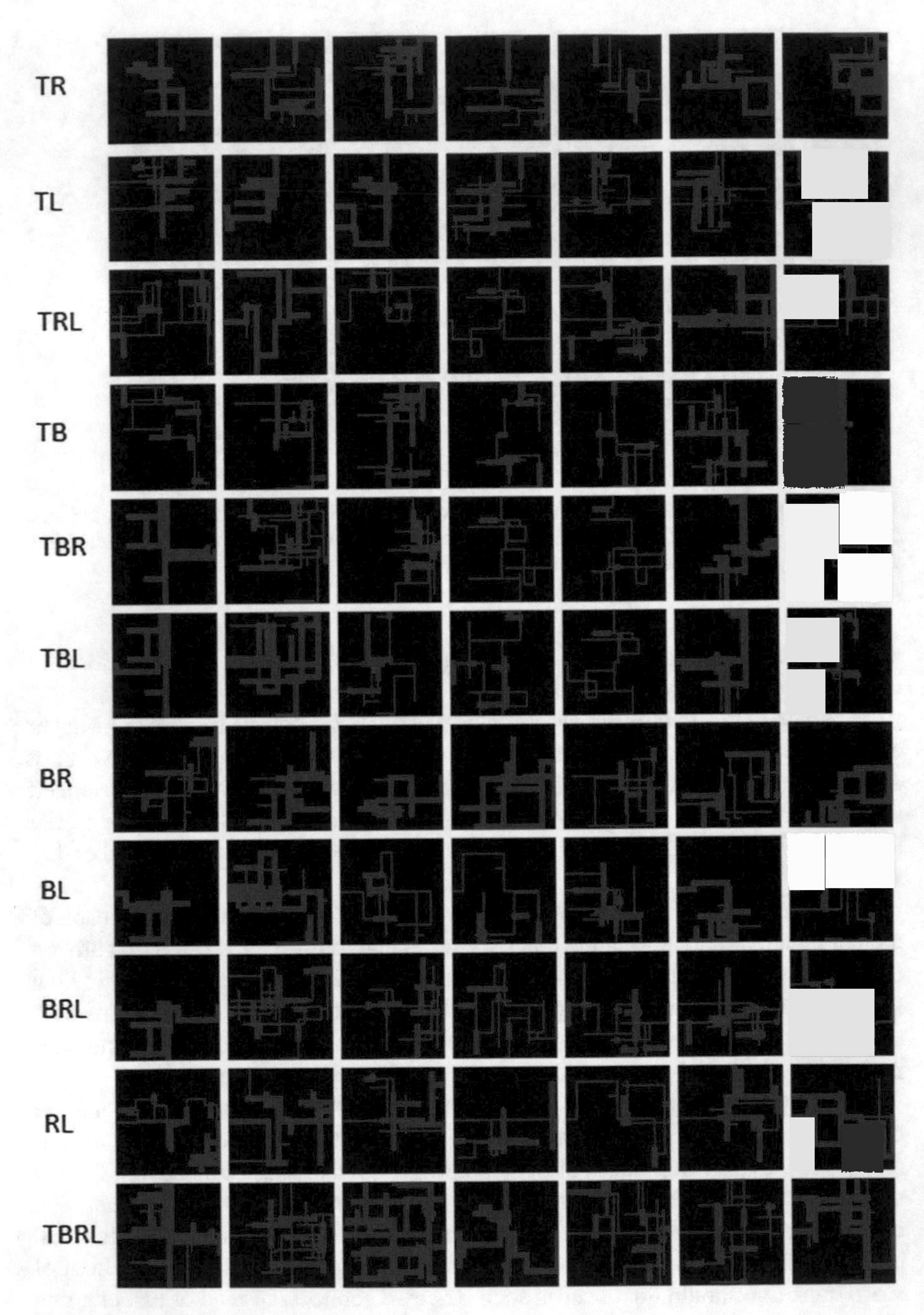

Fig. 5 Example tile shapes for corridors with different connections (T - connection with top tile, B - connection with bottom tile, R - connection with right tile, L - connection with left tile)

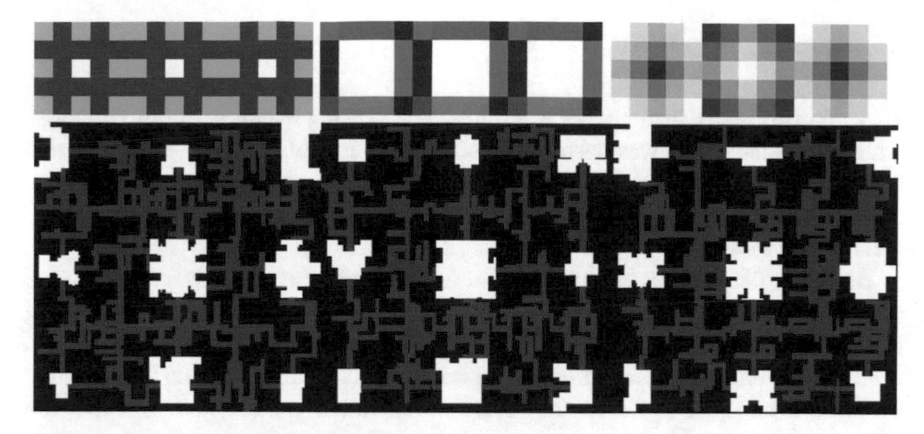

Fig. 6 Example map of dungeon system composed of 3 levels with input maps used for generation: terrain map (top left), connection map for transitions inside each level (top middle) and height map (top right). Each level is divided into 25 tiles, where tile size is set at 31

1. Both tiles contain corridor cells at considered edges. In that case both corridors will be joined at middle and levelled deeper into the tile, either until room cells occur, or depth threshold is achieved.
2. Only one tile contains corridor cells. In that case, corridor tiles will be aligned to second tile containing room, and levelling will also occur only in that tile.
3. Neither one of considered tiles contains corridor cells at considered edges. In that case, we choose part of edges in both tiles, that are classified as room and overlap along the edge, and extrude it to represent sharp transition like ladder or staircase.

We perform additional procedure to mitigate sharper transitions inside corridors, giving them smooth and fine appearance, also distorting some of them, to add detail to our system. At this point we also extrude generated shape to add some volume to resulting terrain instead of representing it as thin plane. Finally, we assign simple texture to our objects, using in-built Blender procedures.

After this step we obtain 3D representation of dungeon system designed in input maps, with each tile represented as separate and editable object. Since our procedure is seed-based, we can easily obtain different variations of final terrain, for any given set of maps. Also, since final results are Blender meshes, they can be not only easily modified, but also readily incorporated in external applications. For overview of our procedure see Algorithm 1.

4 Results and Future Work

Algorithm presented in this paper was implemented using Python scripting language, while 3D visualization was performed using Blender application. Created program can generate multilevel dungeons and represent them as mesh objects. Experiments were performed on a PC with Intel Core i7-4710HQ processor(2,5 GHz per core) and 8 GB of ram.

Since our final objects are intended for further modifications, it is important, that we can obtain different variations of the same terrain in reasonable time. Rendering times of example system, for tiles of different sizes are presented in Table 1. Dungeon used in those experiments is composed of 3 levels each composed of 25 tiles forming a square. While such times do not allow for generation at interactive rate, they are still more than adequate for our purpose, since waiting up to few minutes for example system can be acceptable (especially when modelling such shapes by hand would take much longer, even for smallest tiles). Tiles defined as "empty" were excluded from input terrain map (maps used during experiments are presented at Fig. 6., while second connection map defining transitions to lower levels added connections from large room placed at the middle of level to bottom tile containing corridors).

Algorithm 1. Procedural dungeon generation.

```
program ProceduralDungeonGenerator (Output)
{Input: TerrainMap, HeightMap, ConnectionMapTiles,
ConnectionMapLevels, LevelSize, TileSize, NumberOfLevels,
Seed};
begin
ProcessMaps(TerrainMap, ConnectionMapTiles, ConnectionMapLevels)
for Tile in TerrainMap do:
ConnectTile(ConnectionMapTiles)
GenerateSpaceShape(TerrainMap, Seed)
end for
for all Levels do:
ConnectLevels(ConnectionMapLevels)
end for
PlaceBasicPlanes(HeightMap)
for all Planes do:
UpdateTileData(TileMap)
AllignEdges(TileHeights, TileMaps, TileConnections)
SmoothTransitions(Depth)
ExtrudeRegion()
AssignTexture()
end for
end.
```

Table 1 Generation times for example dungeon [s]. Tile size refers to image size in case of 2D images and initial grid size in 3D object. Dungeon is composed of 3 levels, with size 5×5 tiles.

Tile size	2D shape generation	3D operations	Total time
11×11	4,877	0,712	5,589
31×31	14,316	3,737	18,053
51×51	64,325	19,217	83,542
71×71	177,841	60,663	238,504
91×91	406,751	150,614	557,365

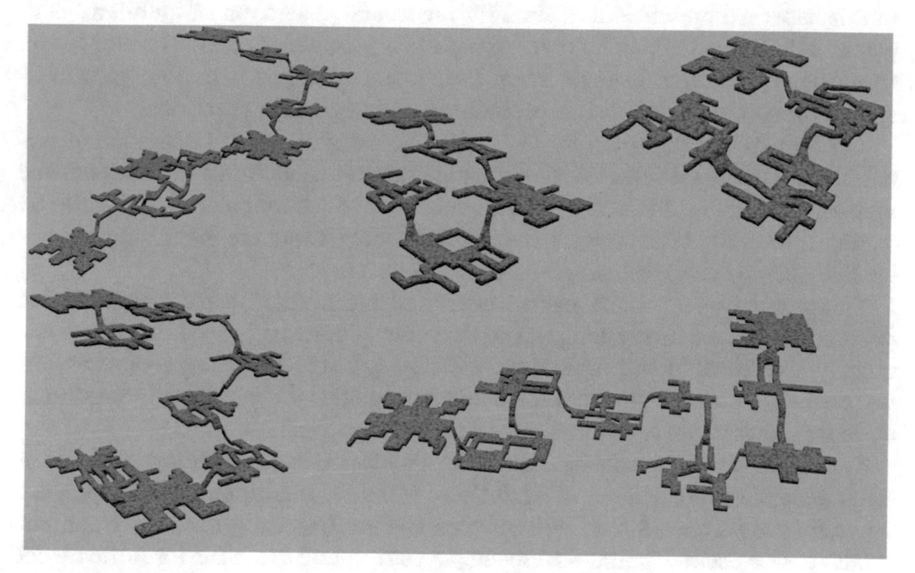

Fig. 7 Example dungeon systems generated by our algorithm

Second important property of our final objects is their ability to represent overall layout provided by user through input files, without limiting possible shapes. We generated different structures, from spiralling sets of corridors, to sets of connected spaces. One huge advantage is that once input maps are created, they can be reused any number of times, resulting in objects with similar layout, but different details. Maximal complexity of output terrain tile in 3D object equals $2 \cdot TileSize^2$ vertices: $TileSize^2$ for initial number of vertices, while second increase comes from final extrusion of terrain, used for adding volume to object (last step can also be omitted, resulting in simpler mesh). Usually final object complexity would be much smaller (especially when concerning corridors, where most of space inside region consists of walls). Fig. 7. shows example dungeons generated during our experiments.

One area, that still is missing from our procedure, is placing additional features (like doors, ladders, stairs, traps and other elements commonly associated with dungeons) through generated system space. We would also like to incorporate additional algorithms for generating room shapes, to further improve variety of elements generated by our algorithm. The same intention can also be extended to part of procedure, that is responsible for generating corridors. We plan to address these elements in future work.

5 Conclusions

In this paper we presented a method for procedural generation of multilevel dungeons, with layout defined by user. To represent general assumptions about terrain arrangement, we use schematic maps. Generation of shapes is performed with L-system for spaces while separate procedure was designed for corridors.

Contrary to existing solutions [5–7, 12, 17], we allow overall definition of layout through our input files. We also generate spaces that are three-dimensional and represent man-made dungeon systems, while most of current procedures focuses on caves. At the same time shapes generated by our algorithm are interesting and can be easily compared to current 2D solutions [3, 9, 11, 25].

One great advantage is that our meshes are not complicated, and entire generation process is not excessively long. Also, since our procedure is seed based, we can quickly generate different variations of the same system. Finally, any created objects can be further edited and adjusted without any additional operations or transitions, since we represent them as meshes inside Blender application.

Our approach still has some drawbacks, like lack of additional layout items, or room generation procedures. Nevertheless, presented algorithm can still produce interesting and large systems with complex layouts. Objects generated by our procedure can be easily incorporated in simple game, provide a basis for further work or be used for visualization purposes during design process.

References

1. Antoniuk, I., Rokita, P.: Generation of complex underground systems for application in computer games with schematic maps and l-systems, In: International Conference on Computer Vision and Graphics, pp. 3–16, Springer, Berlin (2016)
2. Antoniuk, I., Rokita, P.: Procedural generation of adjustable terrain for application in computer games using 2d maps, In: Pattern Recognition and Machine Intelligence, pp. 75–84. Springer, Berlin (2015)
3. Ashlock, D., Lee, C., McGuinness, C.: Search-based procedural generation of maze-like levels. IEEE Trans. Comput. Intell. AI Games **3**(3), 260–273 (2011)
4. Blender application home page: https://www.blender.org/(accesed14.01.2016)
5. Boggus, M., Crawfis, R.: Explicit generation of 3D models of solution caves for virtual environments. CGVR, 85–90 (2009)
6. Boggus, M., Crawfis, R.: Procedural creation of 3d solution cave models, In: Proceedings of IASTED, pp. 180–186 (2009)
7. Cui, J., Chow, Y.W., Zhang, M.: Procedural generation of 3D cave models with stalactites and stalagmites (2011)

8. Ebert, D.S.: Texturing& Modeling: A Procedural Approach. Morgan Kaufmann (2003)
9. Galin, E., Peytavie, A., Marchal, N., Gurin, E.: Procedural generation of roads. Computer Graphics Forum, vol. 29, 2nd edn, pp. 429–438. Blackwell Publishing Ltd., New Jersey (2010)
10. Hendrikx, M., Meijer, S., Van Der Velden, J., Iosup, A.: Procedural content generation for games: A survey. ACM TOMM **9**(1), 1 (2013)
11. Johnson, L., Yannakakis, G.N., Togelius, J.: Cellular automata for real-time generation of infinite cave levels, In: Proceedings of the 2010 Workshop on Procedural Content Generation in Games, p. 10. ACM (2010)
12. Mark, B., Berechet, T., Mahlmann, T., Togelius, T.: Procedural generation of 3d caves for games on the GPU, In: Foundations of Digital Games (2015)
13. Matthews, E., Malloy, B.: Procedural generation of story-driven maps, In: CGAMES, pp. 107–112. IEEE (2011)
14. Merrell, P., Manocha, D.: Model synthesis: a general procedural modeling algorithm. IEEE Trans. Vis. Comput. Gr. **17**(6), 715–728 (2011)
15. Pena, J.M., Viedma, J., Muelas, S., LaTorre, A., Pena, L.: emphDesigner-driven 3D buildings generated using Variable Neighborhood Search, In: 2014 IEEE Conference on Computational Intelligence and Games, pp. 1-8. IEEE (2004)
16. Prusinkiewicz, P., Lindenmayer, A.: The Algorithmic Beauty of Plants. Springer (2012)
17. Santamaria-Ibirika, A., Cantero, X., Huerta, S., Santos, I., Bringas, P.G.: Procedural playable cave systems based on voronoi diagram and delaunay triangulation, In: International Conference on Cyberworlds, pp. 15–22, IEEE (2014)
18. Shaker, N., Togelius, J., Nelson, M.: Procedural Content Generation in Games (2014)
19. N. Shaker, A. Liapis, J. Togelius, R. Lopes, R. Bidara, *Constructive generation methods for dungeons and levels(DRAFT)* Procedural Content Generation in Games, 31-55, 2015
20. Smelik, R., Galka, K., de Kraker, K.J., Kuijper, F., Bidarra, R.: Semantic constraints for procedural generation of virtual worlds, In: Proceedings of the 2nd International Workshop on Procedural Content Generation in Games, p. 9. ACM (2011)
21. Smelik, R.M., Tutenel, T., de Kraker, K.J., Bidarra, R.: A proposal for a procedural terrain modelling framework, EGVE, pp. 39–42 (2008)
22. Smelik, R.M., Tutenel, T., de Kraker, K.J., Bidarra, R.: Declarative terrain modeling for military training games. Int. J. Comput. Games Technol. (2010)
23. Smelik, R.M., Tutenel, T., Bidarra, R., Benes, B.: A survey on procedural modelling for virtual worlds. Comput. Gr. Forum **33**(6), 31–50 (2014)
24. Tutenel, T., Bidarra, R., Smelik, R.M., De Kraker, K.J.: Rule-based layout solving and its application to procedural interior generation, In: CASA Workshop on 3D Advanced Media In Gaming And Simulation (2009)
25. Valtchanov, V., Brown, J.A.: Evolving dungeon crawler levels with relative placement, In: Proceedings of the 5th International C* Conference on Computer Science and Software Engineering, pp. 27-35. ACM (2012)
26. van der Linden, R., Lopes, R., Bidarra, R.: Procedural generation of dungeons. IEEE Trans. Comput. Intell. AI Games **6**(1), 78–89 (2014)

An HMM-Based Framework for Supporting Accurate Classification of Music Datasets

Alfredo Cuzzocrea, Enzo Mumolo and Gianni Vercelli

Abstract In this paper, we use *Hidden Markov Models* (HMM) and *Mel-Frequency Cepstral Coefficients* (MFCC) to build statistical models of classical music composers directly from the music datasets. Several musical pieces are divided by instruments (*String*, *Piano*, *Chorus*, *Orchestra*), and, for each instrument, statistical models of the composers are computed. We selected 19 different composers spanning four centuries by using a total number of 400 musical pieces. Each musical piece is classified as belonging to a composer if the corresponding HMM gives the highest likelihood for that piece. We show that the so-developed models can be used to obtain useful information on the correlation between the composers. Moreover, by using the maximum likelihood approach, we also classified the instrumentation used by the same composer. Besides as an analysis tool, the described approach has been used as a *classifier*. This overall originates an HMM-based framework for supporting accurate classification of music datasets. On a dataset of *String Quartet* movements, we obtained an average composer classification accuracy of more than 96%. As regards instrumentation classification, we obtained an average classification of slightly less than 100% for *Piano*, *Orchestra* and *String Quartet*. In this paper, the most significant results coming from our experimental assessment and analysis are reported and discussed in detail.

A. Cuzzocrea (✉)
DIA Department, University of Trieste and ICAR-CNR, Trieste, Italy
e-mail: alfredo.cuzzocrea@dia.units.it

E. Mumolo
DIA Department, University of Trieste, Trieste, Italy
e-mail: mumolo@units.it

G. Vercelli
DIBRIS Department, University of Genova, Genova, Italy
e-mail: gianni.vercelli@unige.it

R. Bembenik et al. (eds.), *Intelligent Methods and Big Data in Industrial Applications*, Studies in Big Data 40, https://doi.org/10.1007/978-3-319-77604-0_20

1 Introduction

In the field of *music information analysis*, much research has been done in music genre classification, where a model is introduced in order to assign an unknown musical piece a score to a certain class, for example style, period, composer or region of origin. Two main categories of models have been used to do that: *global feature models* (e.g., [16]) and *n-gram models* (e.g., [29]). Global feature models represents every musical piece as a feature vector and use standard machine learning classifiers, whereas n-gram models rely on sequential features and use sequential machine learning algorithms, such as *Hidden Markov Models* (HMM) (e.g., [15]).

In this paper, we apply features derived from the *source-filter model* [12] and HMMs to analyze and classify pieces of classical music, thus combining the two paradigms. This overall originates an HMM-based framework for supporting accurate classification of music datasets. Basically, in this work we adopt the approach that is usually applied to human voice, namely *Mel-Frequency Cepstral Coefficients* (MFCC) (e.g., [12]), and left-to-right HMMs. Indeed, these approaches, in recent initiatives (e.g., [1, 4, 11, 14, 21, 22, 27, 33]), have also been used for musical genre and instrument classification.

We have considered several types of music recording, both choral and instrumental, and we used them to develop statistical models using a classical sequential algorithm. The models have been used to find correlations between composers using the same instrument. Our results shows that this approach can be used both for composer analysis and composer classification. Moreover, the same approach has been used to classify the instrumentation used by a composer, but with some limitation: drums and similar instruments are not well described with the source-filter features and this is verified by our results. For other types of instruments, this method works in a satisfactory way.

Our experimental analysis developed in our research works as follows. We have divided the 400 musical works for composer and instruments. For example, considering *Mozart*, we selected ten pieces of *Piano* (in the following called *Mozart-Piano dataset*), ten pieces of *Orchestra* and ten pieces of *String Quartet* compositions of this author. Of the ten pieces, five pieces are used for the training set and five for the testing set. In addition to this, each piece is then split in two sub-pieces, as follows: only the initial 4 minutes of each piece are used for training the models and the following 4 minutes for testing purposes (of course, the pieces can be longer). By incremental training, we developed an HMM model of *Mozart-Piano*, a model of *Mozart-Orchestra* and a model of *Mozart-String Quartet*, and so forth for each composer. Then, all the available *Piano* compositions are used to compute the likelihood that the musical piece was produced by the *Mozart-Piano* HMM, and so forth for the other compositions. The *confusion matrix* obtained on a maximum likelihood basis describes the relations between the composers using the same instrument. As we show in our experimental assessment and analysis, the obtained results are well-justified on the basis of the known influence among composers. Therefore, a key contribution of this work is to show that left-right HMM and MFCC features can be used to obtain an accurate model of each composer, thus achieving a composite classification framework for music datasets. Indeed, the obtained statistical models can be used to describe the relations among the composers. Our experimental assessment and analysis confirms the benefits coming from our proposal.

The remaining part of this paper is organized as follows. In Sect. 2, previous work on classical music classification are reported and discussed. In Sect. 3, our proposed classification framework is described. In Sect. 4, the experimental results related to the classification of composers and instrumentation are reported. Final remarks and future work analysis conclude the paper in Sect. 5.

2 Related Work

Genre classification has been studied by Yaslan and Cataltepe [32]. They introduce a semi-supervised random feature ensemble method for audio classification that uses labeled and unlabeled data together. Genre classification of classical music has been addressed by Pollastri and Simoncelli [24] by using melody as features and HMM as classifier. They tested the algorithm using four composers. In [17], the MFCCs have been compared to other features for music classification, showing that they are simpler but effective for this purpose. Moreover, approaches based on MFCCs have been used to perform music segmentation (e.g., [3]). In [23], the detection of voice segments in music songs is described. This solution extracts the MFCCs of the sound and uses an HMM to infer if the sound has voice. Marinescu and Ramirez report in [20] an approach that aims at developing models of singers and use them to generate expressive performances similar in voice quality and style with the original singers. Their approach is based on applying machine learning to discover singer-specific timing patterns of expressive singing based on existing performances.

Reference [25] proposes to classify latin music data by exploiting the "cifras" of the songs. Reference [13] focuses on the classification of music data based on lyrics by using non-emotional words. Canonical correlation analysis is instead exploited in [2] to classify Greek folk music. Acoustic and visual features to support music data classification are discussed in [30], whereas global feature models and variable neighborhood search are exploited in [16].

Interesting are also some recent music classification approaches in the context of *deep learning*. For instance, [18] proposes to use deep neural networks to support music genre classification. Convolutional neural networks are instead argued in [34] to be a good classifier for music genre data.

Finally, supervised and semi-supervised methods are studied in [26, 31], respectively.

3 A Framework for Supporting Accurate Classification of Music Datasets Based on HMM

In this Section, we describe our proposed framework for supporting accurate classification of music datasets, by also highlighting the different types of models that we compare in our experimental campaign. Figure 1 provides an overview of our HMM-based classification framework. It essentially comprises two main components: (*i*) *Feature Extraction* and (*ii*) *DHMM*. The first one is the component devoted to extract suitable features from the target music dataset. The second one implements the *Discrete HMM* (DHMM) (e.g., [15]) embedded in our proposal. As we describe in Sect. 3.3, our choice was to select the discrete case as to take advantages from the discrete representation and mapping in data processing.

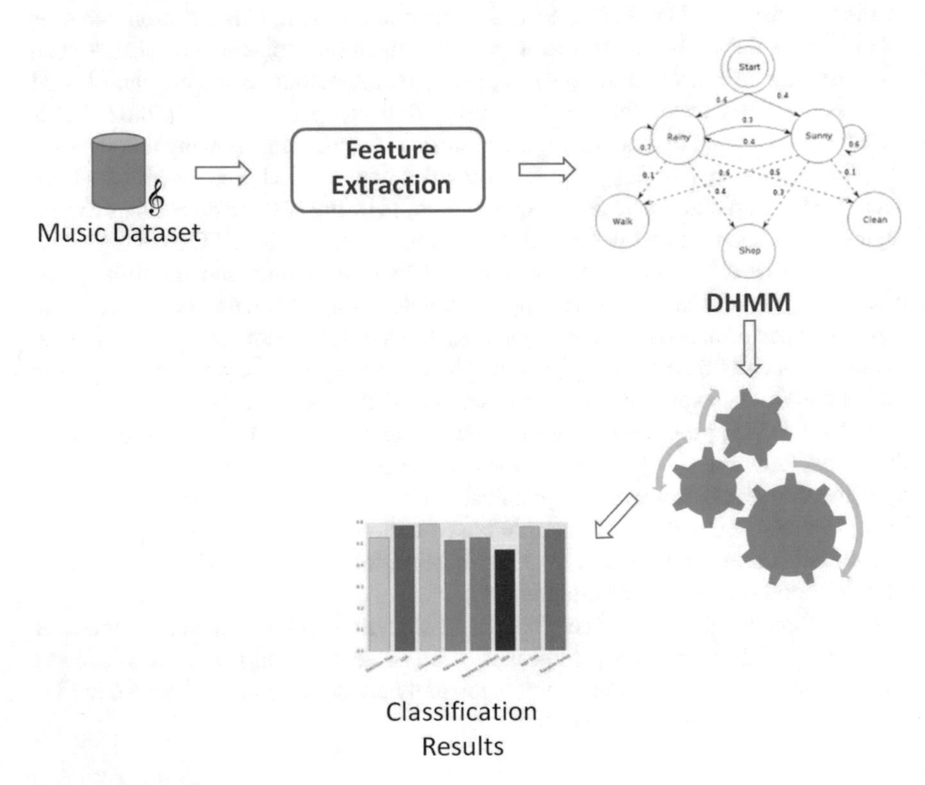

Fig. 1 HMM-based classification framework

3.1 Music Dataset

Our target music dataset comprises a total number of 19 composers, with different styles spanning four centuries, from the 17th to the 20th century, namely: *Bach, Bartok, Beethoven, Berg, Chopin, Debussy, Dvorak, Haydn, Liszt, Mahler, Martinu, Mendelssohn, Mozart, Palestrina, Rihm, Schmittke, Schoenberg, Schumann, Stravinsky*. For the sake of completeness, some musical compositions used in this study are listed, by composer, in the following:

- *Bach*: choral works from BWV225 to BWV229, and the orchestral music from 1046 to 1051;
- *Beethoven*: piano sonatas from 7 to 57, string quartets from Op.127 to Op.135, orchestra symphonies from n.1 to n.9;
- *Chopin*: piano concerto from Op.11 to Op.21, piano suites from Op.22 to Op.52;
- *Mahler*: symphonies from n.1 to n.6;
- *Mozart*: string quartets from KV169 to KV465, orchestra symphonies from n.1 to n.6, piano sonatas from n.10 to n.14;
- *Debussy*: the piano compositions *Arabesque, Ballade, Pagodes, Danse, Images Hommage, Images Reflects, La Boite per Tableau, Nocturne, Preludio, La Cathedrale, Clair de Lune*.

3.2 Feature Extraction

A well-known, simple model of many sounds, like speech and some music, is the source-filter model [12], which we used as fundamental feature extraction model. According to this model, sound is the convolution between the impulse response of a filter with a source signal. For example, in wind instruments, the filter describe the acoustic characteristics of the instrument and the source signal is the signal produced by the reed. The problem is therefore to separate the filter from the source component, which requires ad-hoc deconvolution operations.

One of the approaches to solve this problem is to use *homomorphic deconvolution* [28]. Let $x(n)$ the result of a convolution operation between two signals, the *cepstrum* of $x(n)$ is obtained by taking the logarithm of the spectrum of $x(n)$. Under some circumstances, it is possible to separate the two components. To this end, we use a filter bank, spaced uniformly on a Mel scale, and we considered the output power of these filters in the center of each band-pass filter. We can interpret the output of a single band-pass filter as the kth component of the DFT of the input sequence $x(n)$, denoted by $X(k)$ and defined as follows:

$$X(k) = \sum_{n=0}^{N-1} x(n) e^{j \frac{2\pi}{N} nk} \tag{1}$$

Since we are interested in the center frequencies of the band-pass filters, the kth frequency is moved by $\frac{\pi}{N}$. Therefore:

$$X(k) = \sum_{n=0}^{N-1} x(n) e^{j[\frac{2\pi}{N}k + \frac{\pi}{N}]n} = \sum_{n=0}^{N-1} x(n) e^{j\frac{2\pi}{N}n(k-0.5)} \tag{2}$$

On the basis of this approach, we compute the MFCCs as the $x(n)$ obtained in terms of the inverse filter of Eq. (2), as follows:

$$x(n) = \sum_{k=0}^{M} X(k) \cos(2\pi(k - 0.5)n/N) \tag{3}$$

where $X(k)$ is the logarithm of the output energy for the kth triangular filter, N is the number of filters and M is the number of MFCCs.

3.3 Discrete Hidden Markov Models

In our framework, we make use of DHMM. Markov models describe time series from a stochastic point of view, taking into account the correlations of signals (e.g., [15]).

In HMMs, the output for each state corresponds to an output probability distribution instead of a deterministic event. That is, if the observations are sequences of discrete symbols chosen from a finite alphabet, then, for each state, a corresponding discrete probability distribution describes the stochastic process to be modeled. It is worth noting that, in our framework, we use *discrete observations*, by converting the MFCCs in discrete symbols using standard vector quantization [19], thus achieving the DHMM. The usage of DHMM is motivated by the fact that introducing less symbols makes the model simpler and, at the same, more solid. On the other hand, while it is still possible to convert the continuous observations into discrete ones using vector quantization, some performance degradation due to the quantization process is, however, introduced. Therefore, from a performance point-of-view, it is important to use an overall continuous formulation of algorithms, and consequently introduce ad-hoc optimizations. Our framework adheres to this computational paradigm.

4 Experimental Assessment and Analysis

In this Section, we provide the results of our experimental assessment and analysis devoted to stress the accuracy of the proposed classification framework. In our experimental campaign, we acquired all the sounds at 11025 Hz with 16 bits. The

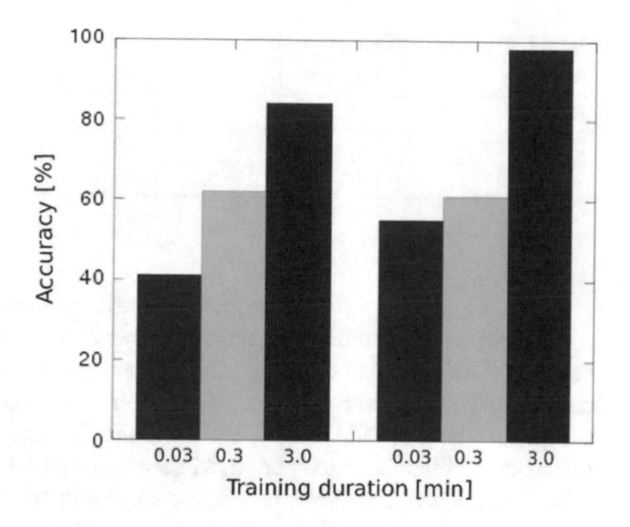

Fig. 2 Average classification accuracy for all the composers (left group of 3 bars) and instrumentation (right group of 3 bars), for different training durations

signal is divided in non overlapping frames of 200 samples, or 18.14 ms, for the subsequent processing. The first problem we explore in this study is the following: what is the best training duration of the DHMM? The second problem is the choice of the number of centroid of the DHMM. The third problem is the choice of the number of states of the DHMM. The last two problems have the following answer: the best results were obtained with 256 centroid and 48 states, so we used these parameters for all the experiments. Regarding the training duration, we computed the average classification accuracy over all the composers and all the instrumentation.

In more details, Fig. 2 reports, by using a bar shape, the average accuracies obtained for the composers and instrumentation, respectively, for three different DHMM training lengths over the 4 minutes basis (see Sect. 1): 0.03, 0.3 and 3.0 min. We observe that the minimum training length to get good enough results is 3.0 min. Furthermore, with more than 3.0 min DHMM training, the accuracy tends to be asymptotic. While this experiment provides average results, it is more interesting to look at the detailed results given as confusion matrixes, as reported in the following Sections.

4.1 Experimental Results for Composer Classification

In this Section, we report the classification results of composers using the same instrumentation, by using the confusion matrix model. The first test is concerned with the choral compositions of *Bach*, *Palestrina* and *Schnittke* (see Table 1). Looking at the confusion matrix, we observe that *Schnittke* is often confused with *Palestrina*. This is probably due to the fact that the Russian composers of the 20th century were influenced by the composition techniques of the Italian and Flemish schools of the 16th century.

Table 1 *Chorus* classification performance

Chorus	*Bach* (%)	*Palestrina* (%)	*Schnittke* (%)
Bach	83.3	16.7	0
Palestrina	0	98.9	0
Schnittke	0	25	75

Then, we study the *String Quartet* compositions (see Table 2).

Looking at the confusion matrix in Table 2, we observe that some confusion between *Beethoven* and *Mozart*, and *Rihm* and *Schoenberg*, respectively, arise. The confusion between *Beethoven* and *Mozart* can be justified by the fact that *Beethoven* was part of the transition of the Classical to the Romantic periods while *Mozart* was a composer of the Classical period. Moreover, we observe some confusion between *Rihm* and *Schoenberg*. Also, this results can be justified by the fact that *Rihm* is a German composer while *Schoenberg* was an Austrian composer, associated with the German expressionist movement, and the former work of *Rihm* was inspired by the *Schoenberg*'s expressionist period.

Next, we make some tests with the *Orchestra* compositions (see Table 3).

From the confusion matrix in Table 3, there are some results to be discussed. The most notably result is probably that *Stravinsky* failed to be recognized as himself, and this can be justified by the fact that he used a lot of percussion in his compositions, and the MFCC, which are derived by a source-filter model of the signal being analyzed, do not describe well this type of sound. Another notably result is that *Stravinsky*'s models are confused with *Chopin*. This result could be justified by the fact that that the musical pieces used for training *Chopin*'s models for orchestral music were concerts for *Piano* and *Orchestra* (mainly Op.11) that inspired *Stravinsky*'s work. On the other hand, *Chopin* is not confused with *Stravinsky* simply because he died before *Stravinsky*. There is also some confusion between *Bach* and *Chopin* while it is not true the opposite. In fact, *Bach*'s model were obtained using *Bach*'s Brandenburg Concertos which did not use instruments typical of *Chopin*'s orchestral music. We also observe an important confusion between *Mahler*'s models and *Berg*'s models, while the opposite does not hold. The linkage between *Mahler* and *Berg* can be explained by the fact that *Mahler* was related with *Schoenberg* as regards for example the instrumental technique, and *Berg* was a *Schoenberg* alumnus.

The classification performances are generally good, with the exception of Chopin, Bach and Stravinsky. Concerning the orchestral compositions of Chopin, it it worth noting that his music considered for training the DHMM are concerts for piano and orchestra. Concerning Bach, his Brandenburg Concertos didn't use percussion or other instruments typical of the orchestral music of following years,as Chopin used. The Stravinsky's poor results can be justified by the fact that he use a lot of percussion, and the source-filter model is not able to model this type of sound.

The last experiments are concerned with the compositions for *Piano* (see Table 4).

Table 2 *String Quartet* classification performance

Quartet	*Bartok* (%)	*Beethoven* (%)	*Dvorak* (%)	*Haydn* (%)	*Martinu* (%)	*Mendelssohn* (%)	*Mozart* (%)	*Rihm* (%)	*Schoenberg* (%)
Bartok	98.7	0	0	0	0	0	0	0	0
Beethoven	0	92	0	0	0	0	8	0	0
Dvorak	0	0	98.5	0	0	0	0	0	0
Haydn	0	0	0	98.6	0	0	0	0	0
Martinu	0	0	0	0	99.1	0	0	0	0
Mendelssohn	0	0	0	0	0	98.9	0	0	0
Mozart	0	0	0	0	0	0	99.2	0	0
Rihm	0	0	0	0	0	0	0	92	8
Schoenberg	0	0	0	0	0	0	0	16	84

Table 3 *Orchestra* classification performance

Orchestra	*Bach* (%)	*Beethoven* (%)	*Berg* (%)	*Chopin* (%)	*Mahler* (%)	*Stravinsky* (%)
Bach	33.3	0	33.3	33.3	0	0
Beethoven	0	83.3	16.7	0	0	0
Berg	0	0	83.3	16.7	0	0
Chopin	0	33.3	0	66.7	0	0
Mahler	0	0	50	16.7	33.3	0
Stravinsky	16.6	0	16.6	50	16.6	0

Table 4 *Piano* classification performance

Piano	*Beethoven* (%)	*Chopin* (%)	*Debussy* (%)	*Liszt* (%)	*Mozart* (%)	*Schumann* (%)
Beethoven	98.7	0	0	0	0	0
Chopin	0	66.6	0	0	16.6	16.6
Debussy	0	16.7	50	0	0	33.3
Liszt	0	33.3	0	33.3	16.6	16.6
Mozart	0	0	0	0	66.7	33.3
Schumann	50	0	0	0	0	50

In this case, an important result is the confusion between *Debussy* and *Schumann*. This may mean that *Debussy* was greatly inspired by *Schumann*. We note also that in some cases *Debussy* is confused with *Chopin*. In fact, *Debussy* was a *Chopin*'s alumnus. Also, there is an important confusion between *Schumann* and *Beethoven*, which may mean that the *Beethoven*'s influence on *Schumann* was quite important. From the results of Table 4, it is worth noting that there is an important confusion between *Liszt* and *Chopin*. In fact, under the influence related with the friendship with *Chopin*, *Liszt*'s developed his poetic and romantic side.

A first conclusion of this set of experiments is that the features derived from the source-filter model are best suitable to model speech (coral compositions), *String* and *Piano* sounds. A second conclusion is that the classification results are well justified in terms of the influence of one composer on another one, thus meaning that the described method can be used to study the influence between composers.

4.2 Experimental Results for Instrumentation Classification

In this Section, we report the classification results of the instrumentation by the same composer, by using the confusion matrix model. In this case we also considered three different instruments: *Piano Solo*, *String Quartet* and *Orchestra*. In Table 5

Table 5 Instrument classification of *Beethoven*'s works

Beethoven	*Piano* (%)	*Orchestra* (%)	*Quartet* (%)
Piano	98.4	0	0
Orchestra	0	98.7	0
Quartet	0	0	98.9

Table 6 Instrument classification of *Mozart*'s works

Mozart	*Piano* (%)	*Orchestra* (%)	*Quartet* (%)
Piano	99.1	0	0
Orchestra	0	98.2	0
Quartet	0	0	98.8

Table 7 Instrument classification of *Haydn*'s works

Haydn	*Piano* (%)	*Orchestra* (%)	*Quartet* (%)
Piano	98.3	0	0
Orchestra	10	90	0
Quartet	0	0	98.8

Table 8 Instrument classification of *Schubert*'s works

Schubert	*Piano* (%)	*Orchestra* (%)	*Quartet* (%)
Piano	99.3	0	0
Orchestra	0	90	10
Quartet	0	0	99.7

and Table 6, we observe that these three settings are successfully recognized for *Beethoven*'s and *Mozart*'s works.

We have also obtained good classification of instrumentation for *Haydn*'s (see Table 7) and *Schubert*'s (see Table 8) works.

Even from the second set of experiments, we can conclude that our proposed framework provides a good accuracy in the case of instrument classification too, thus proving its flexibility and reliability.

5 Final Remarks and Future Work

In this paper, we have proposed a classification framework that exploits the source-filter approach and HMM machine learning techniques for classifying musical recordings. While our framework is rather simple, it proved to ensure good classification accuracy. The experimental results show that these methodologies are well-suited for modeling classical music even when no voices are present in the recordings. A major limitation of this method is due to the weak behaviors in modeling drums, as the feature extraction technique we used removes quite completely such frequencies. This can lead to a modification of the approach to take into account this aspect in future works. Another important line of research to explore concerns with making the proposed framework able to deal with novel challenges posed by emerging *big data trends* (e.g., [5–10].

References

1. Ajmera, J., McCowan, I., Bourlard, H.: Speech/music segmentation using entropy and dynamism features in a HMM classification framework. Speech Commun. **40**(3), 351–363 (2003)
2. Bassiou, N., Kotropoulos, C., Papazoglou-Chalikias, A.: Greek folk music classification into two genres using lyrics and audio via canonical correlation analysis. In: 9th International Symposium on Image and Signal Processing and Analysis, ISPA 2015, Zagreb, Croatia, 7–9 September 2015, pp. 238–243.
3. Bhalke, D.G., Rao, C.B.R., Bormane, D.S.: Automatic musical instrument classification using fractional fourier transform based- MFCC features and counter propagation neural network. J. Intell. Inf. Syst. **46**(3), 425–446 (2016)
4. Cont, A.: Realtime audio to score alignment for polyphonic music instruments, using sparse non-negative constraints and hierarchical HMMS. In: 2006 IEEE international conference on acoustics speech and signal processing, ICASSP 2006, Toulouse, France, 14–19 May 2006, pp. 245–248
5. Cuzzocrea, A.: Accuracy control in compressed multidimensional data cubes for quality of answer-based OLAP tools. In: Proceedings of the 8th International Conference on Scientific and Statistical Database Management, SSDBM 2006, 3–5 July 2006, Vienna, Austria, pp. 301–310
6. Cuzzocrea, A.: Privacy and security of big data: Current challenges and future research perspectives. In: Proceedings of the First International Workshop on Privacy and Secuirty of Big Data, PSBD@CIKM 2014, Shanghai, China, 7 November 2014, pp. 45–47
7. Cuzzocrea, A., Furfaro, F., Saccà, D.: Enabling OLAP in mobile environments via intelligent data cube compression techniques. J. Intell. Inf. Syst. **33**(2), 95–143 (2009)
8. Cuzzocrea, A., Matrangolo, U.: Analytical synopses for approximate query answering in OLAP environments. In: Proceedings of the 15th International Conference Database and Expert Systems Applications, DEXA 2004 Zaragoza, Spain, 30 August, 3 September 2004, pp. 359–370
9. Cuzzocrea, A., Saccà, D., Ullman, J.D.: Big data: a research agenda. In: 17th International Database Engineering & Applications Symposium, IDEAS '13, Barcelona, Spain, 09–11 October 2013, pp. 198–203
10. Cuzzocrea, A., Song, I., Davis, K.C.: Analytics over large-scale multidimensional data: the big data revolution! In: Proceedings of the DOLAP 2011, ACM 14th International Workshop on Data Warehousing and OLAP, Glasgow, United Kingdom, 28 October 2011, pp. 101–104
11. Emiya, V., Badeau, R., David, B.: Automatic transcription of piano music based on HMM tracking of jointly-estimated pitches. In: 2008 16th European Signal Processing Conference, EUSIPCO 2008, Lausanne, Switzerland, 25–29 August 2008, pp. 1–5
12. Fant, G.: Acoustic Theory of Speech Production. Mouton, The Hague (1960)
13. Furuya, M., Oku, K., Kawagoe, K.: Music feeling classification based on lyrics using weighting of non-emotional words. In: Proceedings of the 13th International Conference on Advances in

Mobile Computing and Multimedia, MoMM 2015, Brussels, Belgium, 11–13 December 2015, pp. 380–383
14. Gao, S., Zhu, Y.: A hmm-embedded unsupervised learning to musical event detection. In: Proceedings of the 2005 IEEE International Conference on Multimedia and Expo, ICME 2005, 6-9 July 2005, Amsterdam, The Netherlands, pp. 334–337
15. Ghahramani, Z.: An introduction to hidden markov models and bayesian networks. IJPRAI **15**(1), 9–42 (2001)
16. Herremans, D., Sörensen, K., Martens, D.: Classification and generation of composer-specific music using global feature models and variable neighborhood search. Comput. Music J. **39**(3), 71–91 (2015)
17. J. H. Jensen, M. G. Christensen, M. Murthi, and S. H. Jensen. Evaluation of mfcc estimation techniques for music similarity. In *European Signal Processing Conference, EUSIPCO*, 2006
18. Jeong, I., Lee, K.: Learning temporal features using a deep neural network and its application to music genre classification. In: Proceedings of the 17th International Society for Music Information Retrieval Conference, ISMIR 2016, New York City, United States, 7–11 August 2016, pp. 434–440
19. Linde, Y., Buzo, A., Gray, R.M.: An algorithm for vector quantizer design. IEEE Trans. Commun. 702–710 (1980)
20. M.-C. Marinescu and R. Ramirez. Modeling expressive performances of the singing voice. In *International Workshop on Machine Learning and Music*, 2009
21. Myung, J., Kim, K., Park, J., Koo, M., Kim, J.: Two-pass search strategy using accumulated band energy histogram for hmm-based identification of perceptually identical music. Int. J. Imaging Syst. Technol. **23**(2), 127–132 (2013)
22. Nakamura, E., Yoshii, K., Sagayama, S.: Rhythm transcription of polyphonic piano music based on merged-output HMM for multiple voices. IEEE/ACM Trans. Audio, Speech Lang. Process. **25**(4), 794–806 (2017)
23. R. Nobrega and S. Cavaco. Detecting key features in popular music: case study - singing voice detection. In *International Workshop on Machine Learning and Music*, 2009
24. Pollastri, E., Simoncelli, G.: Classification of melodies by composer with hidden markov models. International Conference on Web Delivering of Music, 0088 (2001)
25. Przybysz, A.L., Corassa, R., dos Santos, C.L., Silla, C.N.: Latin music mood classification using cifras. In: 2015 IEEE International Conference on Systems, Man, and Cybernetics, Kowloon Tong, Hong Kong, 9–12 October 2015, pp. 1682–1686
26. Rajesh, B., Bhalke, D.G.: Automatic genre classification of indian tamil and western music using fractional MFCC. Int. J. Speech Technol. **19**(3), 551–563 (2016)
27. J. C. Ross and J. Samuel. Hierarchical clustering of music database based on HMM and markov chain for search efficiency. In *Speech, Sound and Music Processing: Embracing Research in India - 8th International Symposium, CMMR 2011, 20th International Symposium, FRSM 2011, Bhubaneswar, India, March 9-12, 2011, Revised Selected Papers*, pages 98–103, 2011
28. Tribolet, J.M.: Seismic applications of homomorphic signal processing. Prentice Hall (1979)
29. Wolkowicz, J., Keselj, V.: Evaluation of n-gram-based classification approaches on classical music corpora. In: Proceedings of the Mathematics and Computation in Music - 4th International Conference, MCM 2013, Montreal, QC, Canada, 12–14 June 2013, pp. 213–225
30. Wu, M., Jang, J.R.: Combining acoustic and multilevel visual features for music genre classification. TOMCCAP, **12**(1), 10:1–10:17 (2015)
31. Yang, X., He, L., Qu, D., Zhang, W., Johnson, M.T.: Semi-supervised feature selection for audio classification based on constraint compensated laplacian score. EURASIP J. Audio Speech Music Process. **2016**(9)(2016)
32. Yaslan, Y., Cataltepe, Z.: Audio genre classification with semi-supervised feature ensemble learning. In: International Workshop on Machine Learning and Music (2009)
33. Yeminy, Y.R., Keller, Y., Gannot, S.: Single microphone speech separation by diffusion-based HMM estimation. EURASIP J. Audio Speech Music Process. **2016**(16) (2016)
34. Zhang, W., Lei, W., Xu, X., Xing, X.: Improved music genre classification with convolutional neural networks. In: Interspeech 2016, 17th Annual Conference of the International Speech Communication Association, San Francisco, CA, USA, 8–12 September 2016, pp. 3304–3308

Classification of Music Genres by Means of Listening Tests and Decision Algorithms

Aleksandra Dorochowicz, Piotr Hoffmann, Agata Majdańczuk and Bożena Kostek

Abstract The paper compares the results of audio excerpt assignment to a music genre obtained in listening tests and classification by means of decision algorithms. A short review on music description employing music styles and genres is given. Then, assumptions of listening tests to be carried out along with an online survey for assigning audio samples to selected music genres are presented. A framework for music parametrization is created resulting in feature vectors, which are checked for data redundancy. Finally, the effectiveness of the automatic music genre classification employing two decision algorithms is presented. Conclusions contain the results of the comparative analysis of the results obtained in listening tests and automatic genre classification.

Keywords Music genre classification · Feature extraction · Listening tests

1 Introduction

One of the methods used to categorize music is the systematics based on the music genres. Using the specific features of a track, it allows defining which category or type of music the track belongs to. There are many factors, that have an influence on the assignment of a music piece to the specific music genre, starting with the role of music (film score, religious or orchestral music), through the performance techniques (vocal, piano, orchestral), to the origin of the genre (J-rock, Brit-pop) [3, 17]. It often happens that some features are combined within a track, which results in

A. Dorochowicz · P. Hoffmann (✉) · A. Majdańczuk · B. Kostek
Audio Acoustics Laboratory, Faculty of Electronics, Telecommunications and Informatics, Gdańsk University of Technology, Narutowicza 11/12, 80-233 Gdańsk, Poland
e-mail: phoff@sound.eti.pg.gda.pl

A. Dorochowicz
e-mail: Aleksandra.D@interia.pl

B. Kostek
e-mail: bokostek@audioacoustics.org

R. Bembenik et al. (eds.), *Intelligent Methods and Big Data in Industrial Applications*, Studies in Big Data 40, https://doi.org/10.1007/978-3-319-77604-0_21

genre overlapping. Moreover, music genres are more and more often divided into the sub-genres, so the character of the piece could be described more precisely. Hence, the new genres do not only derive from the main genre (punk rock, new wave, post grunge), but also can be a result of a mixture of genres (symphonic metal, pop rock). It is also worth mentioning that the music genres can also be defined through the audience (adult contemporary, teen pop) or the time, when the piece was created (classic, romantic, modern).

The aim of the study, performed very recently, is to define the extent to which it is possible to correctly assign a track listened to by a subject to the music genre. The results of an online listening survey provided a platform to compare the effectiveness of the performance of the feature vectors created by the authors. Samples used in the study represent the following music genres: pop, rock, rap and hip-hop, classical, jazz, dance and DJ (electronic), hard rock and metal, blues, New Age, country. The music pieces, chosen for the tests, can be divided into two groups: the tracks undoubtedly belonging to the specific genre, and the samples being more ambiguous, where the problems with the classification may appear.

In order to study music assignment to the given music genre, an experiment has been conducted which compared subjective opinion of listeners regarding a given music track and indications of automatic music genre recognition. The goal of the subjective tests performed was to check whether it is possible to assign music excerpts to specific music genres, as well as determining which ones can be considered as uniquely attributed to a particular genre. The unambiguous cases obtained in that way can be parameterized and automatic classification employing decision algorithms can be carried out, resulting in a comparison of its results with subjective tests. The comparison was conducted employing two commonly used classifiers: BayesNet and SMO. It is worth mentioning that the training of the classifiers was based on the set of the tracks contained in the GZTAN database [8] to have a larger representation of parametrized music tracks. The results of the experiment were commented and compared on the basis of the effectiveness measures obtained.

2 Classification of Music Styles and Genres

The concept of music styles and genres was created for the music analysis to describe and specify the music pieces. Divisions created over the centuries worked on different, exclusive planes. With time, more and more division systems appeared, being more and more detailed, so music forms could be described more precisely. Furthermore, the pieces existing inside the given typology can also belong to other categories. To classify them correctly, it is necessary to understand what they are and how they may be characterized. Examples of the categories are style, music form, genre, and also features they are described by tempo, meter, origin, targeted for the specific group of adherents or organizing the diversity of music into some structures.

The concept of the style in music theory exists since the 17th century, but its meaning has changed over the time [14, 18]. In 1649, based on the antique tradition,

Marc Scacchi created the division of the styles into the religious, chamber and scenic (theatrical), and their subtypes. In the next year, Athanasius Kircher expanded it into church style, madrigal and theatrical presuming that the style does not specify the place of the performance, but the affects it triggers, and most importantly, it also predicts the diffusion of the styles. Through the centuries, definitions of the music style have changed, but the music style is the term describing the common features of the compositional technique typical for the specific piece, author, for the national music, historical period [14]. However, the music form is defined according to the vocal or/and music instruments used, type of texture or how many passages it consists of i.e. one-passaged, cyclic, etc. Again, the music forms are typical for a specific group of music pieces schemes, designated by the analysis based on the specific pieces.

There is a large number of types of the music genre division, based on the performers' origin, age of the audience, music ambitions, instruments used, etc. At the beginning of the 2016, the American music magazines Alternative Press [2] and Rock Sound [16] published the research 'What is punk?' created by the Converse company with the Polygraph's analyst [5] about the lack of unequivocal definitions and the difficulties in their differentiation. The research was based on the playlists tagged with the word punk at Spotify and YouTube. Green Day with their songs appears on over the half of them (51%), the next places are Blink-182 (50%), The Offspring (44%), Sum 41 (39%), Rise Against (38%), Fall Out Boy (38%), My Chemical Romance (35%), Bad Religion (30%), Nofx (28%), All Time Low (28%) and A Day To Remember (27%). Music genres the bands belong to are closely related to punk, but they are not precisely punk. They are pop-punk, emo, post-hardcore or metalcore. Following the conclusions of the research cited above, it is not obvious where the specific performer or track belong to. This makes defining the specific track as clearly belonging to a genre a very difficult, if not an impossible task. This may be true both for human listeners and decision algorithms.

2.1 Listening Tests

To obtain statistically significant results, a large group of people should be involved in the subjective listening tests. Even though, one can expect that the listening test results will not fully be homogeneous in such a task as music genre classification. This is due to the individual characteristics of music as well as the ambiguity of some tracks, causing that people will attribute a certain track to different music genres. That is why, it is impossible to obtain 100% accuracy of evaluation of any music sample. Moreover, only certain tendencies can be observed based on listening test results. In addition, there may be different acoustic conditions in which a given music piece can be performed and recorded, such as a studio version may be classified as a different genre than the concert version of the same recording. Another aspect is the fact that different fragments of the same music piece can be identified as different music genres, so that the sample can be attributed to different genres.

Conducting listening tests is affected by the differences in the listeners' perception of the specific features. As already mentioned, obtained results will never overlap completely, which means that the classification will never be unequivocal. The test should be built in a way so that the uncertainty of the features such as personal taste or music sensitivity would be minimized [6]. The ACR method (*Absolute Category Rating*) was applied to conduct subjective tests. It is described in the ITU-T P.910 norm [10], samples with normalized features are the base material. Features of the listening room, as well as test material matter. Samples to be evaluated should be random, appropriately short, but also meaningful. The scale range is from 5 (perfect) to 1 (bad). The numbers chosen for every element have the influence on the score. The test procedure consists in listening to a sample and assigning a rate. The final score is the arithmetic average of the rates given by the listeners and is called *Mean Opinion Score* (MOS) [6].

The material to be assessed was built on 75 high-quality excerpts of the pieces, 10 s long, belonging to the following music genres: pop, rock, rap/hip-hop, classic, jazz, electronic, hard rock/metal, blues, country, R&B, New Age and folk. The samples were chosen to be as clear as possible or ambiguous. Different parts of the piece were used in tests, to check whether they will be classified in the same way or differently. Moreover, same tracks recorded in different acoustic conditions (studio, concert, acoustic versions) or played by other performers, or representing different music genres were in addition chosen. Among the samples, old and new pieces also appeared. The group of classic music consisted of samples from baroque and the 21st century. The same concept was applied to other genre groups, there were samples from the very beginning of the genre creation as well as recent tracks. The samples representing mixed genres were also used (e.g. pop rock).

For the purpose of the study, listening test sessions were conducted with a group of subjects, aged approximately 23 but no older than 30. There were no hearing problems reported among the subjects. Their task was to listen to a series of music excerpts and classify them accordingly to the task to be performed. When conducting listening tests, different types of people participated, with possibly different musical education. This could have an impact on the data collected, however the large number of subjects might minimize the possible error margin.

Because of the huge music material of the test material and constraints imposed by the standard on carrying out listening tests, it was necessary to divide the test into five smaller surveys, containing 15 samples. The survey form with the questions was prepared using the Google Forms. There was an instruction contained on the site created, explaining how to fill the survey in. Samples were added using the HTML, they were played as a list of the files hosted in the cloud. A single test contained 15 music files to play, having only numbers accompanied with 16 questions, 15 related to genre rating and the additional one about the listening equipment (earphones, computer speakers, smartphone/tablet, or external speakers), so it could be verified if there was a big difference in sound, which may affect the rating. The last question was used to verify possible differences in music perception while listening to music through speakers or headphones, and consequently, to determine possible additional factors behind the evaluation.

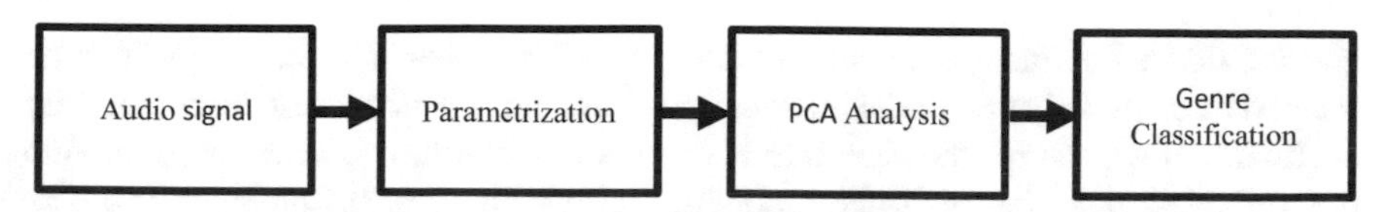

Fig. 1 Simplified diagram of parametrization—classification process

The number of the answers given to each of the 75 samples varies between 36 and 45, and the results were transferred into the spreadsheet and prepared for the analysis.

2.2 *Parametrization and Classification*

The first step to classify the samples with the automatic classification systems was the parametrization of the samples used in subjective tests (Fig. 1). This is the main stage which determines the effectiveness of the classification system. Inaccurate assignment of the parameters at such an early preprocessing stage may result in reduction of the effectiveness of the system.

It is a well-known fact, that the parametrization of music is necessary to gain the high level of the classification efficiency [11]. It helps getting the detailed and precise description of the object, so it could be represented by the finite vectors of numbers, not the linguistic descriptors based on the subjective impressions of the listener. The parametrization process enables to convert such descriptors into the n-dimension feature vectors (FVs), which is necessary in the classification process. Parametrization leads to obtaining the huge number of parameters containing time and spectral information. For decreasing redundancy of the data, the obtained parameters are typically a subject to many operations, including PCA (Principal Component Analysis) [20]. PCA can be defined as a non-parametric method of relevant information extraction from the big databases. It relies on data variances, which makes a basis for a new group of parameters. New descriptors are a linear combination of elements, carrying the most relevant information about the data.

There were two classifiers used for the classification of the parametrized samples: Bayesian Network and SMO (Sequential Minimal Optimization Algorithm). Since both algorithms are well-known, thus only their basic description is to be recalled, further on. Bayesian Networks have a large number of applications in many areas, they owe their popularity due to the ease of resolving problems [7]. The SMO method was created as an extension to the SVN method (Support Vector Machine). The main application area of algorithm is solving the optimization problems and more specifically the quadratic programming problem [4].

The Bayesian Networks method is based on probability theory. Its main feature is independence from events other than its descendants. The model is represented by a graph, not containing the cycles. Every node is a random variable (e.g. parameter), and the ties define the dependencies between the variables [7]. The models can be created by using only part of the information about the dependence of the conditional attributes. However, for the Bayesian Networks to be built, it is necessary to define the variables, links between them, conditional and the a priori probabilities. The data are put into the net sequentially, and then they are upgraded and the posteriori probability is determined. The Bayesian Networks classification enables to predict the assignment of the object to the class [7]. The classification process in this method is carried out through assigning a feature vector to a class, where probability of a posteriori $P(C^{(j)})|a$ is the biggest. When there are more than two classes analyzed, the discriminant functions are calculated. The classifier calculates M discriminant functions for all the classes. As the final choice, the class of Max(M) is determined.

The Sequential Minimal Optimization Algorithm is targeted for the optimization of the quadratic problem solving. The algorithm was created as an extension of the Support Vector Machine. Decomposition of SVM optimization problem as a method of SMO consists in highlighting the two subsets of the parameters: the working set and the passive set [4]. In the SMO method, the working set is the smallest possible two-element set. Because of it, it is not necessary to use complicated numeric methods to solve the sub-problems for any number of parameters. SMO is an iterative algorithm, the method splits the problem into the series of possible smallest sub-problems, which are solved analytically. Because of the Lagrange's requirement of equal linear multipliers, the smallest possible problem affects both multipliers [15]. In the first step of the SMO classifier, the multiplier not fulfilling the conditions KKT (Karush-Kuhn-Tucker) is picked for the optimization. Then, the second multiplier and a pair for optimization are chosen, the step is repeated until the full convergence is achieved. When all the multipliers fulfill the KKT conditions (within the tolerances defined by the user), the problem can be accepted as solved. Although the algorithm ensures convergence, pairs of the multipliers are chosen heuristically to accelerate the gain. It is very important for the huge databases because there are $n(n-1)$ possibilities of the multiplier choice [15].

As mentioned before, the main component of music genres recognition systems is the optimized parametrization block. The prepared feature vectors (FVs) should have a very good separability between parameters. Taking into account these assumptions, the feature vector containing 173 elements, conceived in earlier research studies carried out by the authors [10, 12, 13], was utilized. A collection of 52532 music excerpts described by a set of parameters obtained through the analysis of mp3 recordings was gathered in a database called SYNAT, realized by the Gdansk University of Technology [11]. As it represents a large collection of recordings, that is why it was possible to optimize the FV proposed.

For the recordings included in the database, the analysis band is limited to 8 kHz due to the music excerpt format, this means that the frequency band used for the parameterization is in the range from 63 to 8000 Hz. The prepared feature vector is used to describe parametrically each signal frame. The extracted FVs, in majority are the MPEG 7 standard parameters [13], supplemented with 20 Mel-Frequency Cepstral Coefficients (MFCC), 20 MFCC variances and enlarged by 24 time-related parameters proposed by the authors, which refer to temporal characteristic of the analyzed music excerpt [9].

The parametrization of audio files collected for the purpose of this study included the following steps:

- loading an excerpt of a recording with a duration of approx. 26 s,
- conversion to a 22,050 Hz monophonic signal (right, left channel summing),
- segmentation into 8,192 samples, i.e., 2 to the power of 13, due to the need for the FFT algorithm for spectral analysis,
- calculating Fourier spectra with a sample rate of 4,410 samples (time equal to 0.2 s), and using a Blackman window, hence the overlap size equals to 3,782 samples,
- calculating the power spectrum using a logarithmic scale for each segment.

Due to the sampling frequency of 22,050 Hz, the frequency resolution for Fourier analysis is 2.692 Hz and the analysis band reaches 9986.53 Hz. The entire available frequency band is divided into sub-bands of increasing width directly proportional to the center frequencies. The first sub-band has a center frequency of 100 Hz. The nonlinearity coefficient = 1.194, borrowed from the König scale that defines the ratio of the width of the subsequent sub-bands, was used in the entire analyzed frequency range.

Calculating the spectrograms in the above-described scale, followed by cepstrograms, resulted in cepstral coefficients C_i according to the discrete cosine transform:

$$C_i = \sum_{j=1}^{N} logE_j \cos\left(\frac{\pi i}{N}(j - 0, 5)\right) \tag{1}$$

where:

i- number of a cepstral coefficient,
E_j- energy of the jth sub-band of the filter bank, and
N- number of (filter) channels in the filter bank.

Also, based on the cepstrograms, statistical parameters of individual cepstral coefficients, namely: average value (arithmetic mean), variance and skewness, i.e. moments of the 2nd and 3rd order for the ith cepstral coefficient were also determined:

$$M_i(n) = \frac{1}{K}\sum_{k=0}^{K}[C_i(k) - m_i]^n \tag{2}$$

where:

i- number of a cepstral coefficient,
n- order of the moment (2nd or 3rd).

For cepstral parameters, each sub-vector has a length equal to the number of segments. However, their number is equal to the number of cepstral coefficients (cepstrum order) = 16. The vector length is 2,048 which consists of 16 sub-vectors

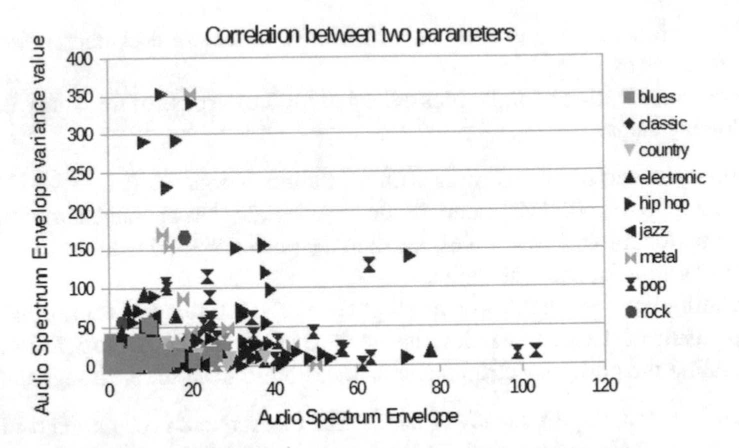

Fig. 2 An example of the correlation analysis of two parameters

with 128 values each. The resulting cepstrogram can be converted into a shorter vector in several ways, such as by delta, trajectory or statistical transformation. It should be noted that the full vector with a length of 2,048 was also tested, but proved to be ineffective in the classification process.

The list of parameters and their definitions were shown in the earlier study [11], thus they are not presented in this paper. However, it is worth noting that the proposed FV was used in the ISMIS 2011 contest in which there were over 120 participants. The best contest result returned almost 88% of accuracy, and later in the authors' own study gained even better effectiveness.

To illustrate difficulties with parameter value separability, an example of the correlation analysis of two parameters is shown in Fig. 2. For decreasing redundancy of the data, authors used the PCA method. Applying PCA allowed for calculation accelerating, because finally the 173 element set was replaced by 33 parameters carrying 90% of information.

3 Result Analysis

3.1 Subjective Tests Results

Below, examples of results analysis and statistics received from the subjective tests are presented. In summary, there were 75 10-s long samples used for the tests, chosen for twelve music genres. Among them there are directly correlated samples: same track, different conditions (Fig. 3); same track, different performers (Fig. 4); same track, different fragments (Fig. 5); same performer, different tracks (Fig. 6).

34 samples received over 50% votes for one music genre, three of them received the result different from the performer's intention. The huge dominance of one of

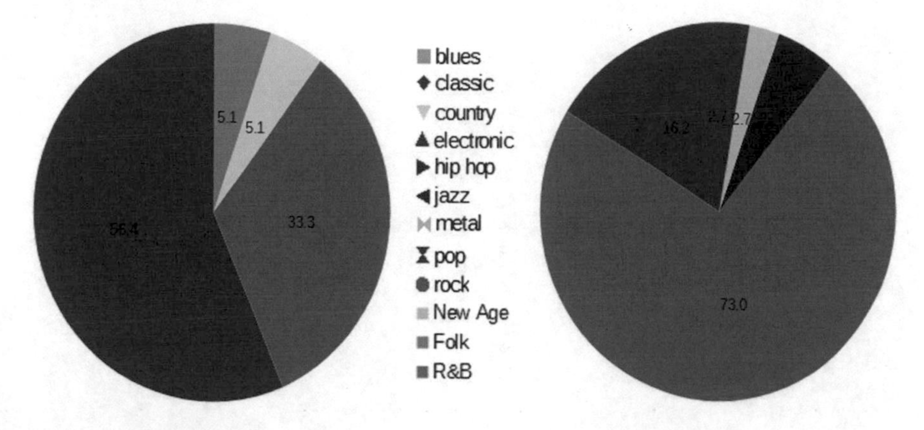

Fig. 3 Two versions of the same track

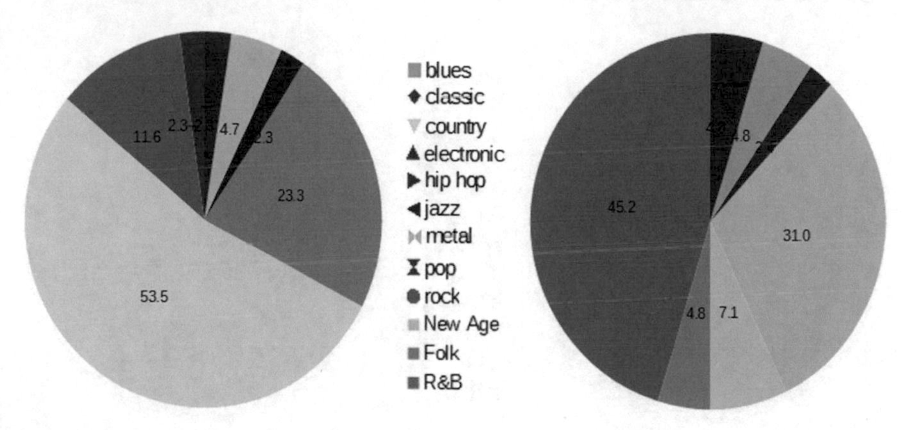

Fig. 4 The same track performed by different artists

the genres is visible in case of 45 samples, three of them were rated differently to what the performer assumed. In case of the two samples, rated differently, their rate was the second most often given rate. Four samples received very similar rates to all music genres. Among wrong rated tracks, in four cases rock was mistaken for pop and in two cases rock was mistaken for New Age. None of the samples received over 50% votes for the following genres: R&B, New Age and folk. There is no influence observed within listeners using different listening equipment.

Additionally, some of the tracks were analyzed more thoroughly. Three samples of the ‘Wherever You Will Go’ performed by The Calling/Alex Band (same band) in different conditions (concert, studio and acoustic version) received very similar results and all were classified as rock (in sequence 60.9, 51.1, 58.5%). Two fragments of the ‘Remembering Sunday’ of All Time Low feat. Juliet Simms were rated differently: the first one was rated as rock (37.8%, pop on the second position with 31.1%), and the second one as pop (34.1%, then New Age 25%, rock 22.7%). The

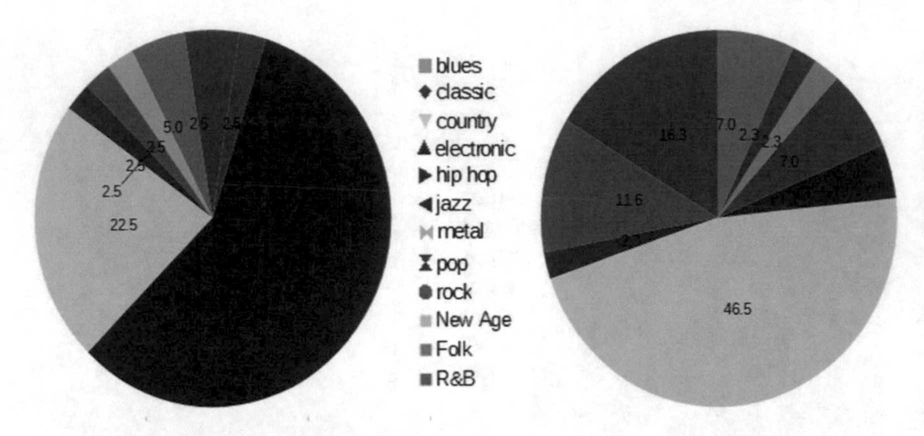

Fig. 5 Different fragments of the same track

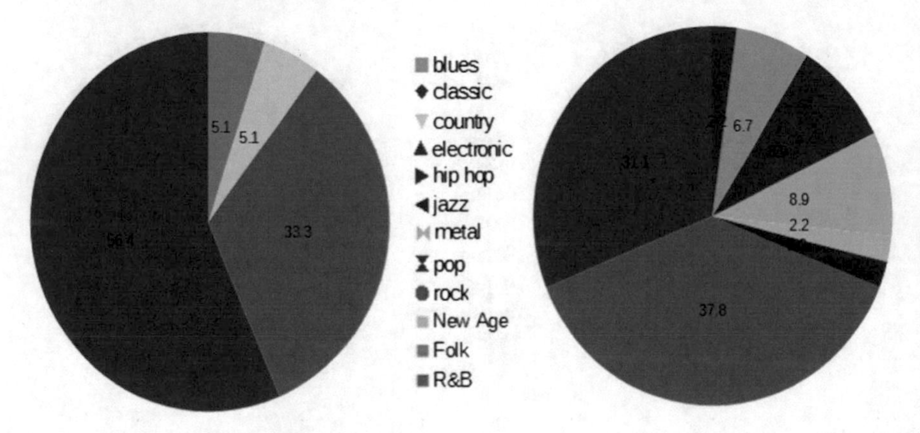

Fig. 6 Different track performed by the same artist

piece 'Heute Hier, Morgen Dort' was rated differently depending on the performer. The Hannes Wader's folk version was classified as country (64.5%) and the rock version made by Die Toten Hosen 45.2% of listeners rated as rock (31% as hard rock/metal). Two fragments of 'Seize The Night' by art rock musician Meat Loaf, depending on the fragment, were rated differently (classic 57.5%, New Age 46.5%), see Fig. 4.

3.2 Classification Results

In order to verify the subjective test results, i.e. the assignment of tracks to music genres performed by listeners, a comparison between listening test results and information provided by an automatic music genre recognition system has been executed.

As mentioned before, for that purpose two classifiers, i.e. BayesNet and SMO from the WEKA environment were employed. The experiment results have been analyzed on the basis of precision, recall and f-measure values.

Two stages should be clearly discerned: training and testing, as different databases were utilized. For the purpose of the experiment, the system of automatic genre recognition, has been trained on the basis of GZTAN [8]. The system training set included 1000 music excerpts, equally divided among the previously mentioned genres.

The second stage, namely testing was performed on the same collection of songs as utilized in the listening tests. Before the classification, the songs were parametrized with the system described in Sect. 2.2.

Carrying out an automatic classification of genres and comparing its results with the subjective perception of listeners, required selecting samples to be listened to in the experiment. As already mentioned, the selection consisted in choosing only those samples which were unequivocally assigned to one of the 10 music genres recognizable by the system: classical, blues, country, disco, hip-hop, jazz, metal, pop, reggae, rock. The level of 50% has been set as the acccptable threshold for clearly recognizing a music genre. Based on these assumptions, 30 excerpts were utilized as test FVs on previously taught classifiers. Each of the 30 samples had the minimal length of 30 s, required by the automatic genre recognition process.

After checking out the assignment of the object to specific classes (music genres), the following parameters were compared (Eqs. 3–6):

- The number of correctly classified entities in a form of number and percentage;
- The number of incorrectly classified entities in a form of number and percentage;
- The Kappa statistics, measuring compatibility between the proposed allocation instance to a class, and an actual one, which describes general accuracy of the model [12];
- The mean absolute error measures the mean size of errors in the data set, not considering their direction. The mean absolute error can be described as a mean of the absolute differences between the proposal (result) and the analogous period of observation [19];
- Root mean square error describes the accuracy of the classification compared to the analogous period of observation [19];
- Relative absolute error describes, how big deviation are the results of the classification process from the expected values [1];
- Root relative squared error [1].

The next part of the results analyzes in detail data in reference to all the classes and shows True Positive Rate (TP), which says, whether the genre x was classified correctly as the genre x. While the False Positive Rate (FP) gives the opposite result, i.e. information if the genre x was classified as a genre y. Furthermore, a number of factors related to the precision of the algorithm are calculated. The values are given in relation to a particular object, and averaged. These measures are as follows:

- Precision describes how many of the results returned by a classifier as True is objectively true. The parameter is described by a formula:

$$Precision = PPV = \frac{TP}{TP + FP} \tag{3}$$

- Recall describes how many of the results were detected by a classifier:

$$Recall = \frac{TP}{TP + FN} \tag{4}$$

- F-Measure is a measure of unbalanced sets (significantly different numbers of the learning and testing sets)

$$F = \frac{2 \cdot Precision \cdot Recall}{(Precision + Recall)} \tag{5}$$

- MCC (*Matthews Correlation Coefficient*) is the measure of the binary classification:

$$MCC = \frac{TP \cdot TN - FP \cdot FN}{\sqrt{(TP + FP)\,(TP + FN)\,(TN + FP)\,(TN + FN)}} \tag{6}$$

- ROC Area is described as an area under the ROC curve. The ROC curve shows the relationship between the TP and FP results.

In the case of the Bayesian Networks and the SMO classifiers, the number of correctly and incorrectly rated samples is the same, but the error values differ. For the Bayesian Networks classifier, the mean absolute error is four times smaller than for the SMO classifier, while in the case of the root mean square error, the difference is twice smaller. The relative absolute error is ten times bigger for the SMO classifier, and the root relative squared error twice smaller. It should be mentioned that the training of the classifiers was based on the set of the tracks contained in the GZTAN database [8].

The data obtained were analyzed in detail in reference to all classes and some other factors were calculated, such as: True Positive Rate, False Positive Rate, Precision, Recall, F-Measure, MCC (*Matthews Correlation Coefficient*) and finally ROC (Receiver Operating Characteristics) and PRC (Precision/Recall). For the classification of blues, country, hip-hop and jazz, the result is very similar, but there are differences for the classic music in the ROC parameter: ROC Area (0.910 and 0.889) and PRC Area (0.861 and 0.736). The differences are also visible for the electronic music (0.655 and 0.172; 0.091 and 0.033), pop (0.992 and 0.980; 0.833 and 0.967) and rock (0.949 and 0.966; 0.934 and 0.942). Because of this, the differences also appeared in the mean values (0.955 and 0.938; 0.919 and 0.872). Tables 1 and 2 show the summary of results produced by the Bayesian Networks and SMO classifiers.

The results obtained in the listening tests and the learning algorithms show that only three samples in both cases were classified incorrectly. For the Belief Networks

Table 1 Summary of results obtained for both classifiers

	Bayesian Net	SMO
Correctly classified instances (%)	27/90	27/90
Incorrectly classified instances (%)	3/10	3/10
Kappa statistic	0.8808	0.8808
Mean absolute error	0.0201	0.1631
Relative absolute error (%)	11.14	90.62
Root relative squared error (%)	47.14	92.44
Total number of instances	30	30

Table 2 Summary of measure results obtained for both classifiers

Precision		Reacall		F-Measure		Class
BN	SMO	BN	SMO	BN	SMO	
1	1	1	1	1	1	Blucs
0.833	0.833	0.833	0.833	0.833	0.833	Classical
1	1	1	1	1	1	Country
0	0	1	1	0	0	Disco
1	1	1	1	1	1	Hip-hop
1	1	1	1	1	1	Jazz
1	1	1	1	1	1	Metal
0.833	0.833	1	1	0.909	0.909	Pop
1	1	1	1	1	1	Reggae
1	1	0.875	0.875	0.933	0.933	Rock
0.906	0.906	0.9	0.9	0.9	0. 9	Avg.

classifier Bach's 'Largo' was classified as electronic, and the Jarre's 'Equinoxe' as classic. 'Problem Child' by AC/DC was classified as pop. Moreover, listeners rated the samples of 'Jasey Rae' as two different genres, and so did the classifier. In the case of SMO classifier, it incorrectly classified 'Unchained Melody' by Henry Mancini (jazz), rating it as blues.

4 Conclusions

The aim of the study was to compare the efficiency of the listeners music classification compared to the automatic music genre classification experiment designed by the authors. First, data resulting from the music genre assignment the listeners were gathered. In the second part of the experiment, the same music samples were used in the automatic genre classification process. Conducted tests showed to what extent the listeners' assignment and the results of automatic classification system are correlated.

Automatic classification show that three songs have been classified differently in comparison to the listeners' perception in subjective tests. Incorrect recognition has occurred for music genres, such as: classical, disco, pop, rock. The efficiency of using classifiers was at the same level (90%). The F-Measure, Precision, and Recall measures also match that level. Based on the experiment, it is impossible to clearly confirm the superiority of any of the classifiers. It means that the instruments contained in the track and the performance techniques affect similarly the way both listeners and the classifiers evaluate music.

Samples of the different part of the track, different types of recordings, different artists, that were rated differently show that it is not always possible to assign a piece to the particular music genre.

References

1. Abramowitz, M., Stegun, I.A.: Handbook of Mathematical Functions with Formulas, Graphs, and Mathematical Tables, 9th printing. Dover, New York (1972)
2. Alternative Press. http://www.altpress.com/index.php/news/entry/what_is_punk_this_new_infographic_can_tell_you. Accessed Jan 2017
3. Benward, B., Saker, M.: Music: In Theory and Practice, 7th ed., vol. I, p. 12 (2003)
4. Candel, D., Nanculef, R., Concha, C., Allende, H.: A Sequential Minimal Optimization Algorithm for the All-Distances Support Vector Machine, CIARP 2010, LNCS, vol. 6419, pp. 484–491. Springer, Berlin, 2010
5. Definition of Punk. http://poly-graph.co/punk/. Accessed Jan 2017
6. Dorochowicz, A., Majdańczuk, A.: Conducting subjective listening tests of an audio graphic equalizer with automatic music genre recognition. M.Sc., Faculty of ETI, Gdansk University of Technology, Gdańsk, 2016 (in Polish)
7. Friedman, N., Geiger, D., Goldszmidt, M.: Bayesian network classifiers. Mach. Learn. **29**, 139–164
8. GZTAN Database. http://labrosa.ee.columbia.edu/millionsong/blog/11-2-28-deriving-genre-dataset. Accessed Jan 2017
9. Hoffmann, P., Kostek, B.: Bass enhancement settings in portable devices based on music genre recognition. J. Audio Eng. Soc. **63**(12), 980–989 (2015). http://dx.doi.org/10.17743/jaes.2015.0087
10. ITU P.910 (04/08) Standard. https://www.itu.int/rec/T-REC-P.910-200804-I/en
11. Kostek, B., Hoffmann, P., Kaczmarek, A., Spaleniak, P.: Creating a Reliable Music Discovery and Recommendation System, pp. 107–130. Springer (2013)
12. McHugh, M.L.: Interrater reliability: the kappa statistic. Biochem. Med. **22**, 276–282 (2012). https://doi.org/10.11613/BM.2012.031
13. MPEG 7 Standard. http://mpeg.chiariglione.org/standards/mpeg-7
14. Pascall, R.: The New Grove Dictionary of Music and Musicians, red. Stanley Sadie, 24, 2/London, pp. 638–642 (2001)
15. Platt, J.: Sequential Minimal Optimization: A Fast Algorithm for Training Support Vector Machines, Microsoft Research MSR-TR-98-14 (1998)
16. RockSound. http://www.rocksound.tv/news/read/study-green-day-blink-182-are-punk-my-chemical-romance-are-emo. Accessed Jan 2017

17. Rosner, A., Kostek, B.: Classification of music genres based on music separation into harmonic and drum components. Arch. Acoust. **39**(4), 629–638 (2014). https://doi.org/10.2478/aoa-2014-0068
18. Seidel, W., Leisinger, U.: Die Musik in Geschichte und Gegenwart, ed. Ludwig Finscher, Sachteil, 8, Kassel-Basel-etc., pp. 1740–1759 (1998)
19. Tofallis, A.: A better measure of relative prediction accuracy for model selection and model estimation. J. Oper. Res. Soc. (2015)
20. Williams, L.J., Abdi, H.: Principal component analysis. Wiley Interdiscip. Rev.: Comput. Stat. **2** (2010)

Handwritten Signature Verification System Employing Wireless Biometric Pen

Michał Lech and Andrzej Czyżewski

Abstract The handwritten signature verification system being a part of the developed multimodal biometric banking stand is presented. The hardware component of the solution is described with a focus on the signature acquisition and on verification procedures. The signature is acquired employing an accelerometer and a gyroscope built-in the biometric pen plus pressure sensors for the assessment of the proper pen grip and then the signature verification method based on adapted Dynamic Time Warping (DTW) method is applied. Hitherto achieved FRR and FAR measures for the verification based exclusively on the biometric pen sensors and for the comparison on the parameters retrieved from the signature scanning pad are compared.

Keywords Signature verification · Biometric pen · Dynamic time warping

1 Introduction

The handwritten signature verification system presented in the paper has been developed in the scope of the project which aim is to create multimodal biometric system for bank client identity verification. The modalities based on which the verification is performed are: handwritten signature put down using the developed biometric pen, voice, face image, face contour, and hand veins pattern. It is assumed that depending on the degree of required security of the banking operation the particular modalities are chosen separately or jointly [1].

The main purpose of the undertaken task is to develop improved algorithms and methods to verify identity based on a dynamic analysis of the electronic handwritten signature. In the course of the work, methods of obtaining individual characteristics of electronic handwritten signature are developed and tested for their suitability in the

M. Lech (✉) · A. Czyżewski
Faculty of Electronics, Telecommunication and Informatics, Multimedia Systems Department, Gdansk University of Technology, Gdansk, Poland
e-mail: mlech@sound.eti.pg.gda.pl

R. Bembenik et al. (eds.), *Intelligent Methods and Big Data in Industrial Applications*, Studies in Big Data 40, https://doi.org/10.1007/978-3-319-77604-0_22

verification process. First, the electronic handwritten signature should be examined with the use of the device previously developed at the Gdansk University of Technology in the form of an experimental electronic pen. Subsequently, a new technology of intelligent pen for mobile terminals is to be developed. Dynamic parameters of the handwritten signature as opposed to the static ones, describe the process of signature giving, which enables its fuller and more stable representation. The group of dynamic parameters, among others, include the time length of giving signature, the pen pressure on the surface, the tilt of the pen and others. These parameters are variable during the signature giving, making it possible to extract the individual features. Owing to the continuous recording of parameters of the signature, it is much less likely to impersonate another subject than in the case of using the signature representation to verify the signature graphically. The main research problem associated with this task is to develop the optimal dynamic parameter vector with a choice of optimal classifier in terms of reliability and efficiency. In the course of the research, various ways of acquiring and representing the dynamic signature are to be tested, including the registration of the signature making process. This paper presents results of the hitherto achieved results in this domain, namely an application of the DTW (Dynamic Time Warping) method for discerning original signatures from forged ones.

The aim of the conducted research is also to examine the technologies developed in the framework of this project under close-to-real conditions. During the tests, the observations are conducted to verify the assumptions made. The overall research methodology is based on finding a representative group of users (at least 10,000), which will take part in biometric verification tests using developed methods. The results will be used to optimize the developed methods. In addition to the statistical data, the result of this phase will help to determine a subjective assessment of satisfaction of persons undergoing experimental verification of identity. Based on the results, the conclusions will be developed and published in the future. A large Polish bank as an institution supporting the project takes an active part in the implementation of this task by organizing the research group and test environment. So far laboratory tests were organized, described in this paper, employing an experimental biometric stand located in Gdansk University of Technology, however, the first setups of this type were installed in real bank outlets, at the time of this paper writing (see Fig. 1).

2 Signature Acquisition and Verification Method

The hardware of the handwritten signature verification subsystem consists of a wireless biometric pen (Fig. 1) and a 7″ resistive touch screen mounted on a LCD display. The biometric pen is equipped with a 3-axis accelerometer, a 3-axis gyroscope, a surface pressure sensor and 2 touch pressure sensors. The resistive touch screen and the LCD display are connected to the computer responsible for acquiring the biometric features from all the modalities (called Biometric Hub) via USB cable and HDMI cable, respectively. The biometric pen communicates with the Biometric Hub using Bluetooth 4.0 + LE. Power is supplied by a rechargeable battery mounted inside

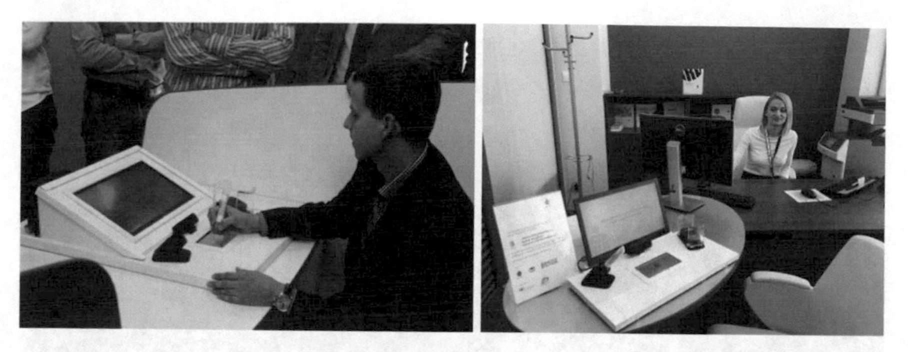

Fig. 1 Experimental biometric stands installed in the laboratory of Gdansk University of Technology (left) and in the real bank outlet (right)

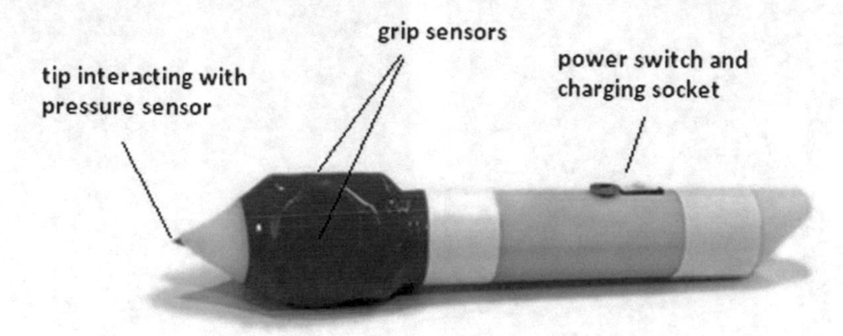

Fig. 2 The wireless biometric pen prototype

the pen. The battery can be easily charged using a mini-jack cable connected to the socket located near the on/off switch (Fig. 2).

In order to provide repeatable signals from the biometric pen during the authentication sessions the grip manner and firmness are monitored based on the values of pressure acquired from touch pressure sensors. Moreover, the relations between the values of acceleration in 3 axes from the accelerometer are transmitted via a Bluetooth radio built-into the biometric pen. If any of the touch sensors is not pressed or the pen is rotated improperly, the information on how to hold the pen properly is displayed to the user on the screen. The signature is acquired only if the grip is firm and the position of the held pen is proper. The resistive computer screen is used only for the purpose of providing a graphic feedback to the user while putting down the signature, whereas the verification is based solely on the measurements from the sensors mounted in the pen (transmitted to the computer wirelessly).

The values of acceleration and angular positions representing the given signature excluding the pen movements before and after putting down the signature are retrieved based on the monitor surface pressure sensor indications. Each time a user puts his hand off the screen (the pressure sensor indicates values below empirically set threshold) a software timer is started. The timer is reset when the user puts his

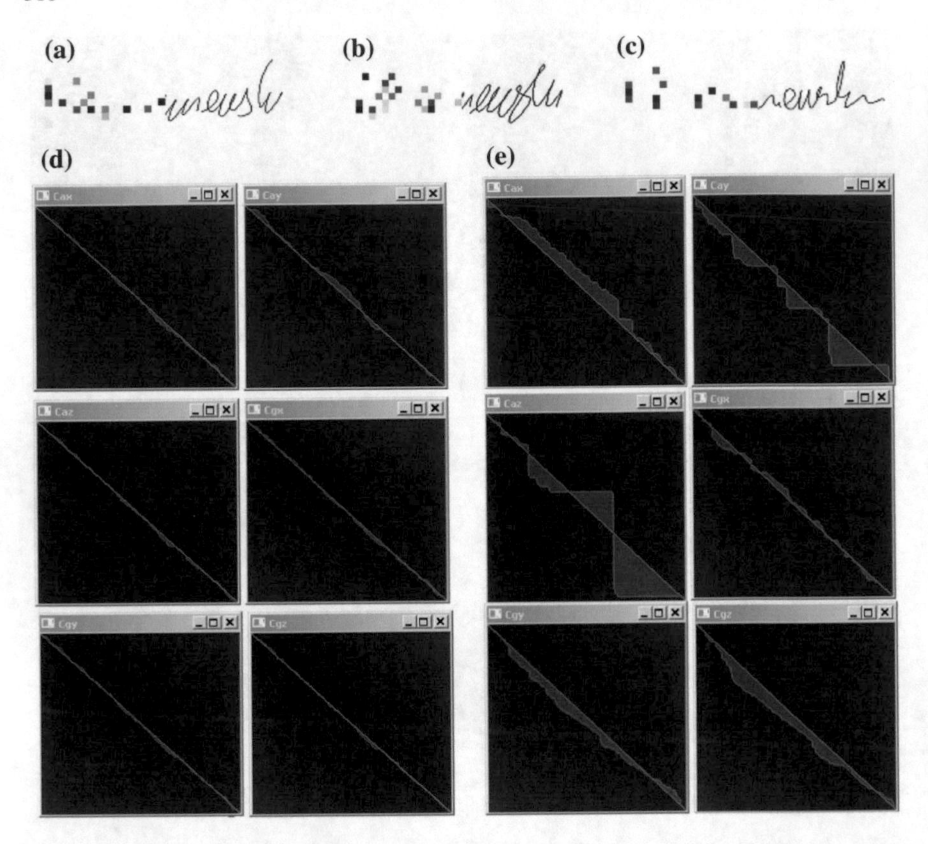

Fig. 3 The anonymized model signature (**a**), another authentic signature (**b**) and their forgery (**c**), and DTW accumulated cost matrices for the 3-axis acceleration (c_{ax}, c_{ay}, c_{az}) and angular position (c_{gx}, c_{gy}, c_{gz}), both for the authentic signature (**d**) and its forgery (**e**)

hand down again to write the successive part of the signature. If the timer reaches 1500 ms it is assumed that the signature is ended and the verification phase begins.

The signature verification has been based on 6 dynamic measures, which will be described later on in the paper, created employing the results of the dynamic time warping (DTW) algorithm. The method is based on the assumption that time-domain functions of two arbitrary authentic signatures entail less warping than the functions of the authentic and the forged signature (Fig. 3f, g). Therefore, the information derived from the DTW method and then used for the verification process represents the convergence of a "diagonal" and an optimal cost path in the accumulated cost matrix. The matrix is created in the manner given by Eq. 1, where $\gamma_{i,j}$ is the accumulated cost in cell (i, j), d is the distance (Eq. 2) between the elements of functions F and G (Eq. 3) representing values of a particular parameter, such as acceleration or angular position, of two arbitrary signatures of lengths m and n:

$$\gamma_{i,j} = d\left(f_i, g_j\right) + \min\left(\gamma_{i-1,j-1}; \gamma_{i-1,j}; \gamma_{i,j-1}\right) \tag{1}$$

$$d\left(f_i, g_j\right) = \left|f_i - g_j\right| \tag{2}$$

$$F = f_1, f_2, \ldots, f_i, \ldots, f_n \tag{3}$$

$$G = g_1, g_2, \ldots, g_j, \ldots, g_m \tag{4}$$

The distance metric given by Eq. 2 has been chosen empirically and it turned out that it outperformed other popular metrics, providing the best EER (Equal Error Rate) measure during experiments. The standard back tracing manner [2] of finding the optimal path **w**, given by Eqs. 5 and 6, in the accumulated cost matrix has been used:

$$\mathbf{w} = \{w_k, w_{k-1}, \ldots, w_0\} \quad \max(m; n) \leq k < m + n - 1 \tag{5}$$

$$w_k = \begin{cases} (i-1, j-1) & \gamma_{i-1,j-1} = \min\left(\gamma_{i-1,j-1}; \gamma_{i-1,j}; \gamma_{i,j-1}\right) \\ (i-1, j) & \gamma_{i-1,j} = \min\left(\gamma_{i-1,j-1}; \gamma_{i-1,j}; \gamma_{i,j-1}\right) \\ (i, j-1) & \gamma_{i,j-1} = \min\left(\gamma_{i-1,j-1}; \gamma_{i-1,j}; \gamma_{i,j-1}\right) \end{cases} \tag{6}$$

In the previous work by the authors [3, 4] another path tracing method had been proposed which performed well with the measures representing in time domain the shape of the signature put down on the graphic tablet using an ordinary tablet stylus.

In the work presented herein, for the 3-axis acceleration (ax, ay, az) and 3-axis angular position from a gyroscope (gx, gy, gz), 6 measures have been defined, denoted respectively by $c_{ax}, c_{ay}, c_{az}, c_{gx}, c_{gy}, c_{gz}$, representing a degree of convergence of DTW matrix "diagonal" and the optimal path. The convergence is defined as the sum of absolute differences of y positions of pixel belonging to the matrix "diagonal" and pixel belonging to the optimal path cost, for the same x position.

The process of verification involves a comparison of model signatures with those obtained in the current authentication session. The assessment of the degree of signature authenticity p, within a range [0; 1], using the 6 measures involves a comparison of their values p'_s with threshold values p_{THR}, according to Eq. 7:

$$p = \begin{cases} 1 & p'_s < p_{THR} \\ \frac{p_{THR}}{p'_s} & p'_s > p_{THR} \end{cases} \tag{7}$$

where p'_s is a value of measure p_s obtained from DTW method after rescaling, according to Eq. 8, where n and m define the size of the accumulated cost matrix:

$$p'_s = 10000 \frac{p_s}{\mathrm{nm}} \tag{8}$$

Thus, the thresholds p_{THR} could be set empirically to fixed values, and for convergence of the path and the diagonal equaled 300.

The global similarity ratio value is the average value of all the p values.

3 Assessment of FRR and FAR Measures

In order to compare the performance of the verification based on the developed biometric pen with the performance of the verification based on a typical modern banking approach employing a signature pad plus an unequipped pen, the FRR and FAR measures assessment was performed. The assessment was also made for the system with verification based on the DTW measures for parameters obtained exclusively from the resistive surface. These parameters were acceleration **a** (Eqs. 9–11) retrieved from the (x, y) movement of the pen tip on the resistive surface over time t and a signature trajectory $\boldsymbol{\alpha}$ over time t represented in the polar coordinate system, according to Eqs. 12 and 13:

$$\mathbf{a} = \sum_{i=1}^{N-1} a_i \tag{9}$$

$$a_i = \frac{\upsilon_i - \upsilon_{i-1}}{t_i - t_{i-1}} \tag{10}$$

$$\upsilon_i = \frac{\sqrt{(x_i - x_{i-1})^2 + (y_i - y_{i-1})^2}}{t_i - t_{i-1}} \tag{11}$$

$$\boldsymbol{\alpha} = \sum_{i=1}^{N-1} \alpha_i \tag{12}$$

$$\alpha_i = \arctan \frac{|y_i - y_{i-1}|}{|x_i - x_{i-1}|} \cdot \frac{180}{\pi} \tag{13}$$

23 persons took part in the experiments. From each person 3 model signatures were collected. The FRR (*False Rejection Rate*) measure, i.e. the ratio of the number of authentic signatures rejected by the system to the number of all authentic signatures and the FAR (*False Acceptance Rate*) measure, i.e. the ratio of the number of forged signatures accepted by the system divided by the number of all forgeries, have been assessed. In an ideal system both measures should be equal to 0. Considering the probability of a signature authenticity (further understood as the similarity ratio), the acceptance thresholds in the resistive surface-based verification and the biometric pen-based verification had been set based on the results of the assessment and it turned out that they equaled 0.84 and 0.82, respectively. Such thresholds were considered optimal, as they provided simultaneously the smallest possible value of FRR and FAR measures.

3.1 *Assessment of FRR Measure*

In order to check the short-term invariability related to putting down the signature, the FRR assessment has been performed twice, i.e. directly after obtaining the model signatures and after five more days. In both series from 21 persons 10 samples of

Table 1 Distribution of true positives (T) and false negatives (F) in FRR measure assessment of the resistive surface-based verification attempts made directly after obtaining the models and after 5 days since obtaining the models

	direct attempts										*postponed attempts*									
P1	T	T	T	T	T	T	T	T	T	T	T	T	T	T	T	T	T	T	T	T
P2	T	T	T	T	T	T	T	T	T	T	T	T	T	F	T	T	F	T	T	T
P3	T	T	T	T	T	T	T	T	T	T	T	T	T	F	T	T	F	T	T	T
P4	F	F	F	F	F	T	F	T	F	T	x	x	x	x	x	x	x	x	x	x
P5	T	T	T	T	T	T	T	T	T	T	T	T	T	T	T	T	T	T	T	T
P6	F	F	T	T	T	F	T	F	F	F	T	F	T	F	F	F	F	F	T	F
P7	T	T	T	T	T	T	T	T	T	T	T	T	T	T	T	T	T	T	T	T
P8	T	T	T	T	T	T	T	F	T	T	T	T	T	T	T	T	T	F	T	T
P9	T	T	F	T	T	T	F	T	T	T	T	T	T	T	T	F	F	T	T	T
P10	T	T	T	F	F	T	T	T	F	T	T	T	T	T	T	T	T	T	T	T
P11	T	T	T	T	T	T	T	T	T	T	F	F	F	F	F	F	F	F	T	T
P12	T	T	T	T	T	T	T	T	T	T	T	T	T	T	T	T	T	T	T	T
P13	T	T	T	T	T	T	T	T	T	T	T	T	T	F	T	T	T	T	F	T
P14	T	T	T	T	T	T	T	T	T	T	T	T	T	T	T	T	T	T	T	T
P15	T	T	T	T	T	T	T	T	T	T	T	T	T	T	T	F	T	T	T	T
P16	T	T	T	T	F	T	F	F	T	T	T	T	T	T	T	F	F	F	F	F
P17	T	T	T	T	T	T	T	T	T	T	F	T	T	T	T	F	T	F	T	T
P18	T	T	T	T	T	T	T	T	T	T	T	F	T	T	T	F	F	T	T	T
P19	T	T	T	T	T	T	T	T	T	T	F	T	T	F	T	F	F	F	F	T
P20	T	T	T	T	T	T	T	T	T	T	F	T	T	T	F	F	T	T	T	T
P21	T	T	T	F	T	T	T	T	T	T	T	F	T	F	T	T	T	F	T	F
	FRR = 0.110										FRR = 0.245									

his/her signature have been collected. A sample including 440 authentic signatures was obtained in this way, including 3 model signatures obtained from each person. The person P4 was not able to participate in the delayed verification (marked 'x' in tables).

In Tables 1 and 2 the distributions of true positives and false negatives for the assessment of the FRR measure, for the resistive surface-based verification and the biometric pen-based verification, have been presented. The FRR measure for the resistive surface-based verification attempts, occurring directly after obtaining the signature models equaled 0.110. The FRR measure for the resistive surface-based verification attempts occurring after 5 more days increased to 0.245. The FRR measure for the biometric pen-based verification attempts, occurring directly after obtaining the signature models equaled 0.129. The FRR measure for the biometric pen-based verification attempts occurring after 5 more days increased to 0.305.

In Figs. 4 and 5 the box-and-whisker plots for the resistive surface-based verification and the biometric pen-based verification, have been presented. The median value of the similarity ratio for the resistive surface-based verification attempts, occurring

Table 2 Distribution of true positives (T) and false negatives (F) in FRR measure assessment for biometric pen-based verification attempts made directly after obtaining the models and after 5 days since obtaining the models

	direct attempts										*postponed attempts*									
P1	T	T	T	T	T	T	T	T	T	T	T	T	T	T	T	T	T	T	T	T
P2	T	T	T	T	T	F	T	T	T	T	T	T	T	T	T	T	T	T	T	F
P3	T	T	T	F	T	T	T	T	T	T	T	F	F	T	T	F	T	T	F	F
P4	F	T	T	F	F	T	F	F	T	T	x	x	x	x	x	x	x	x	x	x
P5	T	T	T	T	T	T	T	T	T	T	T	T	T	T	T	T	T	T	T	T
P6	T	T	T	T	T	T	T	F	T	T	T	T	T	T	T	T	F	T	T	T
P7	T	T	T	T	T	T	T	T	T	T	T	T	T	T	T	T	T	T	T	T
P8	T	T	T	T	T	T	T	T	T	T	F	F	T	T	T	T	T	T	T	T
P9	T	T	T	T	T	T	T	T	T	T	T	F	T	T	T	F	T	T	T	T
P10	T	T	T	T	T	T	T	T	T	T	F	T	T	T	T	F	F	T	T	T
P11	T	T	T	T	T	T	T	T	T	T	T	T	T	T	T	T	T	T	T	T
P12	T	T	F	T	T	T	F	T	T	T	F	F	T	T	T	T	T	T	T	T
P13	F	T	T	F	F	F	T	F	F	F	F	F	F	F	F	F	F	F	F	F
P14	T	T	T	T	T	T	T	T	T	T	T	T	T	T	T	F	F	T	T	T
P15	T	T	T	T	T	T	F	T	T	T	F	T	F	F	F	F	F	T	F	F
P16	T	T	T	T	F	T	F	T	T	T	F	T	F	T	T	T	F	F	F	F
P17	T	T	T	T	T	T	T	T	T	T	F	T	T	T	F	T	T	F	T	F
P18	T	F	T	T	T	T	T	T	T	T	F	T	T	T	T	F	F	F	F	F
P19	T	T	F	T	T	T	T	F	T	F	F	T	F	T	F	F	F	F	F	T
P20	T	T	T	T	T	T	T	F	T	T	F	T	F	T	T	T	F	F	F	F
P21	T	T	T	T	T	F	F	F	T	T	F	F	F	T	F	T	T	T	T	T
	FRR = 0.129										FRR = 0.305									

directly after obtaining the signature models equaled to 0.948, whereas after 5 more days it decreased to 0.905. The median of the similarity ratio for the biometric pen-based verification attempts, occurring directly after obtaining the models equaled to 0.919, and after 5 more days it decreased to 0.855.

3.2 *Assessment of FAR Measure*

The assessment of the FAR measure was performed in such a way that each person's signature was forged by another person among the test group. The first 10 persons among the group were skilled in forging the signatures and the remaining 13 persons performed only simple (non-skilled) forgeries. The first 10 persons were skilled in forging both visual and dynamic aspects of signatures by looking how the signature owner is putting down the signature. Such an approach is much more restrictive in comparison with the approach commonly used in which in skilled forgeries only the

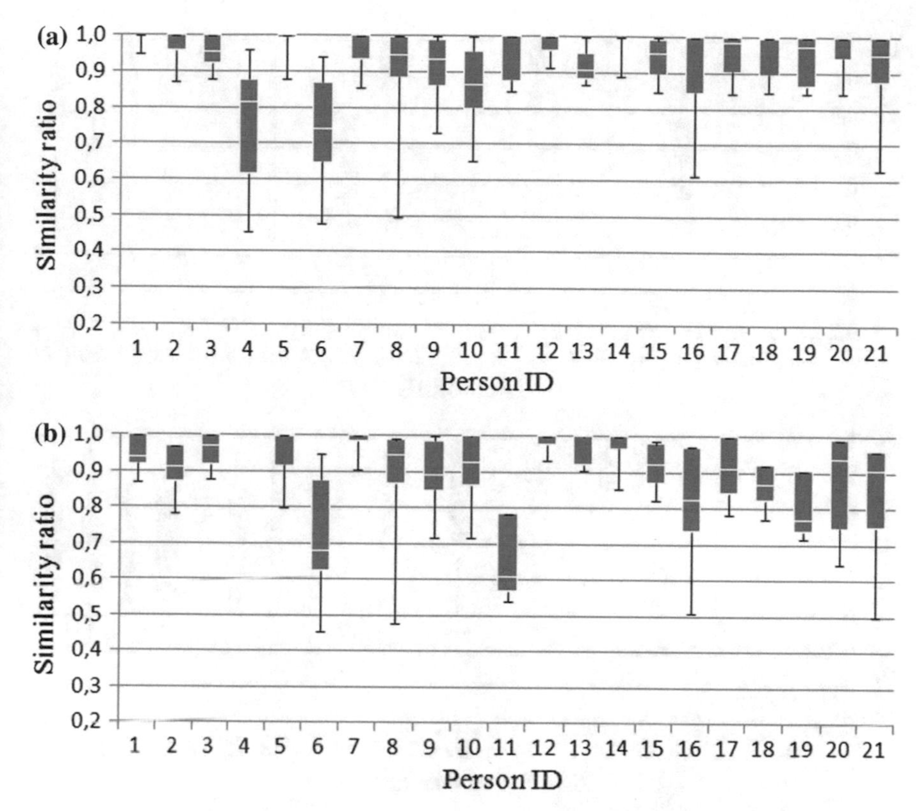

Fig. 4 FRR measure assessment box-and-whisker plots for the resistive pad-based verification, for attempts occurring directly after obtaining the models (**a**) and after 5 days since obtaining the models (**b**)

signature image is imitated and the dynamic features not necessarily reflect the ones produced by the signature owner [5]. The remaining 13 persons tried to forge another person's signature by imitating the signature's image.

In Table 3 and Fig. 6 the results of FAR measure assessment for skilled forgeries have been presented. The FAR measure of the resistive pad-based verification was unacceptably high, since it equaled to 0.38 with the median value of the similarity ratio equal to 0.760. The FAR measure of the biometric pen-based verification equaled to 0.09 with the median of the similarity ratio equal to 0.628. The significant impact on the FAR measure has the signature of the person P5 which was well-calligraphed, turning out to be the most easy to forge. Basing the verification on measurements from sensors mounted in the biometric pen enabled to globally decrease the difference between the 1st and the 3rd quartiles, leading to more reliable detection of forgeries.

In Table 4 and Fig. 7 the results of FAR measure assessment for simple forgeries have been presented. The FAR measure of the resistive pad-based verification was

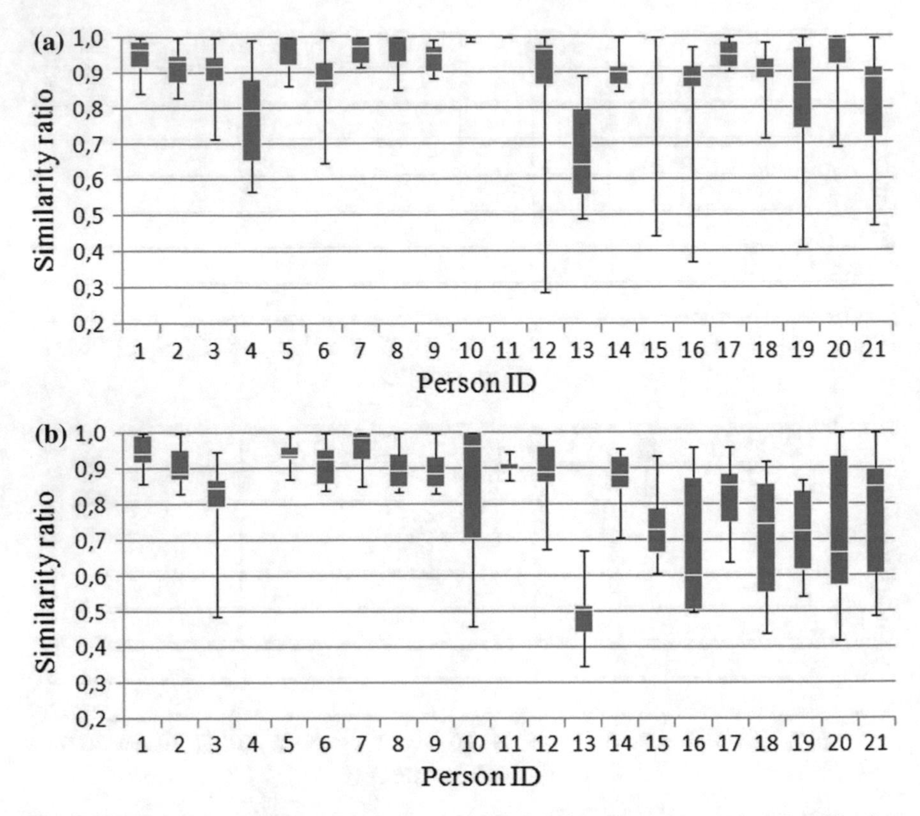

Fig. 5 FRR measure assessment box-and-whisker plots for the biometric pen-based verification, for attempts occurring directly after obtaining the models (**a**) and after 5 days since obtaining the models (**b**)

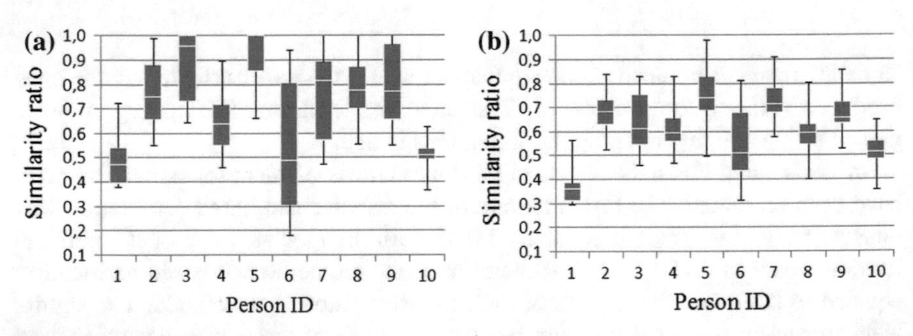

Fig. 6 FAR measure assessment box-and-whisker plots for the resistive pad-based verification (**a**) and biometric pen-based verification (**b**), for skilled forgeries

still high, and equaled to 0.15 with the median value of the similarity ratio equal to 0.635. The FAR measure of the biometric pen-based verification equaled to 0.02 with the median of the similarity ratio equal to 0.482.

Table 3 Distribution of true negatives (N) and false positives (P) in the FAR measure assessment of skilled forgeries verification based exclusively on the resistive pad and exclusively on the biometric pen

	resistive pad-based verification										*biometric pen-based verification*									
P1	N	N	N	N	P	N	N	N	N	N	N	N	N	N	N	N	N	N	N	N
P2	P	N	N	N	P	P	N	N	P	N	N	N	N	N	N	N	N	N	N	N
P3	N	N	P	P	P	P	P	P	P	N	N	N	N	N	N	N	N	N	N	N
P4	N	N	N	N	P	N	N	N	N	N	N	N	N	N	N	N	N	N	N	N
P5	N	P	P	P	P	P	P	P	N	P	N	N	N	N	N	N	P	N	P	P
P6	N	N	N	N	P	N	N	N	P	P	N	N	N	N	N	N	N	N	N	N
P7	N	N	P	P	N	N	P	P	N	P	N	N	N	N	N	N	N	N	N	P
P8	P	N	N	P	N	N	N	P	N	P	N	N	N	N	N	N	N	N	N	N
P9	N	P	N	P	P	P	N	N	N	P	N	P	N	N	N	N	N	N	N	N
P10	N	N	N	N	N	N	N	N	N	N	N	N	N	N	N	N	N	N	N	N
	FAR = 0.38										FAR = 0.09									

Table 4 Distribution of true negatives (N) and false positives (P) in the FAR measure assessment of simple forgeries verification based exclusively on the resistive pad and exclusively on the biometric pen

	resistive pad-based verification										*biometric pen-based verification*									
P11	N	N	N	N	N	N	N	N	N	N	P	P	N	N	N	N	P	N	N	N
P12	N	P	P	P	P	P	P	P	P	P	N	N	N	N	N	N	N	N	N	N
P13	N	N	P	N	N	N	P	N	N	N	N	N	N	N	N	N	N	N	N	N
P14	N	N	N	N	N	N	N	N	N	N	N	N	N	N	N	N	N	N	N	N
P15	P	N	N	P	P	N	N	P	N	P	N	N	N	N	N	N	N	N	N	N
P16	N	N	N	N	N	N	N	N	N	N	N	N	N	N	N	N	N	N	N	N
P17	N	N	N	P	N	N	N	P	P	P	N	N	N	N	N	N	N	N	N	N
P18	N	N	N	N	N	N	N	N	N	N	N	N	N	N	N	N	N	N	N	N
P19	N	N	N	N	P	N	N	P	N	N	N	N	N	N	N	N	N	N	N	N
P20	N	P	N	N	N	N	N	N	N	N	N	N	N	N	N	N	N	N	N	N
P21	N	N	N	N	N	N	N	N	N	N	N	N	N	N	N	N	N	N	N	N
P22	N	N	N	N	N	N	N	N	N	N	N	N	N	N	N	N	N	N	N	N
P23	N	N	N	N	N	N	N	N	N	N	N	N	N	N	N	N	N	N	N	N
	FAR = 0.15										FAR = 0.02									

4 Conclusions

The accelerometer and gyroscope mounted in the biometric pen enabled to create a solution providing reliable verification independent from the surface on which the signature is written. The signature verification employing 3-axis acceleration instead using of the acceleration measure derived from the signature put down on a computer screen with resistive touch screen technology, plus considering 3-axis

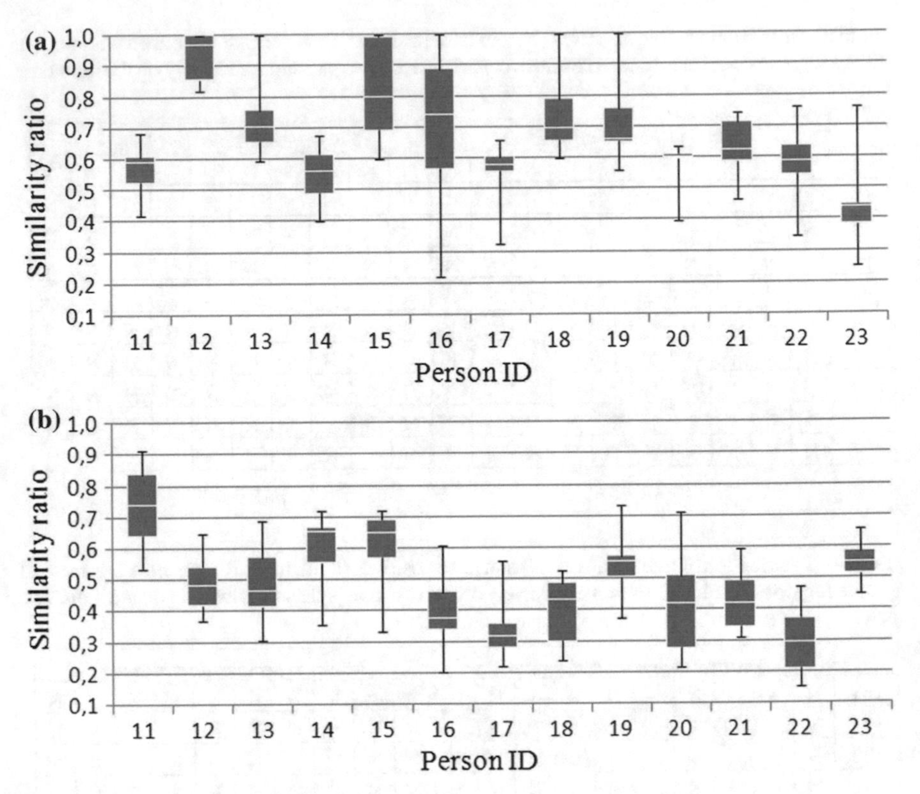

Fig. 7 FAR measure assessment box-and-whisker plots for the resistive pad-based verification (**a**) and biometric pen-based verification (**b**), for simple forgeries

angular positions instead of the angular two-dimensional positions of the signature pattern, enabled a substantial decrease of FAR measure from 0.38 to 0.09 for skilled forgeries, and from 0.15 to 0.02 for simple forgeries. Simultaneously, the median value of the similarity ratio in FAR measure decreased from 0.760 to 0.628 for skilled forgeries and from 0.635 to 0.482 for simple forgeries. The FRR measures were slightly lower in case of the resistive surface-based verification. This may be due to the novelty factor of such a biometric technology as the users kept pointing out that they are unfamiliar with a shape of the biometric pen and that their resulting signatures look differently to the ones produced using a standard pen.

In the developed multimodal biometric system for bank client identity verification the tip of the biometric pen is shaped identically to the ball pen tip, except it does not leave the ink trace on the paper. Exchanging the tip with its ink writing version (so called "wet signature") would enable to perform additionally FRR and FAR measures assessment of the signatures put down on the paper. In this way it would be possible to check if the FRR measure stability depends on the novelty factor (e.g. resistive screen, untypical shape of the pen) or it is solely associated with the behavioral aspect of putting down the signature (a signature shape and dynamics depending on a current person disposition, their mood etc.).

Hitherto achieved results employing the DTW method for discerning original signatures from forged ones will be broadened up in future research employing machine learning algorithms to mine data in a large repository of signatures collected using the developed biometric pen in 100 cashier stands at bank outlets.

Acknowledgements This work was supported by the grant No. PBS3/B3/26/2015 entitled "Multimodal biometric system for bank client identity verification" co-financed by the Polish National Centre for Research and Development.

References

1. Czyżewski, A., Bogdanis, G., Kostek, B., Lech, M., Bratoszewski, P., Hoffmann, P.: Automatic verification of banking clients based on multimodal biometrics. Biometrics (London) (2016)
2. Piyush Shanker, A., Rajagopalan, A.N.: Off-line signature verification using DTW. Pattern Recognit. Lett. **28**, 1407–1414 (2007)
3. Lech, M., Bratoszewski, P., Czyżewski, A.: System weryfikacji autentyczności podpisu odręcznego (Handwritten signature authenticity verification System). Przegląd Telekomunikacyjny+Wiadomości Telekomunikacyjne (8–9), 1145–1148 (2016) (in Polish)
4. Lech, M., Czyżewski, A.: A handwritten signature verification method employing a tablet. In: Signal Processing: Algorithms, Architectures, Arrangements, and Applications (SPA), pp. 45–50, Poznań, Poland (2016)
5. Nautsch, A., Rathgeb, C., Busch, C.: Bridging gaps: an application of feature warping to online signature verification. In: International Carnahan Conference on Security Technology (ICCST), Rome, pp. 1–6 (2014)

Part VI
Text Processing

Towards Entity Timeline Analysis in Polish Political News

Katarzyna Baraniak and Marcin Sydow

Abstract Our work presents a simple method of analysing occurrences of entities in news articles. We demonstrate that frequency of named entities in news articles is a reflection of events in real world related to these entities. Occurrences and co-occurrences of entities between portals were compared. We made visualisation of entities frequency in a timeline which can be used to analyse the history of entity occurrences.

1 Introduction

The availability of textual data in web portals gives many opportunities of information extraction. News media creates articles about politics and events which should reflect the real world. The amount of information makes it difficult for a human to know and identify all important and objective information. We found it important to extract information from web portals such as news articles and create methods that help to identify most important events.

Analysis of the time series based of news articles or user generated content can give us answers for questions about real life events. We may find out what caused stock market crashes or drop of support of a candidate in president elections. Comparing news texts and user sentiments we can find what topics were the most important in an election campaign.

In our research, we want to show how timeline of entities frequency can reflect occurrences of real events and can be used to compare occurrences of entities in different web portals. This work presents an early stage of research on how to use named entity occurrences in news articles and comments in order to analyse real-world events, their presentation in the media and reception by the on-line users.

K. Baraniak (✉) · M. Sydow
Polish-Japanese Academy of Information Technology, Warsaw, Poland
e-mail: katarzyna.baraniak1@pjwstk.edu.pl

M. Sydow
Institute of Computer Science, Polish Academy of Sciences, Warsaw, Poland
e-mail: msyd@ipipan.waw.pl

R. Bembenik et al. (eds.), *Intelligent Methods and Big Data in Industrial Applications*, Studies in Big Data 40, https://doi.org/10.1007/978-3-319-77604-0_23

At this early stage we present proof-of-concept demo on gathering text from news media, using basic summaries and visualisation of entities that result in promising results to be further developed.

2 Problem Specification

The goal of this research is an analysis of entities appearing in political news articles published on leading on-line Polish web portals.

More precisely, the goal is to find if there are any patterns related to named entities (persons, organisations, places, events) encountered in the news articles in time. We want to make a visual analysis of differences in entity distributions among different time periods.

Additionally, we want to identify if there are differences between news portals in a number of mentions of certain entities in their articles. We also want to analyse co-occurrences of entities.

3 Experiments

Our approach was to crawl the data from 2 popular web portals and find all entities appearing in each article. We choose portals that are known as representing diverse political views. We recognised named entities in each article and present an analysis of them.

3.1 Data

A simple crawler developed in Python programming language was used to collect data concerning articles and comments. We have collected data from 01.01.2017 to 22.04.2017. The language of the textual data is Polish.

We collected the following data: entities in articles, dates of articles, comments to articles, authors of comments and dates of comments.

Web news portals that we consider in the current status of the research are the following:

- `wpolityce.pl`
  - 5519 articles
  - 255427 comments
- `gazeta.pl`
  - 521 articles

– 83832 comments

At the current stage of our research we focus on the contents of articles. The analysis of comments is planned to be included in the continuation of this research.

All the analysed articles are in the category "*politics*".

One of the analysed web portals is commonly known as an example of moderate conservative and another as liberal.

3.2 Entity Detection

Named entity recognition was done using tool `liner2` [2]. This tool is available by REST API and is a part of Clarin-PL project. This tool recognises named entities and assigns categories to each of them. There are several category models available for `liner2`, but we have noticed that using 4 models increase chances of finding all entities. These models are:

- names - recognizing borders of named entities boundaries.
- 5nam - recognizing 5 categories: first name, last name, country, city, street.
- top9 - recognising 9 categories: adjective, event, facility, living, location, numex, organization, other and product.
- n82 - recognising 9 categories as in top9 and their 82 subcategories.

We excluded two models for temporal entities: timex1, timex4, as we do not use them in our research. We found that sometimes phrases recognised as entities do not represent true entities or sometimes entities are merged when they appear one by one in a text. Despite this, we are able to find the most frequently occurring entities.

We collected entities appearing in the text of articles and article titles. Because of declension of words in Polish, after recognising an entity we take its base form to count its occurrences. At this stage of research, for simplicity, we do not focus on entity linking and co-reference resolution i.e. we assume that two different base form entities are always treated separately even if they have the same disambiguation meaning. For example, we collected separate entities 'Tusk' and 'Donald Tusk' even if they most probably represent the same person (a former prime minister of Poland). Despite this, the number of entities appearing as most frequent was sufficiently representative to observe a relation between real world events and higher occurrence frequencies of entities. We envisage to include more advanced named entity recognition (NER) techniques in the continuation of this research as it is a challenging issue itself, especially for a highly inflectional language as Polish.

We have recognised 24207 unique entities in `wpolityce.pl` articles and 3542 unique entities in `gazeta.pl` articles.

3.3 Analysis of Entities Occurrence

In this part of experiments, we analysed the frequency of entities in news portals. Firstly, we compared data from gazeta.pl and wpolityce.pl and entity occurrences among different months. Entity frequency is defined as the number of occurrences of each recognised entity in all documents in a month. As a next step, we visualise frequency of entities in time using area plots. This time in visualisation, we use frequency of entities appearing in all articles from one day.

We also analyse co-occurrence of entities in articles from the whole period of time of data we have collected. We can use it to investigate which entities relate to each other. Based on data representing entities in articles we created a matrix elements of which represent the number of articles where each pair of entities co-occurs.

For example if entity 'Tusk' appears twice and entity 'Kaczyński' (the president of "PiS" ("Law and Justice") the major party in the Polish parliament) appears three times in one article we count it as one co-occurrence. The number of the articles where entity 'Kaczyński' and 'Tusk' appears together is the number of their co-occurrences.

4 Results

As the results of our preliminary experiments we study the following characteristics:

- the most frequently occurring entities
- entity timelines, i.e. area plots representing occurrence intensity of entities over time
- co-occurrence statistics of entities

The results presented in this work concern some example entities to illustrate our approach.

Tables 1 and 2 present the most frequent entities among the first 4 months in 2017. Some entities are the most frequent ones in both portals other appears just in one of them. We can observe that some entities are encountered with similar frequency in every month. These are entities which give us less information about current popular topics in the news. Not surprisingly for each month for both portals 'Polska' (Poland) entity is among the most common entities. Similar popularity is observed for the entity 'PiS' the name of major party in the Polish parliament.

One can notice that some entities appear more often in one web portal than in another. We observed the following phenomenon: in the first month of analysis, gazeta.pl more often uses entities related to the currently ruling party, when wpolityce.pl represents the higher frequency of entities related to opposition.

Figures 1 and 2 present intensity timelines of three example entities from the most frequent ones: 'Sala kolumnowa' ('Column Hall') - the name of the room in the Polish parliament where the voting on the Polish budget had to be exceptionally moved in December 2016, 'MON' (abbreviation of the name of the Polish Ministry

Table 1 Most common entities in web portal gazeta.pl (most frequent entity is at the bottom)

January		February		March		April	
Entity	Freq.	Entity	Freq.	Entity	Freq.	Entity	Freq.
Sejm	315	PiS	132	Polska	247	PiS	160
PiS	242	MON	95	PiS	225	Polska	56
Polska	141	Polska	89	Donald Tusk	177	Macierewicz	44
Kaczyński	96	Kaczyński	85	Tusk	163	MON	44
Sala kolumnowa	93	Beata Szydło	75	Rada Europejska	142	Sejm	38
Jaroslaw Kaczyński	69	Sejm	60	UE	105	Jaroslaw Kaczyński	35
Beata Szydło	51	Warszawa	58	Sejm	98	PO	30
KOD	45	Szydło	54	Kaczyński	93	Donald Tusk	29
Macierewicz	44	BOR	53	polski	88	Warszawa	29
senat	44	Macierewicz	53	MON	75	Antoni Macierewicz	23

Table 2 Most common entities in web portal wpolityce.pl (most frequent entity is at the bottom)

January		February		March		April	
Entity	Freq.	Entity	Freq.	Entity	Freq.	Entity	Freq.
Polska	2712	Polska	2383	Polska	3551	Polska	1264
Sejm	2293	PiS	1466	PiS	1818	PiS	1210
PiS	1927	Warszawa	1092	UE	1417	Polak	519
KOD	895	polski	645	Donald Tusk	1194	Donald Tusk	438
Polak	758	UE	638	Tusk	1088	Tusk	429
Kijowski	594	Sejm	532	polski	1015	Sejm	390
Nowoczesna	581	Polak	499	Europa	996	MON	373
zloty	537	Niemcy	496	Polak	801	zloty	332
polski	537	zloty	493	Rada Europejska	722	polski	311
TK	515	Europa	481	Unia Europejska	629	UE	293

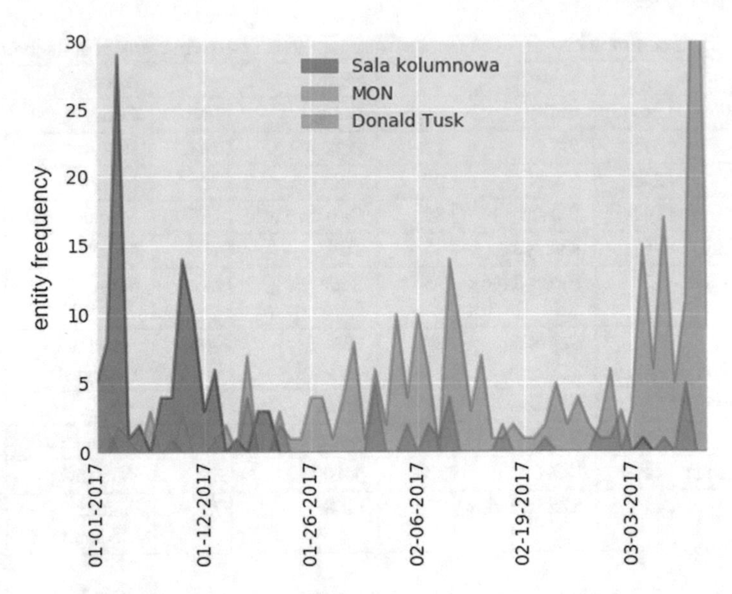

Fig. 1 Example of timeline for entities: Sala Kolumnowa, MON, Donald Tusk encountered in news articles at gazeta.pl from 01.01.2017 to 10.03.2017

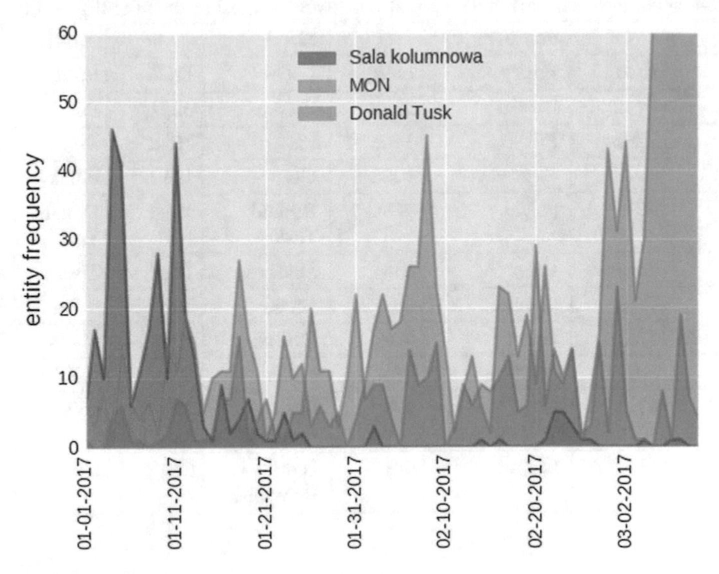

Fig. 2 Example of timeline for entities: Sala Kolumnowa, MON, Donald Tusk encountered in news articles at wpolityce.pl from 01.01.2017 to 10.03.2017

of Defense) and 'Donald Tusk' (a former prime minister of Poland, who resigned in 2014 for taking a position of a president of the European council). It can be noticed that entity distributions in these examples seem to be strongly associated with some real events of high importance in Polish politics that actually took place at that time.

First of all, we can observe a high frequency of the entity 'sala kolumnowa' at the beginning of the year. 'Sala kolumnowa'. The timeline shows that this place was often mentioned in Polish news in January. After some time this entity is almost not mentioned in articles of both portals anymore. With a background knowledge of the political situation in Poland, we can draw a suggestion that this represents particular event: Sejm meeting unexpectedly took place in Column Hall due to the illegal blocking of the main voting room by the opposition. This event provokes discussion about the legality of the blocking or validity of the voting between the ruling party and the opposition what had a reflection in the news from that time.

Also for entity 'Donald Tusk' which represents Polish politician, former prime minister, we can find some interesting pattern. Just before a day of his election for President of the European Council the number of occurrencies rapidly grow for both portals. Again it was a controversial event in the Polish political scene.

The third entity 'MON' is related to the Polish Ministry of National Defence. Higher frequency of this entity observed in February 2017 seems to be naturally related to an affair that actually took place at this institution that time.

Except for mentioned similarities, we can observe some differences. For example 'MON' entity is mentioned rather regularly at portal 'wpolityce.pl', and at 'gazeta.pl' mainly at February. 'Donald Tusk' entity occurs more often at 'wpolityce.pl' portal than in 'gazeta.pl' in the first two months. Entity 'MON' is more popular than entity 'Donald Tusk' at 'gazeta.pl' but 'Donald Tusk' entity is, in general, more popular entity than 'MON' at 'wpolityce.pl'.

Co-occurrence of entities to exemplary entities are shown in Table 3 for portal gazeta.pl and in Table 4 for wpolityce.pl. Each examplary entity has its co-occurring entities. Entities are considered as co-occuring if they encounter in the same article at least once. In each case entities 'Polska', 'PiS' are repeated because they are among most frequent ones and relate to the country and government. For entity 'Donald Tusk' the most interesting one is 'Rada Europejska' (EU Council) which is strongly connected to his current function. 'MON' occurs most often with entity 'Macierewicz' and 'Antoni Macierewicz' which relates to the person of the current Minister of National Defense. For entity 'Sala kolumnowa' there is some difference between both portals. Both use entities 'sejm' (parliament), 'PiS' (the ruling party), 'Polska' but only in one the phrase 'Sala kolumnowa' is mentioned with 'PO Michał Szczerba' (a politician from the current opposition known for initialising the illegal process of blocking the work of the Polish parliament what resulted in moving the work into the 'Sala kolumnowa' room) and 'PO i Nowoczesna' (the names of the current opposition parties in Poland) but at 'wpolityce.pl' it co-occurs with 'senat' (the higher chamber of the Polish parliament). Further improvements are needed like entity linking and entity cleaning.

Table 3 Co-occurrence entities in web portal gazeta.pl (most frequent co-occurring entity is at the bottom)

Donald tusk		MON		Sala kolumnowa	
Entity	Freq.	Entity	Freq.	Entity	Freq.
Rada Europejska	66	Macierewicz	58	sejm	30
PiS	64	Antoni Macierewicz	50	PiS	21
Tusk	63	PiS	31	PO i Nowoczesna	16
Polska	63	Polska	29	PO Michał Szczerba	13
polski	42	NATO	26	PO	12

Table 4 Co-occurrence entities in web portal wpolityce.pl (most frequent co-occurring entity is at the bottom)

Donald tusk		MON		Sala kolumnowa	
Entity	Freq.	Entity	Freq.	Entity	Freq.
Polska	729	Polska	249	sejm	134
Tusk	606	Antoni Macierewicz	206	PiS	95
PiS	526	Macierewicz	199	Polska	72
Rada Europejska	463	PiS	177	PO	52
polski	155	polski	155	senat	46

5 Related Work

Entity extraction and summarisation are problems which appear in many disciplines. The concept of event or entity timelines has been used in some previous research, but we are not aware of any research publication that focused on Polish political news.

In the work [1] the authors present their method called 'Timemachine' which automatically generates timelines of events based on the knowledge base. Authors focus on three quality criteria: relevance of events to entities, temporal diversity, and content diversity.

In article [3] authors describe their method of creating and analysing timelines of named entities. Their research is focused on entity evolution. As a part of their work, they analyse also political entities.

Recent work [4] presents a timemachine for Portuguese news. The results of their work are presented as an interactive web tool allowing searching and visualising news stories. Besides timelines, they present co-occurrences with an entity as an egocentric network.

An interesting approach is presented in a paper [5] where authors describe a problem of connecting events described in news articles. They introduced a method using timelines to automatically find coherent chain linking events and topics described in articles.

Most of the current research on automatically generated timelines were focused on news and knowledge bases in English or other popular languages. In our research, we use only news in Polish, which has not been studied as carefully as other languages. Additionally, we are using the timeline to compare entities occurrence between news portals.

6 Conclusions and Future Work

The results of our work show that frequency of entity occurrences in political news has a potential for representing real life events, even for quite simple techniques used. We showed that it is possible to track important and unusual events based just on the frequency of entities in the news. Our work presents promising results and creates many ways of extension. As our analysis presents still ongoing research we envisage to extend our research in several directions, for example:

- named entity detection can be improved by introducing disambiguation and co-reference resolution which is challenging task in politics domain, especially for highly-inflective languages as Polish
- extend the timeline for a longer period of time and collect data from a larger number of web news portals.
- extend our research and timeline visualisation to be able to recognise events relevant to entities.
- interesting would be to analyse entities in real-time and publish visualisations online.
- add analysis of comments: which entities are most commonly commented by users and how
- introduce sentiment analysis of entities in the timeline based on comments and other sources of user generated content. It seems it will be useful information in determining how news articles influence public opinions and how public opinion is changing over time and under the influence of events.
- extend the work towards comparative analysis of particular on-line media.

Acknowledgements The work is partially supported by the Polish National Science Centre grant 2012/07/B/ST6/01239.

References

1. Althoff, T., Dong, X.L., Murphy, K., Alai, S., Dang, V., Zhang, W.: Timemachine: timeline generation for knowledge-base entities. CoRR (2015). arXiv:1502.04662, http://dblp.uni-trier.de/db/journals/corr/corr1502.html#AlthoffDMADZ15
2. Marcinczuk, M., Kocon, J., Janicki, M.: Liner2 - a customizable framework for proper names recognition for polish. In: Bembenik, R., Skonieczny, L., Rybinski, H., Kryszkiewicz, M., Niezgodka, M. (eds.) Intelligent Tools for Building a Scientific Information Platform. Studies in Computational Intelligence, pp. 231–253. Springer, Berlin (2013). http://dblp.uni-trier.de/db/series/sci/sci467.html#MarcinczukKJ13
3. Mazeika, A., Tylenda, T., Weikum, G.: Entity timelines: visual analytics and named entity evolution. In: Macdonald, C., Ounis, I., Ruthven, I. (eds.) CIKM. pp. 2585–2588. ACM (2011). http://dblp.uni-trier.de/db/conf/cikm/cikm2011.html#MazeikaTW11
4. Saleiro, P., Teixeira, J., Soares, C., Oliveira, E.: Timemachine: Entity-centric search and visualization of news archives. In: European Conference on Information Retrieval. pp. 845–848. Springer (2016)
5. Shahaf, D., Guestrin, C.: Connecting the dots between news articles. In: Proceedings of the 16th ACM SIGKDD International Conference on Knowledge Discovery and Data Mining. pp. 623–632. KDD '10, ACM, New York (2010). http://doi.acm.org/10.1145/1835804.1835884

Automatic Legal Document Analysis: Improving the Results of Information Extraction Processes Using an Ontology

María G. Buey, Cristian Roman, Angel Luis Garrido, Carlos Bobed and Eduardo Mena

Abstract Information Extraction (IE) is a pervasive task in the industry that allows to obtain automatically structured data from documents in natural language. Current software systems focused on this activity are able to extract a large percentage of the required information, but they do not usually focus on the quality of the extracted data. In this paper we present an approach focused on validating and improving the quality of the results of an IE system. Our proposal is based on the use of ontologies which store domain knowledge, and which we leverage to detect and solve consistency errors in the extracted data. We have implemented our approach to run against the output of the AIS system, an IE system specialized in analyzing legal documents and we have tested it using a real dataset. Preliminary results confirm the interest of our approach.

Keywords Information extraction · Natural language processing · Ontologies Data curation · Legal document analysis

M. G. Buey (✉) · C. Roman
InSynergy Consulting S.A., Madrid, Spain
e-mail: mgbuey@unizar.es

C. Roman
e-mail: croman@isyc.com

A. L. Garrido · C. Bobed · E. Mena
Department of Computer Science and System Engineering, University of Zaragoza, Zaragoza, Spain
e-mail: garrido@unizar.es

C. Bobed
e-mail: cbobed@unizar.es

E. Mena
e-mail: emena@unizar.es

R. Bembenik et al. (eds.), *Intelligent Methods and Big Data in Industrial Applications*, Studies in Big Data 40, https://doi.org/10.1007/978-3-319-77604-0_24

1 Introduction

Nowadays, many organizations are fully engaged in the development of software systems focused on extracting information from documents. Information Extraction (IE) is an area of Natural Language Processing (NLP) devoted to obtain specific information from unstructured or semi-structured texts. When this activity is performed by humans, it involves a high cost in time, staff, and effort. Therefore, automation represents great savings for businesses. In order to face this problem, different algorithms and approaches have been developed for over 40 years, and there are many software companies that offer adaptable solutions to different use cases. However, they are still far from finding efficient and comprehensive solutions applicable to any context. On the other hand, although these systems make the work much easier, most of them extract data without checking their quality and searching for possible errors. In this context, the most common errors [1] are: the appearance of empty or duplicate data, the presence of poorly structured data, the existence of entities with data distributed among other entities, and the appearance of sets of entities whose data correspond to a single one. The task in charge of checking and ensuring the quality of the data is well known as *Data Curation* [2].

In this work, we propose an ontology-based approach to improve the quality of the outcomes from automatic IE systems, and so perform *Data Curation* on the extracted data. Once the extraction process is completed, our proposal leverages the knowledge stored in a domain ontology (an *ontology*, as defined by Gruber, is a formal and explicit specification of a shared conceptualization [3]) to: (1) perform a review of the extracted data, curating possible errors, and (2) improve substantially the quality of the results by correcting and enriching them exploiting the available knowledge.

Our approach is implemented on top of the AIS system [4]. AIS is an Ontology-Based Information Extraction (OBIE) system [5] which aims at extracting relevant information from documents in natural language [6, 7]. While AIS is specialized in legal documents, its extraction process is guided by an ontology which stores the knowledge about the structure and the content of different types of documents, as well as appropriate extracting operations. This decoupling makes it possible to adapt it easily to other domains, reusing much part of the modeling efforts. Currently, AIS is being used by *OnCustomer*,[1] a Customer Relationship Management (CRM) platform, developed by the InSynergy Consulting (ISYC) company,[2] whose goal is to anticipate, meet, and respond to the needs and expectations of potential customers. Regarding the company, ISYC is focused on the innovation and the development of new technologies, and offers national and international integral solutions and services. ISYC is part of Tessi,[3] a multinational leader in Business Process Outsourcing (BPO) in multiple countries, and it has its headquarters placed in Madrid (Spain), having several offices in other Spanish cities, and in Italy.

[1] https://www.isyc.com/es/soluciones/oncustomer.html.

[2] https://www.isyc.com.

[3] https://www.tessi.fr.

Liquidadora en los inmediatos treinta días hábiles siguientes, responsabilidades derivadas del incumplimiento de la legalidad vigente. -----

Esta escritura ha sido redactada conforme a minuta facilitada por la Entidad acreedora y contiene condiciones generales de su contratación. En consecuencia, yo, la Notario, advierto a los otorgantes de la posible aplicación de Ley 7/1998, de 13 de Abril, sobre Condiciones Generales de la Contratación. -------

De acuerdo con lo previsto en la Ley Orgánica 15/1999, de 13 de Diciembre, de Protección de Datos de carácter Personal, los comparecientes quedan informados y aceptan que los datos recabados y que en esta escritura constan, han quedado incorporados a los ficheros automatizados de esta Notaría, cuya exclusiva finalidad es la formalización de este instrumento y su seguimiento posterior, sin perjuicio de su utilización en las comunicaciones que legalmente proceda cumplimentar, y que podrán ejercer los derechos de acceso, rectificación, cancelación y oposición correspondientes. ---------------------

Permito a los comparecientes conforme

68

Fig. 1 Sample page (in Spanish) of a notarial deed of constitution of a mortgage, the type of document used in our dataset. Content has been partially blurred or deleted for privacy reasons

Hence, to show the feasibility and the benefits of our approach, we have performed a set of preliminary tests with data extracted by AIS from a real legal document dataset, composed of notarial deeds of constitution of mortgages (in Spanish). These type of documents are usually very large, verbose, and use a very formal language that makes it difficult to find the relevant information to extract, even for a human reader. A sample page can be seen in Fig. 1. The results of such experiments are promising and they encourage us to continue working in this line.

The paper is structured as follows. Section 2 gives an overview of the state of the art concerning our approach. Section 3 describes our system, including the kind of knowledge described in the ontology and the data curation process performed. Section 4 presents the results of the empirical study conducted to assess our proposal. And finally, Sect. 5 offers some concluding remarks and directions for future work.

2 Related Work

In this section, we present the main works related to our approach from the point of view of the two main tasks performed by our system: Data Curation, and Data Enrichment.

Data Curation is a set of processes responsible for ensuring the quality of data. Most of the developed approaches are focused on assist manual or semi-automatic processes [2]. In our case, we propose an automatic approach to handle this goal after IE processes. Related to the field are the *Data Cleansing* [8] processes whose aim is to detect and correct (or remove) corrupt or inaccurate records from a dataset. They identify incomplete, incorrect, inaccurate, or irrelevant parts of the data and then replace, modify, or delete the dirty or coarse data. However, Data Curation and Data Cleansing processes are both focused on working over databases or data warehouses [9]. In these scenarios, IE processes have been already perform and the extracted data is already stored. The most similar approach that we have found is the xCurator Project.[4] This project is focused on analyzing large volumes of user-generated data on the Web. For that purpose, the authors enhance the quality of the data by extracting entities; identifying their types and their relationships to other entities; merging duplicate entities; linking related entities (internally and to external sources); and publishing the results on the Web as high-quality Linked Data [10]. While we focus also on managing the information of extracted entities, our approach performs the data curation over the results returned by IE processes which manage documents in natural language and before storing the extracted information.

Data Enrichment is another related field which comprises activities that allow adding extra information to the extracted data. In this scope, we can find works as "Linking Tweets to News" [11], a framework for enriching short texts from the social network *Twitter*[5] with long texts from event news, or works as [12] where the authors

[4]http://dblab.cs.toronto.edu/project/xcurator/.

[5]https://twitter.com/.

enrich a corpus to improve word sense induction. This kind of systems usually focus on recognizing Named Entities (NE) [13] in the text,[6] which serve as entry points to consult ontologies which are populated with such NEs, their properties, and the relationships between them [14, 15]. Our approach follows a similar strategy, but also leans on other external services in case the ontology used does not store related information.

3 System Overview

Our approach has been implemented as an independent module on top of the IE process of the AIS system. As above mentioned, AIS extracts information from text-based documents in a flexible way to adapt to different domains; however, it currently lacks a data post-processing module to clean, correct, and enrich the extracted data. Our proposal starts there, after the extraction process, taking the data extracted by AIS as input, and curating such data to improve the outcome of the overall IE process. In this pipeline, the whole IE process is guided by the information stored in the domain ontology, i.e., AIS uses the ontology for the extraction, and this approach uses it for data curation purposes. In this section, first, we will detail the contents of the domain ontology used, and, then, we move onto the details of our data curation process.

3.1 The Ontology

The ontology which guides both the extraction process of the AIS system, and our data curation process, stores the following information (see Fig. 2):

- *Document taxonomy and structure*: The different types of documents are classified hierarchically in a taxonomy. Each document class contains information about the sections that shape their kind of documents, and each of these sections includes further information about which properties and entities appear within them.
- *Entities and management operations*: The ontology also stores which entities and their attributes (modeled as properties) have to be identified, how they should be processed, and how they relate to other entities in the context that they appear. Besides, the key attributes (i.e., the set of attributes which define an entity uniquely) for each of the entities are marked as such (e.g., the name, the surname, and the national identity document of an entity *Person*). This decoupled knowledge makes it possible to reuse and adapt easily both the extraction operations, and the access points to data curation methods developed for each entity among different types of documents.

[6]A Named Entity is a unique identifier of an entity in a text, e.g.,'Marie Curie' is a NE of a person.

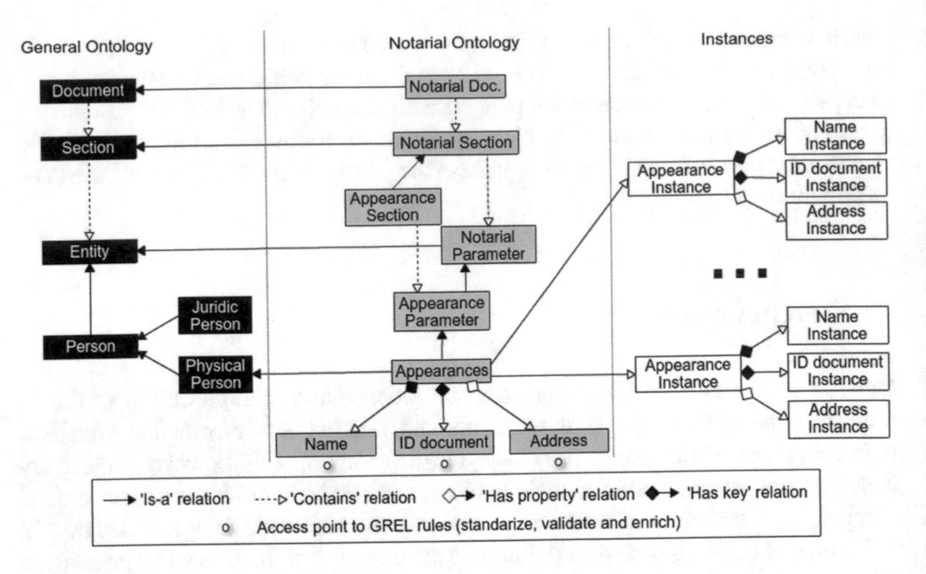

Fig. 2 Ontology of a notarial document

- *Entity instances*: Finally, the ontology also stores information about previously extracted and curated data as instances of the different entities defined in the model. This information is the extensional knowledge that our proposed approach uses.

We have designed and implemented this ontology as an ontology network [16]. We chose this methodology because most documents are classified following a hierarchy with different features. This model allows us to define a modular knowledge-guided architecture, so new document types can be added dynamically.

Figure 2 shows an excerpt of our ontology that defines a notarial document. Black boxes represent high level definitions that are used regardless of the document typology: document, sections, and entities. Gray boxes represent data that are shared among notarial documents regardless of their particular purpose (i.e., the introductory section and the appearances). White boxes represent instances stored for this kind of typology. Regarding to the relations defined in the ontology, straight lines represent the 'is-a' relation, while dashed lines represent the 'contains' relation. The white circles represent the access points to the standardization and enriching operations which are modeled in the ontology as annotations of the corresponding classes. Here we can also see how each 'Appearance' entity contains two key attributes (modeled with 'Has key' and 'Has property' relations respectively): 'Name' and 'ID document', and the 'Address' attribute.

The ontology is populated with the information stored in the repositories of real environments where AIS is currently being used. This information corresponds to previously extracted data from different processed documents. We included a periodic process in charge of recollecting these data, and populating the ontology. This periodic update of the information in the ontology is done by using an R2RML mapping,[7] which basically maps each of the entities and the properties in the domain ontology to a set of SQL queries which accesses the appropriate data, and allows to format the data according to our ontology, updating the knowledge leveraged by our system.

3.2 The Data Curation Process

The data curation process proposed in this work analyzes the structured data returned by AIS. These data include the information about the entities and their attributes extracted from the documents. An example of the results returned by the system after processing a document is as follows[8]:

```
<Data>
   <Parameter name="DOC_Title"
    value="Declaration of loan with mortgage guarantee and
           constitution of exchange, granted by Bank S.A.
           in favor of Marketing Services S.L."/>
   <Parameter name="DOC_Signing_Date"
    value="May 15th, two thousand fourteen"/>
   <Parameter name="DOC_Notary"
    value="Vicent Rogers"/>
   <Entity name="DOC_Appearing">
      <Attributes>
        <Parameter name="PROP_Person_Name" value="Julia"/>
        <Parameter name="PROP_Person_Last_Name" value="Pearson"/>
        <Parameter name="PROP_Id_Number" value="012345678G"/>
        <Parameter name="PROP_Marital_State" value="single"/>
        <Parameter name="PROP_Address" value="street Gran Vía"/>
        <Parameter name="PROP_City" value="Madrid"/>
        <Parameter name="PROP_Country" value="Spain"/>
        <Parameter name="PROP_Location" value="Madrid"/>
      </Attributes>
   </Entity>
</Data>
```

The data curation process takes these results and curates and enriches them following six steps (see Fig. 3):

1. *Basic cleaning*: First, the process cleans empty fields and duplicated information. We dedicate the first step to these two type of errors due to their frequency and their particularly nocive effects in subsequent steps.
2. *Data Refinement*: This step is focused on refining those entity attributes whose content is corrupted, e.g., the first name of a person contains the full name of one

[7] R2RML: RDB to RDF Mapping Language, https://www.w3.org/TR/r2rml/.

[8] This example is directly taken from the experiments dataset. Proper names and specific data have been altered for reasons of privacy.

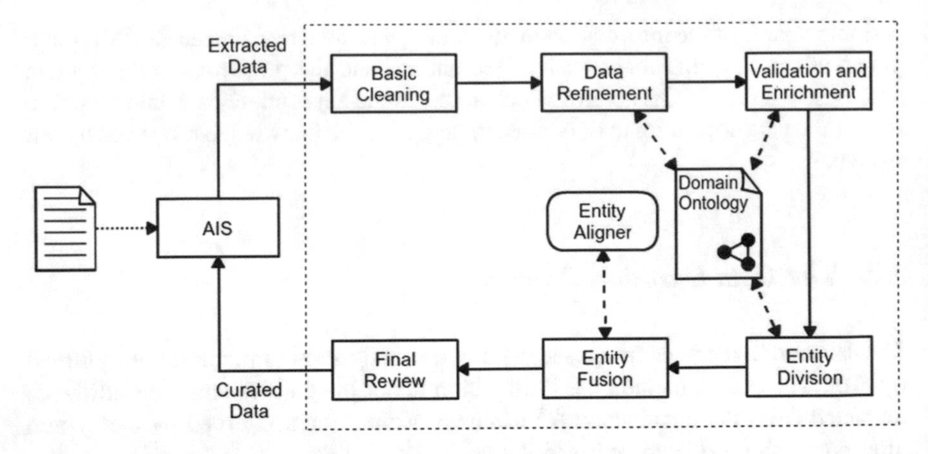

Fig. 3 Steps of the data curation process

or more different persons. For each attribute of each entity, this step searches instances stored into the ontology, if it detects that an attribute is not valid, it modifies the attribute with the appropriate value stored in the ontology. If it does not find an equivalent value, the process does not change it. For example, AIS returns the following results for an 'Appearing' entity:

```
<Parameter name="PROP_Person_Name" value="Julia Pearson"/>
<Parameter name="PROP_Person_Last_Name" value="Julia Pearson"/>
<Parameter name="PROP_Id_Number" value="012345678G"/>
```

So, this step identifies the correct value for each attribute corrupted (i.e., the 'Person Name' and the 'Person Last Name') and changes them to:

```
<Parameter name="PROP_Person_Name" value="Julia"/>
<Parameter name="PROP_Person_Last_Name" value="Pearson"/>
<Parameter name="PROP_Id_Number" value="012345678G"/>
```

3. *Validation and Enrichment*: In this step, the data is standardized, validated, and enriched exploiting different sources of information. In our current proposal, these sources mainly include: (1) previously extracted and verified data included as instances in the ontology, and (2) available external information services (e.g., *Google Maps API*[9]). We apply this step to standardize and validate the value of the attributes of the entities, with the goal of solving later duplicities. For example, the type of a street can be extracted as 'St.' or 'Street', but they are representing the same concept; or the system could extract from a certain postal address only the street and the city, but the zip code could be obtained from Google Maps.

[9]https://developers.google.com/maps/documentation/geocoding/start.

4. *Entity Division*: Once the data is validated and enriched, this step: (1) detects attributes which belong to different entities, but are assigned to a single extracted entity, and (2) separates such attributes assigning them to the appropriate ones. The detection is focused on the set of attributes that uniquely define an entity (i.e., their key attributes), e.g., the data extracted for a person could include two driving license numbers. In order to detect potentially unified entities, the process searches entities in the ontology using combined key attributes. In case of success, it creates a new entity with such data, and deletes those attributes from the rest of entities. If no entity is found or the attributes of other entity remain in the result, the process creates a number of entities equals to the maximum number of occurrences of a key attribute, and each attribute is assigned to an entity. For example, AIS could return the following results for an 'Appearing' entity:

```
<Entity name="DOC_Appearing">
   <Attributes>
      <Parameter name="PROP_Person_Name" value="Julia"/>
      <Parameter name="PROP_Person_Last_Name" value="Pearson"/>
      <Parameter name="PROP_Id_Number" value="012345678G"/>
      <Parameter name="PROP_Person_Name" value="Richard"/>
      <Parameter name="PROP_Person_Last_Name" value="Smith"/>
      <Parameter name="PROP_Id_Number" value="11111111H"/>
      <Parameter name="PROP_City" value="Madrid"/>
      <Parameter name="PROP_Country" value="Spain"/>
      <Parameter name="PROP_Location" value="Madrid"/>
   </Attributes>
</Entity>
```

We can see that two actual entities have been returned as a single one by the system. Therefore, this step divides it into two different entities in the following way:

```
<Entity name="DOC_Appearing">
   <Attributes>
      <Parameter name="PROP_Person_Name" value="Julia"/>
      <Parameter name="PROP_Person_Last_Name" value="Pearson"/>
      <Parameter name="PROP_Id_Number" value="012345678G"/>
      <Parameter name="PROP_City" value="Madrid"/>
      <Parameter name="PROP_Country" value="Spain"/>
      <Parameter name="PROP_Location" value="Madrid"/>
   </Attributes>
</Entity>
<Entity name="DOC_Appearing">
   <Attributes>
      <Parameter name="PROP_Person_Name" value="Richard"/>
      <Parameter name="PROP_Person_Last_Name" value="Smith"/>
      <Parameter name="PROP_Id_Number" value="11111111H"/>
      <Parameter name="PROP_City" value="Madrid"/>
      <Parameter name="PROP_Country" value="Spain"/>
      <Parameter name="PROP_Location" value="Madrid"/>
   </Attributes>
</Entity>
```

5. *Entity Fusion*: This step is just the opposite to the previous one: It finds extracted entities which share key attributes and merge their attributes into a single one. As an example, consider that the information of a particular person might be distributed along the document: An straight-forward extraction process might return the data as belonging to different people. AIS could return the following entities which correspond with the same person:

```
<Entity name="DOC_Appearing">
   <Attributes>
      <Parameter name="PROP_Person_Name" value="Richard"/>
      <Parameter name="PROP_Person_Last_Name" value="Smith"/>
      <Parameter name="PROP_Id_Number" value="11111111H"/>
      <Parameter name="PROP_City" value="Madrid"/>
   </Attributes>
</Entity>
<Entity name="DOC_Appearing">
   <Attributes>
      <Parameter name="PROP_Person_Name" value="Richard"/>
      <Parameter name="PROP_Id_Number" value="11111111H"/>
      <Parameter name="PROP_Country" value="Spain"/>
   </Attributes>
</Entity>
```

At this step the process identifies that these data correspond with the same 'Appearing' entity and merges them:

```
<Entity name="DOC_Appearing">
   <Attributes>
      <Parameter name="PROP_Person_Name" value="Richard"/>
      <Parameter name="PROP_Person_Last_Name" value="Smith"/>
      <Parameter name="PROP_Id_Number" value="11111111H"/>
      <Parameter name="PROP_City" value="Madrid"/>
      <Parameter name="PROP_Country" value="Spain"/>
   </Attributes>
</Entity>
```

For this purpose, our approach uses an *Entity Aligner* module which includes different functions and methods to measure the similarity between two entities [17–19]. More details about this module are described below in this section.

6. *Final Review*: We include this last step to remove those attributes or information which remain at the end of the process and which have not been assigned to any entity or validated, or do not cover the minimum cardinality specified in the ontology.

Our approach uses intensively the information stored into the domain ontology to support the data curation process in different steps, relying on its already validated information and trusted information services to improve the quality of the extracted data. Note how this approach could be easily adapted to other domains whenever there is available curated information to be leveraged, which is usually the case of general CRM systems where users review and validate data.

Entity Aligner As we have seen, in the fusion step, our proposal needs to assess whether two different entities are the same one or not. The Entity Aligner (see Fig. 3) centralizes all the methods to compare different elements during the curation process (e.g., it compares how much two entities resemble each other), and determines whether two entities must remain separated, or two different attributes must be merged. In particular, this module offers the following functionalities:

- *Comparison between attributes.* Two attributes which correspond to different types are considered totally different. Otherwise, we calculate a similarity distance between the values of both attributes, which are extracted in string format. For this purpose, we use the Levenstein distance [20] which is a a string metric for measuring the difference between two sequences. It calculates the minimum number of single-character edits (insertions, deletions, or substitutions) required to change one sequences into the other, and it returns a value in the range [0, 1], 0 meaning that they are completely different and 1 meaning that they are totally equal.
  We have also to take into account that two attributes which share the same type, despite being extracted correctly, can be extracted in a different format. For example, a monetary value can be represented as 'forty-five euro' or '45 euro'. To solve this, this module transforms values into their canonical form before performing any comparison.
- *Comparison between entities.* In order to compare two entities, they must fit the same type. The module calculates a similarity measure between them which depends on the Levenstein distance between the attributes they contain. It takes into account the distance of all the key attributes (or non-key attributes if at least one of the entities which are being compared not contain attributes of this type), and averages the measures obtained. If the average score overcomes a configurable threshold, the entities are considered that they reference to the same entity. When all the attributes of an extracted entity are equal to all the attributes of a different extracted entity, the entities are considered equal.

4 Experimental Evaluation

We have implemented our proposal on top of AIS. We have used Open Refine[10] for standardizing, validating, and enriching the extracted data in the *Validation and Enrichment* step of our approach. Open Refine is an open source framework that enables processing large amounts of datasets. It allows to apply over the data a cleansing process to standardize fields and eliminate duplications. This framework also includes rules called General Refine Expression Language (GREL)[11] which make it possible to clean and enrich the data through Web services (we have used Google Maps). The framework only processes tables, and it is unable to solve all

[10]http://openrefine.org/.

[11]https://github.com/OpenRefine/OpenRefine/wiki/General-Refine-Expression-Language.

detected errors; however, it offers interesting functions that we have wrapped and incorporated into our solution.

Experimental Settings and Results

To evaluate the improvement of the results offered by our data curation process, we have prepared a dataset of 100 real notarial documents in Spanish which have been manually processed to obtain the actual data which have to be extracted. These documents have been processed by the AIS system and then the extracted data obtained have been handled by our data curation process.

To assess the quality of our approach, we have used the well-known metrics used to evaluate the performance of information extraction systems: *Precision (P)*, *Recall (R)*, and *F-Measure (F)* [21]. As we wanted the evaluation to focus on the extraction of well-structured entities, we tested our preliminary evaluation on the data extracted about person entities. Person entities are the most complex to extract from notarial documents, and they usually suffer more extraction errors. A person entity has been defined with a set of attributes that corresponds to its first name, last name, national identity document, marital status, and address, being the first name, last name, and national identity document the key attributes. In our evaluation, we have assumed that a extracted person has been processed correctly if it contains at least all the key attributes (i.e., name, surname, and ID) and they are completely equal to the expected ones. An example of the information extracted can be found in Fig. 4, where we have highlighted the data corresponding to the names of the person entities and the seller; it can be seen that first and middle names, being both compound in this case, can easily be confused with the two last names when they are extracted.[12]

Figure 5 shows the results obtained in our experiments. Applying our proposal allows the system to improve its F-Measure from 0.5 to a value of 0.61 (we have to bear in mind that we were dealing with the most complex entities in the domain). Analyzing the intermediate steps, we have seen that both the *Basic Cleaning* and the *Data Refinement* steps participates almost in every extraction preparing the values of the attributes before the following steps. The GREL standardization rules helps to remove successfully all the duplicate appearances. On the other hand, the GREL validation rules do not offer a great improvement because the invalid attributes are deleted if a valid attribute exists in the ontology to replace it.

However, whenever GREL enrichment rules and Google Maps were used (e.g., to obtain zip codes), we have obtained a great increase of the recall. Also, the enrichment of entities with the ontology fixes many structure mistakes. Finally, in general, the steps that further improve the results are the *Entity division* and *Entity fusion*, as they re-structure those person entities which have been extracted wrongly.

---

[12]In Spain, people have a first name, an optional middle name, and *two* mandatory last names, the first one if the father family name and the second one is the mother family name, although legally this order can be interchanged.

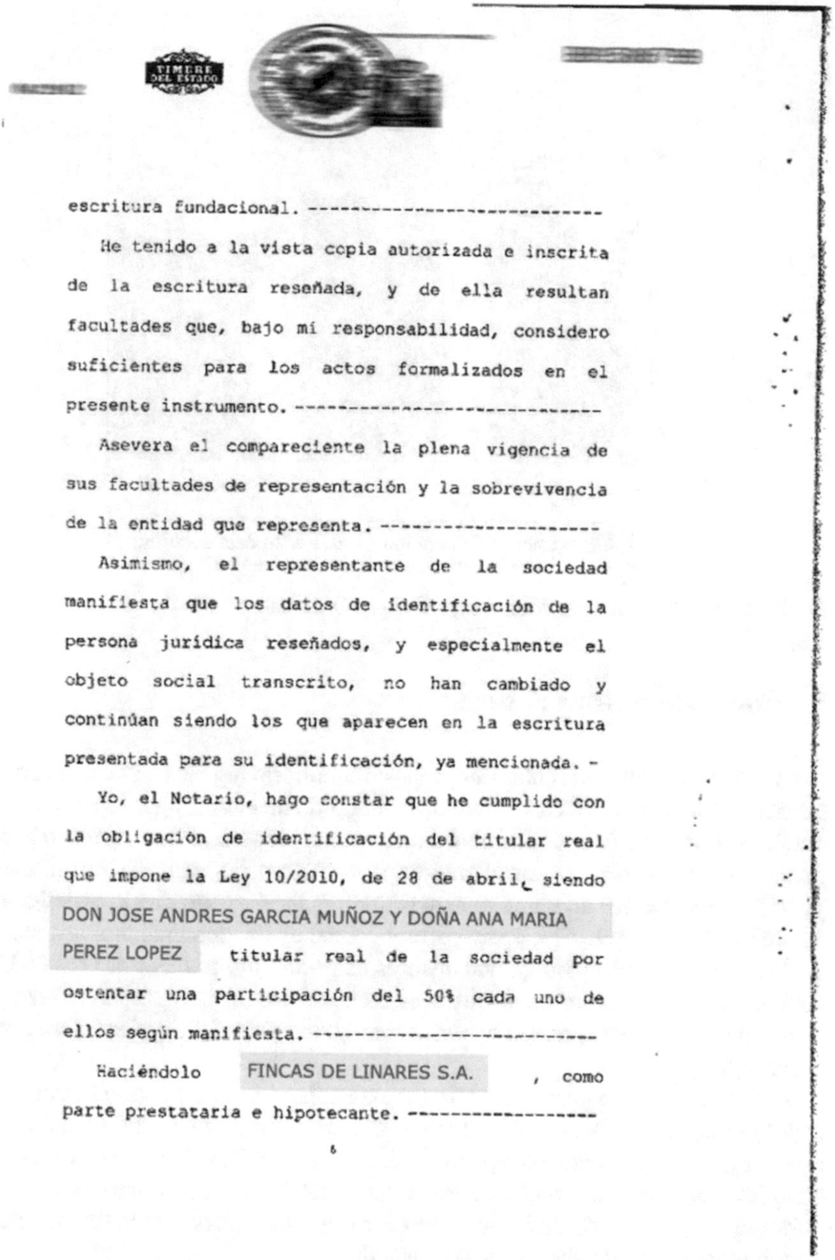

TIMBRE DEL ESTADO

escritura fundacional.---------------------------

He tenido a la vista copia autorizada e inscrita de la escritura reseñada, y de ella resultan facultades que, bajo mí responsabilidad, considero suficientes para los actos formalizados en el presente instrumento.----------------------------

Asevera el compareciente la plena vigencia de sus facultades de representación y la sobrevivencia de la entidad que representa.--------------------

Asimismo, el representante de la sociedad manifiesta que los datos de identificación de la persona jurídica reseñados, y especialmente el objeto social transcrito, no han cambiado y continúan siendo los que aparecen en la escritura presentada para su identificación, ya mencionada.-

Yo, el Notario, hago constar que he cumplido con la obligación de identificación del titular real que impone la Ley 10/2010, de 28 de abril, siendo DON JOSE ANDRES GARCIA MUÑOZ Y DOÑA ANA MARIA PEREZ LOPEZ titular real de la sociedad por ostentar una participación del 50% cada uno de ellos según manifiesta.--------------------------

Haciéndolo FINCAS DE LINARES S.A. , como parte prestataria e hipotecante.-----------------

6

Fig. 4 Sample page (in Spanish) of a notarial deed of constitution of mortgage. Content has been partially blurred or deleted, and original names have been replaced for privacy reasons

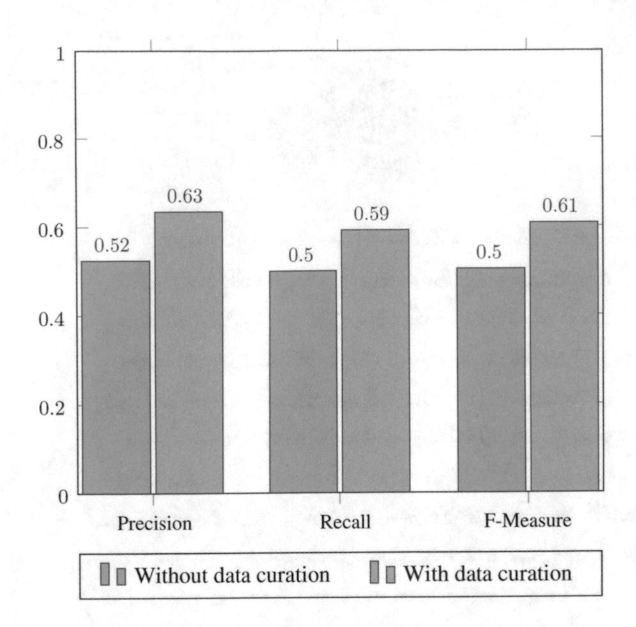

Fig. 5 Results obtained without applying and applying the data curation

5 Conclusions and Future Work

In this paper, we have presented our proposal for improving the results of information extraction processes, which is based on leveraging the knowledge stored in a domain ontology with the aim of performing data curation processes. Our approach has been implemented in a real industrial environment (automatic legal document analysis) yielding good results: an improvement of 11% in F-Measure. The knowledge-based nature of our approach, as it is completely guided by the domain ontology, enhances its flexibility and adaptability, and ensures its portability to other use cases. In particular, its application to other business contexts would only require to adapt the data and knowledge expressed by the ontology; the rest of the system would remain unchanged.

As future work we plan to work in (1) expanding the tests in order to improve the quality data extracted, and (2) detecting and extracting other kind of information, including relevant document-type individual data (see Figs. 6 and 7) as well as more complex data when different individual data must be combined into more complex information (see Fig. 8). Moreover, another important work line is the detection of co-references between the entities extracted.

PACTO PRIMERO. Capital del préstamo. -------

La PARTE DEUDORA recibe de [illegible], en este acto, a su entera satisfacción, mediante abono en cuenta, en concepto de préstamo mutuo, la cantidad de (38.000,00 €) TREINTA Y OCHO MIL **EUROS.** --

Cada uno de los prestatarios que integran la PARTE DEUDORA responde solidariamente del cumplimiento de las obligaciones contraidas en esta escritura. --

La PARTE DEUDORA consiente en que de la suma prestada se deduzcan las cantidades precisas para cancelar los gravámenes que afecten a la finca que en esta escritura se hipoteca. --------------

PACTO SEGUNDO. Amortización. ---------------

A) Vencimiento final y devolución del préstamo. --

El plazo de vencimiento final del préstamo viene determinado por los pagos convenidos en este pacto. --

Fig. 6 Example of fragment of a real notarial document (in Spanish) with key information to extract: "`The total amount of the mortgage is €38.000,00`". Content has been partially blurred or deleted for privacy reasons

timiento. Así mismo, el Banco presta su consentimiento al conocer la aceptación del prestatario al firmar esta escritura, lo que unido a la entrega efectiva del importe del préstamo reconocida anteriormente implica el perfeccionamiento jurídico del contrato de préstamo, de acuerdo con lo previsto en el artículo 1.262 del Código Civil. ---------------------------

2ª.- DURACION. VENCIMIENTOS. REEMBOLSO ANTICIPADO. ---

2.1. Duración. --------------------------------

El préstamo se ha pactado por un plazo de 360 meses, contados a partir del día 30 de Junio próximo, más el denominado "período de ajuste", integrado por los días comprendidos desde la fecha de formalización de esta escritura y el día antes citado, ambos inclusive. ---------------------------------------

Se entiende que los años, plazos y períodos en los que, en su caso, se divide el préstamo, son siempre sucesivos, sin solución de continuidad, y que el día inicial que en cada caso se indica está incluido en el cómputo. ---------------------------------

2.2. Vencimientos. ---------------------------

- 4 -

Fig. 7 Example of fragment of real notarial document (in Spanish) with relevant information to extract: "`The mortgage duration is 30 years beginning on next June 10`". Content has been partially blurred or deleted for privacy reasons

les. **Compensación por desistimiento por amortización anticipada subrogatoria y no subrogatoria.** ----------

La parte prestataria tendrá la facultad de amortizar anticipadamente la totalidad o parte del capital del préstamo con las siguientes condiciones: ---

-Que dé aviso por escrito al Banco con un mes de antelación a la fecha de pago, indicando el importe de capital que desea reembolsar, -------------------

-Que dicho importe no sea inferior a **TRESCIENTOS EUROS CON CINCUENTA Y UN CÉNTIMOS DE EURO (300,51 €)**.

-Que abone también los débitos vencidos, que en su caso existieran, y los intereses que devengue el capital anticipadamente reembolsado hasta la fecha de pago. Estos intereses se calcularán por días al "tipo de interés vigente" en la citada fecha. ------------

En la fecha de pago, el Banco tendrá el derecho a percibir una compensación por desistimiento total y parcial equivalente: -----------------------------

a) al 0,50% del capital amortizado anticipadamente cuando la amortización anticipada se produzca dentro de los cinco primeros años de vida del préstamo, o --

- 7 -

Fig. 8 Example of fragment of real notarial document (in Spanish) with relevant and combined information to extract: "`The mortgage can be partially refunded in amounts not lower than € 300.51, paying a fee of 0.50% when done within the first 5 years`". Content has been partially blurred or deleted for privacy reasons

Acknowledgements This research work has been supported by projects TIN2013-46238-C4-4-R, TIN2016-78011-C4-3-R (AEI/FEDER, UE), and DGA/FEDER.

References

1. Rahm, E., Do, H.H.: Data cleaning: problems and current approaches. IEEE Data Eng. Bull. **23**(4), 3–13 (2000)
2. Curry, E., Freitas, A., ORiáin, S.: The role of community-driven data curation for enterprises. In: Linking Enterprise Data, pp. 25–47 (2010)
3. Gruber, T.R.: Toward principles for the design of ontologies used for knowledge sharing. Int. J. Human Comput. Stud. **43**(5–6), 907–928 (1995)
4. Buey, M.G., Garrido, A.L., Bobed, C., Ilarri, S.: The AIS project: boosting information extraction from legal documents by using ontologies. In: Proceedings of the 8th International Conference on Agents and Artificial Intelligence (ICAART 2016), pp. 438–445 (2016)
5. Wimalasuriya, D.C., Dou, D.: Ontology-based information extraction: an introduction and a survey of current approaches. J. Inf. Sci. **36**(3), 306–323 (2010)
6. Borobia, J.R., Bobed, C., Garrido, A.L., Mena, E.: SIWAM: using social data to semantically assess the difficulties in mountain activities. In: Proceedings of 10th International Conference on Web Information Systems and Technologies (WEBIST'14), pp. 41–48 (2014)
7. Garrido, A.L., Buey, M.G., Muñoz, G., Casado-Rubio, J.L.: Information extraction on weather forecasts with semantic technologies. In: International Conference on Applications of Natural Language to Information Systems (NLDB 2016), pp. 140–151. Springer International Publishing, Berlin (2016)
8. Maletic, J.I., Marcus, A.: Data cleansing. In: Data Mining and Knowledge Discovery Handbook, pp. 21–36. Springer, Boston, MA (2005)
9. Sarpong, K.A.M., Arthur, J.K.: Analysis of data cleansing approaches regarding dirty data-a comparative study. Int. J. Comput. Appl. **76**(7) (2013)
10. Yeganeh, S., Hassanzadeh, O., Miller, R. J.: Linking semistructured data on the web. In: Interface (2011)
11. Guo, W., Li, H., Ji, H., Diab, M.T.: Linking tweets to news: a framework to enrich short text data in social media. In: Proceedings of the 51st Annual Meeting of the Association for Computational Linguistics (ACL 2013), pp. 239–249 (2013)
12. Wang, J., Bansal, M., Gimpel, K., Ziebart, B.D., Clement, T.Y.: A sense-topic model for word sense induction with unsupervised data enrichment. Trans. Assoc. Comput. Linguist. **3**, 59–71 (2015)
13. Sekine, S., Ranchhod, E.: Named Entities: Recognition, Classification and Use. John Benjamins Publishing Company (2009)
14. Hu, Y., McKenzie, G., Yang, J.A., Gao, S., Abdalla, A., Janowicz, K.: A linked-data-driven web portal for learning analytics: data enrichment, interactive visualization, and knowledge discovery. In: LAK Workshops (2014)
15. Yosef, M.A.: U-AIDA: a customizable system for named entity recognition, classification, and disambiguation. Ph.D thesis, Saarland University (2016)
16. Suárez-Figueroa, M. C., Gómez-Pérez, A., Motta, E., Gangemi, A. Ontology engineering in a networked world. Springer Science and Business Media (2012)
17. Euzenat, J., Valtchev, P.: Similarity-based ontology alignment in owl-lite. In: Proceedings of the 16th European Conference on Artificial Intelligence (ECAI 2004), pp. 323–327. IOS Press, Amsterdam (2004)
18. Bollegala, D., Matsuo, Y., Ishizuka, M.: Measuring semantic similarity between words using web search engines. In: Proceedings of the 16th International World Wide Web Conference (WWW'07), pp. 757–766 (2007)

19. Jiang, Y., Wang, X., Zheng, H.T.: A semantic similarity measure based on information distance for ontology alignment. Inf. Sci. **278**, 76–87 (2014)
20. Yujian, L., Bo, L.: A normalized levenshtein distance metric. IEEE Trans. Pattern Anal. Mach. Intell. **29**(6), 1091–1095 (2007)
21. van Rijsbergen, C.J.: Information Retrieval, 2nd. edn. Butterworth-Heinemann (1979). ISBN 0408709294

To Improve, or Not to Improve; How Changes in Corpora Influence the Results of Machine Learning Tasks on the Example of Datasets Used for Paraphrase Identification

Krystyna Chodorowska, Barbara Rychalska, Katarzyna Pakulska and Piotr Andruszkiewicz

Abstract In this paper we attempt to verify the influence of data quality improvements on results of machine learning tasks. We focus on measuring semantic similarity and use the SemEval 2016 datasets. To achieve consistent annotations, we made all sentences grammatically and lexically correct, and developed formal semantic similarity criteria. The similarity detector used in this research was designed for the SemEval English Semantic Textual Similarity (STS) task. This paper addresses two fundamental issues: first, how each characteristic of the chosen sets affects performance of similarity detection software, and second, which improvement techniques are most effective for provided sets and which are not. Having analyzed these points, we present and explain the not obvious results we obtained.

Keywords Corpora quality · Annotation rules · SemEval · Semantic textual similarity

1 Introduction

A high quality corpus is needed in many NLP tasks. Often no publicly available corpus suits our needs and we have to create a new one. As we collect data and annotate it, there arises a question whether to further improve the corpus or just to start training our model.

K. Chodorowska · B. Rychalska · K. Pakulska · P. Andruszkiewicz (✉)
Samsung R&D Institute Poland, Plac Europejski 1, Warsaw, Poland
e-mail: p.andruszki2@samsung.com

K. Chodorowska
e-mail: k.chodorowsk@samsung.com

B. Rychalska
e-mail: b.rychalska@samsung.com

K. Pakulska
e-mail: k.pakulska@samsung.com

R. Bembenik et al. (eds.), *Intelligent Methods and Big Data in Industrial Applications*, Studies in Big Data 40, https://doi.org/10.1007/978-3-319-77604-0_25

We focused on semantic similarity measuring which is a standard NLP task: given a pair of sentences, a system assesses their similarity degree (on a predefined semantic scale). We decided to research the issue of corpora refinements because we have noticed that the datasests used for building models for semantic similarity measuring within SemEval competition [3] lack quality in both grammatical correctness and annotations on semantic similarity.

Early work on sentence similarity established the basic operation framework: sentence similarity is usually computed as a mean of word similarities across the two input sentences [11]. SemEval (Semantic Evaluation) is currently the most important competition aiming to select the best solutions in this area. In the training phase participants are asked to train their software using the dataset provided (an unsupervised approach is also possible). For the purpose of this task, a paraphrase is defined as a "restatement of an utterance in another form" [8], or a sentence that is "believed to maintain the same idea or semantic meaning" [13].

In the testing phase, participants are provided with an evaluation set to perform a similarity-measuring subtask using their software. The paraphrase detection software is trained on a monolingual corpus of parallel sentence with similarity score labels assigned. The training datasets provided by SemEval contain pairs of English sentences and are drawn from publicly available sources. They include, among others: MSR-Paraphrase [5], MSR-Vid [6], both from Microsoft Research, as well as the SMTeuroparl: WMT2008 and SMTnews [1] and some of the sets provided in the preceding years.

In this paper we present and analyze different types of errors in the above-mentioned datasets (Sect. 4). Then we discuss detailed annotation guidelines (Sect. 5) which might be helpful for SemEval organizers and participants. Moreover, the guidelines facilitate refinement of datasets with respect to consistency in annotation and grammatical correctness of sentences. To conclude our study, we verify the effect our improvements had on final results of the semantic similarity system (Sect. 6). To this end we use the software we developed for SemEval competition [9]. The system won 2016 SemEval English Semantic Textual Similarity task. We also separatelly used aligner [11] that won or was at the top in previous editions of SemEval.

2 Related Work

Quality of a corpus is really important, though only a few works can be found that investigate the influence of the quality of a corpus on the final results.

Labadié and Prince [4] verifies the influence of the quality of corpora by comparing the results of topic based text segmentation on cleaned and raw corpora. The authors claim that deterioration of corpora leads to poorer results. The influence has a slightly different pattern regarding the drop in precision or recall for two analyzed algorithms. Unfortunately, the authors compare the results on two different corpora: one cleaned and the second raw. The corpora have significantly different content. This is not the same corpus that has been cleaned but two different corpora. It is the same for two sets of corpora: natural and artificial. In total they used four different French corpora. Though the conclusions are reasonable, they may be questioned due to comparison of results obtained on different corpora.

The corpus quality influence on Chinese-English statistical machine translation has been investigated in [14]. The authors filter the noise and then compare results on original corpus and denoised ones. They use three methods, the third is a combination of the first two methods, that filter noise based on the length ration and the translation ratio. The first method achieves better results, second worse, and third only slightly better.

In [12], the quality, size and vocabulary coverage of a corpus in corpus-based cross-language information retrieval have been verified. The experiments show that topical nearness of the corpus and the translated queries is the most important from these tree factors. High quality of a corpus is also important, as the authors obtained better results for high quality corpus than for noisy one. The size of a corpus influence was not so clear. The experiments seem to show that to some threshold there is improvement and further the accuracy does not increase significantly.

In the paper, we investigate the influence of the quality of corpora on a different task - semantic similarity. We improve quality of semantic similarity corpora by performing both grammar and label corrections. Then we compare the results for the original and improved corpora.

3 Our Software: Short Review

In our research we intended to develop software that could detect both the similarity between single words and longer phrases. It employs two important components: the unfolding recursive autoencoder (RAE) [10] and the penalty-award weight system based on WordNet [7]. For maximum efficiency, we combine RAE and several other solutions, including a monolingual word aligner [11], in an ensemble classifier. Please refer to [9] for more detailed information about the system.

The complete pipeline included a RAE module, a WordNet module, a normalization module and a sentence similarity matrices computing module.

After parsing the sentences into dependency trees, RAE is used to learn word embeddings in the form of vectors, until a whole sentence vector is computed. Then sentence similarity matrices are computed to generate similarity scores for two candidate sentences. The WordNet module adjusts the Euclidean distance between RAE vectors with awards and penalties based on the semantic similarity of word pairs. For maximum efficiency, we combined RAE and several other solutions in an ensemble classifier. An important part of this combination, is a monolingual word aligner made of two algorithms. The basic algorithm aligns identical word sequences, named entities and content words (using dependencies and surrounding words, respectively) in the two sentences, outputs the alignment score and gives the correlation between the alignment score and the labels. Alignment score is calculated according to [11]:

$$score(S^1, S^2) = \frac{na(S^1) + na(S^2)}{n(S^1) + n(S^2)},$$

where $n(S^i)$ and $na(S^i)$ are the number of content words and aligned content words in a sentence Si, respectively. The corrected aligner is a Linear Support Vector Regression using the modified basic aligner, a Bag of Words and a number of other features.

4 SemEval Datasets: Data Characteristics

SemEval datasets were compiled using several popular methods used for development of paraphrase corpora. One of the most common approaches involves asking participants to provide one-sentence descriptions of pictures and then using their statements as paraphrases. The Images 2014 set [2] was most likely developed with this method and clearly not edited afterwards, as it contained a number of errors that could affect learning processes. We analyze those errors in this section.

4.1 Lexical Errors (LEX)

We noticed that some of the sentences provided in the Images 2014 dataset contain misleading lexical errors, such as: *A potted plant with only a few sprouting*. A human annotator would easily interpret this utterance as: *A potted plant with only a few sprouts*. However, paraphrase detection software relying on properties of words (e.g., their role as verbs or nouns) would not be able to process it properly.

4.2 Grammatical Errors (GRAM)

A common difficulty within these sets concerns pervasive lack of definite or indefinite articles (a typical phenomenon in written informal speech), for instance:

Black cow walking under trees in pasture versus
A cow walking under the tree in the pasture.

Such errors may affect work of the module aligning identical word sequences and context words which also relies on dependency relations.

4.3 Spelling Errors (SPELL)

A number of typing errors affected performance of our software, e.g.,

A longed-haired cat with it's eyes closed versus
A close up of a cat with its eyes closed.

The spelling error in the former sentence affects tokenization ("its" vs. "it" + "s" - two tokens instead of one), structure of the dependency tree, part-of-speech tagging as well as alignment, hindering proper processing. There were also sentences where spelling errors affected work of all RAE, Wordnet and aligner, making some words unrecognizable, e.g.: *People sitting on acouch* versus *People sitting on the porch.*

4.4 Unclear Scoring Procedure

The scores assigned to the sentence pairs do not seem to be obtained simply from calculating the average of five scores provided by five annotators, as the README file purports. If each sentence pair was assessed five times with a whole number, integer and the gold standard was the average of those five scores, it would not be possible for scores such as 4.667 or 4.714 to appear in the corpus. Moreover, the linguists who evaluated the sets probably had not been instructed on the purpose of the evaluation.

After applying grammatical, lexical and spelling corrections to faulty sentences, we learned the model on a split of a corrected Images 2014 set, then ran a test on another split, getting visibly improved results (more on this in Sect. 7). We attempted to apply these principles to other problematic sets, i.e. SMTeuroparl and SMTnews (2012) datasets, where we encountered other types of errors.

4.5 Mixed Languages (MIX)

The SMTeuroparl and SMTnews corpora were most likely compiled with another popular method, this time relying on machine translation. It involves either translating an utterance into another language and back again to get its paraphrase, or using multiple translations of the same sentence from different languages. These two corpora contain multiple traces of other languages, especially French, including both French vocabulary ("regresseront") and syntax ("A businessman Chinese influential, who share his time..."). Consider the following SMTeuroparl sentence pair with a score of 4.75:

Consumers will lose out, employees will lose out, Europe will lose competitive strength and growth.
The users are the losers, employees, and European competitiveness and growth regresseront.

The score indicates that all annotators but one considered those two pairs "completely equivalent." While Images 2014 dataset only features single, isolated mistakes, SMTeuroparl contains whole sentences that border on unintelligible, yet still received a high score (4.6): *Mr Gayssot, Greek ships have sunk without warning, when there were no storms in the vicinity.; Vessels, without which there is the slightest, Mr Gayssot.* An NLP system trained on corpora of grammatically correct sentences would not classify such a pair as high-level paraphrases.

4.6 External Rules (ERUL) Not Followed

The scoring is sometimes inconsistent with SemEval's rules. The following pair of sentences received a score of 5.00, despite the fact that one main verb ("dismissed") is not the synonym of the other ("disrupt"), and neither are the nouns ("demands" and "applications"):

Only a month ago, Mubarak dismissed demands for constitutional reform as 'futile'
Only a month ago, disrupted Muybarak applications of constitutional reform were a futile.

4.7 No Internal Consistency in Scoring (INC)

The gold standard score for sentences grouped in a single set is often internally inconsistent. Suppose we examined the following two pairs, where the first sentence is identical in both cases and the other two are its paraphrases (both ranked 5.00):

Only a month ago, Mubarak dismissed demands for constitutional reform as 'futile'
Only a month before, muybarak refused demands of constitutional reform by taxing them with 'futile'
Only a month ago, disrupted Muybarak applications of constitutional reform were a futile.

Table 1 Distribution of errors in SemEval datasets

Dataset	MIX	GRAM	ERUL	INC
SMTnews 399 total	92 **23%**	96 **24%**	124 **31%**	198 **50%**
SMTeuroparl training	73 **10%**	450 **61%**	409 **55%**	550 **75%**
SMTeuroparl test	21 **4%**	141 **30%**	141 **30%**	307 **66%**

While the first of the paraphrases contains a misspelling of the proper name and an incorrectly used preposition, the other one is incorrect in many more aspects. Importantly, sentence pairs with the same mistakes sometimes received different scoring, while differences in scoring sometimes did not account for the scope of actual disparity. Table 1 presents error distribution in the most problematic datasets. Error rate was counted as a proportion of sentences with an error of a given type to all sentences. Some sentences contained more than one type of error.

5 Annotation Guidelines

In response to the above problems (mostly to INC and ERUL-type errors), we devised a set of rules for proper annotation of the faulty datasets.

Prior to creating our scoring guidelines, we had conducted an in-depth review of all the published corpora and scrutinized their score labels. Our goal was to establish the following: how do annotators perceive the importance of additional or missing information with respect to the core semantic content of a sentence. We adhered to the 6-grade (0–5) scoring policy. While annotating a sentence pair, we first tried to match it to the rules of the highest reasonable score. If these rules were not satisfied, we moved to the lower levels until there was a match. Highest and lowest similarity ranges were easier to distinguish, while the middle ranges were more vague and posed difficulties.

5.1 Range 5 *Rules*

According to the SemEval Gold Standard: "The two sentences are completely equivalent, as they mean the same thing." From detailed analyses we concluded that a *range 5* paraphrase allows one of the following variations:

(I) The same elements are put in a different order:

(a) *If convicted, Paracha could face at least 17 years in prison.*
(b) *Paracha faces at least 17 years in prison if convicted.*

(II) Information content is the same, but synonyms are used.

(a) Sendmail said the **software** can even be set up to permit business-only usage.
(b) The **product** can be instructed to permit business-only use, according to Sendmail.

(III) Some modifiers are omitted that do not change the information content of a sentence:

(a) *FBI agents arrested a former partner of Big Four accounting firm Ernst & Young on* ***criminal charges*** *of obstructing federal investigations.*
(b) *A former partner of accountancy firm Ernst & Young was arrested by FBI agents on* ***charges*** *of obstructing federal investigations.*

(IV) Measuring units are converted ("about 3 km" vs. "two miles").

(a) A second shooting linked to the spree was a November 11 shooting at Hamilton Central Elementary School in Obetz, **about 3 km** from the freeway.
(b) Another shooting linked to the spree occurred Nov. 11 at Hamilton Central Elementary in Obetz, about **two miles** from the freeway.

5.2 Range 4 *Rules*

According to the SemEval Gold Standard: "The two sentences are mostly equivalent, but some unimportant details differ". If a paraphrase is ranked 4, it means both of its sentences contain almost the same pieces of information and differ only in small, relatively unimportant details, such as:

(I) Numbers appearing in the paraphrased sentence were rounded up or down. Rounding could be done up to full hundreds, provided that an "over", "more than", "under", "less than" modifier was used, e.g., more than 2,000 versus 2,100, such as in:

(a) A rebel who was captured said **more than 2,000** insurgents were involved in the attack.
(b) A captured rebel said **2,100** combatants had been involved in the offensive.

(II) When a precise date was modified, leaving a year and removing a day and/or month: e.g., "2001" versus "August 2001."

(a) DeVries was voluntarily castrated in **2001**, an operation he contends removed his ability to become sexually aroused.
(b) DeVries, who was voluntarily castrated in **August 2001**, has said the surgery took away his ability to become sexually aroused.

(III) When a person's first name was omitted, leaving just the last name, e.g., "Calvin Hollins Jr." versus "Hollins Jr."

(a) **Calvin Hollins Jr.**'s attorney, Thomas Royce, has repeatedly said his client had no link with the company that owned and operated E2.
(b) The lawyer for **Hollins Jr.**, Thomas Royce, has said his client had no link to the company that owned and operated E2.

(IV) When the sentences were identical but for the reported speech marker, such as "he said", "she said", "according to the paper", etc.

(V) When constituents of reported speech indicator other than a full sentence (including descriptions of place, time, manner, etc.), were missing:

(a) *"We condemn the Governing Council headed by the United States, "Muqtada al Sadr said* ***in a sermon at a mosque****.*
(b) *"We condemn and denounce the Governing Council, which is headed by the United States, "Moqtada al-Sadr said.*

(VI) When some modifiers or intensifiers providing important additional information were missing:

(a) *Earlier Thursday, PeopleSoft formally rejected the* ***unsolicited bid*** *from Oracle.*
(b) *Thursday morning, PeopleSoft's board rejected the Oracle takeover* ***offer****.*

5.3 Range 3 *Rules*

According to the SemEval gold standard: "The two sentences are roughly equivalent, but some important information differs/is missing". Creating an exhaustive definition for *Range 3* paraphrases was a huge challenge. Our own guidelines rely heavily on the "difference in information content", sometimes collapsing on imprecise terms (such as "multiple discrepancies" or "important sentence elements") which, however, tended to rise similar interpretations among linguists. The guidelines for *range 3* are as follows:

(I) Important sentence elements change to more general, e.g.: "Ben" versus "the boy."

(a) A passer-by found **Ben** hiding along a dirt road in Spanish Fork Canyon about 7 p.m., with his hands still taped together but his feet free.
(b) A passer-by found **the boy** hiding along a dirt road in Spanish Fork Canyon about 7 p.m. Thursday, with his feet free but his hands still taped.

(II) There occur tense differences in the main verb (apart from the specific cases covered in the Sect. 4.1), e.g.: "were published" versus "are being published."

(a) The findings **were published** in the July 1 issue of the Annals of Internal Medicine.
(b) The findings **are being published** today in the Annals of Internal Medicine.

(III) There are medium-level differences in information content, e.g.:

(1) part of a conjunct phrase (ABC and XYZ) is missing:
(a) *We need to change old habits and seriously rethink business-as-usual.*
(b) *He urged employees to* **"avoid complacency"** *and to "change old habits and seriously rethink business-as-usual.*

(2) a whole dependent clause (starting with "but," "so," "although", etc.) is missing:
(a) *The US Senate judiciary committee overcame a significant hurdle yesterday in the battle to create a trust fund to pay victims of asbestos exposure.*
(b) *The Senate judiciary committee on Tuesday overcame a significant hurdle in the battle to create a trust fund to pay victims of asbestos exposure* ***but the most difficult obstacles remain****.*

(3) a subordinate clause is missing:
(a) *I like apple pie.*
(b) *I like the apple pie* ***you bought****.*

(4) there are multiple discrepancies in constituent structure, i.e., when some constituents (modifiers, some dependent elements) are missing or were replaced by something else in a paraphrase. Typically two or three differences of this kind entail a *3 range* score:
(a) *Toll,* ***Australia's second-largest transport company****, last week offered NZ75 a share for Tranz Rail.*
(b) *Toll last week offered to buy the company for NZ75c a share,* ***or 158 million dollars****.*

5.4 Range 2 *Rules*

According to the SemEval Gold Standard: "The two sentences are not equivalent, but share some details". The difference between *range 2* and *range 3* paraphrases was doubly difficult to grasp and express in terms of guidelines. SemEval organizers even acknowledged this in the 2015 edition of the Spanish task and eliminated one tier of score entirely, arguing that the difference in information content for tiers 2 and 3 was not clear-cut and too vague to easily translate into Spanish (Agirre et al., 2015). Our specifics for level 2 paraphrases were as follows:

(I) Sentences contain different numbers (apart from rounding up/down, as mentioned in Sect. 4.1).

(II) There are major differences in information content. In order to establish what constituted a major difference lowering the score to 2, we concentrated on the "key" sentence elements:

(1) different objects (often occurring in short paraphrases), e.g.: "flute" versus "guitar".

(a) A man is playing a **flute**.
(b) A man is playing **guitar**.

(2) more than 3 pairs of different constituents:
(a) *The* ***dead*** *woman was also wearing a* ***ring*** *and a Cartier watch.*
(b) *"It's a* ***blond-haired*** *woman wearing a Cartier watch on her wrist,"* ***the source said.***

(3) missing main clause (with subordinate clause still present):
(a) ***He will replace*** *Ron Dittemore, who announced his resignation April 23.*
(b) *Dittemore announced his plans to resign on April 23.*

(4) considerable changes in reported phrases with speakers and circumstances left unchanged:
(a) ***"We don't know if any will be SARS,"*** *said Dr. James Young, Ontario's commissioner of public safety.*
(b) ***"We're being hyper-vigilant,"*** *said Dr. James Young, Ontario's commissioner of public safety.*

5.5 Range 1 *Rules*

According to the SemEval Gold Standard: "If a paraphrase is ranked as 1, it means that the two sentences are very different in meaning, but are on a similar topic". Our rule of thumb guidelines for this range are as follows:

(I) There are major differences in the subject/main verb/object group, such as:

(a) different Subject + same Verb + different Object
(b) same Subject + different Verb + different Object
(c) same Subject + different Verb + same Object

(II) Names or other details concerning speakers are the same, but reported parts of a sentence are completely different:

(a) ***"This child was literally neglected to death,"*** *Armstrong County District Attorney Scott Andreassi said.*
(b) *Armstrong County District Attorney Scott Andreassi said* ***the many family photos in the home did not include Kristen.***

5.6 Range 0 *Rules*

If a paraphrase is ranked 0, it means that the two sentences deal with entirely different topics (in other words, they are not related to each other to the extent required for paraphrases).

6 Results

We split the Images 2014 dataset into a training dataset of 750 pairs and a test set of 151 pairs, and ran them through the basic detection software (RAE + Wordnet) to establish a baseline for the uncorrected set. The results for original and corrected files can be found in Table 2. The fist column indicates the dataset and its state: corrected or original. The second and third column show Pearson correlation for RAE and aligner. The last column is a mean score generated by aligner.

Contrary to SMTeuroparl and SMTnews (please refer to Table 1), most of the labels for this set seemed to reflect the SemEval Golden Standard; that is, were properly assigned, so we did not change them. The Pearson correlation between the original Gold Standard and the result we got constituted 79%. We used the Pearson correlation as this measure was used in SemEval competition. After applying grammatical, lexical and spelling corrections, we performed another run, this time getting a result of 82%. We also tested the corrected Images 2014 on the sole aligner, where it got a correlation of 76% with the original dataset and 77% with the corrected one.

Before applying the same procedure to the SMTeuroparl dataset, we had to apply our guidelines to the paraphrases with questionable scores (see Sect. 5). We adopted SemEval's practice of labelling sentence pairs with an average of several annotations. The observed proportional agreement of our annotators amounted to 0.42 on the training set and 0.62 on the test set, and the disparity between scores rarely exceeded 1, so we did not make any further adjustments to the score. Then we used the original SMTeuroparl training (750 pairs) and test (459 pairs) datasets on a basic RAE + Wordnet detector, achieving a correlation of 46.8%. As we applied corrections and used a modified set of labels, the correlation dropped to 41.9%.

There could be several reasons for this discrepancy. Tests on the aligner showed a 49% correlation for the original SMTeuroparl set labels and the alignment score. On the other hand, the correlation for the alignment and the corrected SMTeuroparl was as low as 38%, even though the average of the alignment score itself for the corrected version was slightly higher (1.81 vs. 1.80 for the uncorrected version).

Table 2 RAE and aligner results of original and corrected datasets

Dataset	RAE (%)	Aligner (%)	Aligner score
Images 2014 (original)	79	76	1.64
Images 2014 (corrected)	82	77	1.65
SMTeuroparl (original)	46.8	49	1.80
SMTeuroparl (corrected)	41.9	42	1.81

We attempted to work out why the seemingly reasonable adjustments which improved performance for the Images 2014 dataset not only failed to improve detection on the SMTeuroparl dataset, but also impaired it. First, the SMTeuroparl dataset is visibly unbalanced. The average score for the original label set is 4.31, with *range 4* paraphrases making up as much as 68% of the dataset. SemEval does not publish data on annotator agreements for the individual sets. The pairs seem to rarely be assessed as 5-score, while ranges granted by our annotators (following the annotation guidelines) often included 5. Moreover, annotators working with us could only use full numbers, while the software operates on fractions. Often the results from the detector approached the previous uncorrected label (for instance, 4.8 and a unanimous 5 from our annotators). Further, the set (being a transcript of EU parliamentary proceedings) was incredibly context-specific. Besides, most other sets penalized time difference. For instance, the pair [6]:

The cat played with a watermelon.
The cat is playing with a watermelon

is assessed as a *4 range* pair in the MSRvid dataset, and we decided to keep that principle in the annotation guidelines (see Sect. 5.3). However, in political speeches, a difference in grammatical tense is not always considered a difference in meaning, because a sentence still expresses the same idea. It is likely that for this reason SemEval's annotators did not always choose to penalize tense differences and our annotators complained about having to do so. Ironically enough, adhering to our guidelines and using score penalty for tense differences provided additional handicap to our software, as the RAE uses lemmas with no account of tense, aspect, mood, or voice. The same holds for the aligner which is tuned to reflect SemEval's way of annotation and did not acquire new rules, as it is unsupervised.

7 Conclusions and Future Work

The quality of SemEval datasets is not high enough. We analyzed these sets, summarized errors and provided precise, unified annotation guidelines. Moreover, we verified the influence of corrections on the results obtained by our semantic similarity systems. Considering our findings, two conclusions arise. The first and most obvious one is that using impure texts (e.g., with spelling mistakes or traces of other languages) can be confusing for software relying on properties of, and tools designed for a specific language, such as word embeddings trained on monolingual corpora. Second, proceedings of the European Parliament are extremely difficult to process semantically even for human speakers, due to their high level of abstraction, figurative character, and complex synonymous expressions whose link to the respective synonyms is difficult to establish. These texts do not easily conform to annotation guidelines that seem to do for other sets, making it harder to devise a unified scoring procedure. Third, choosing a machine-translated, unedited text severely complicates scoring, as it becomes unknown how to treat a sentence whose phrasing is deeply incorrect but whose meaning can be inferred from the other sentence. Another question is whether making world-knowledge part of a benchmark for measuring computer software's performance is reasonable. The datasets contain pairs that are awarded high score by annotators who relied on their "knowledge of the world". Correct recognition of such paraphrases would be rather tough for a system. We plan to research this in our future work.

References

1. ACL 2008 Third Workshop on Statistical Machine Translation: WMT2008 Development Dataset (2008). http://www.statmt.org/wmt08/shared-evaluation-task.html/
2. Agirre, E., Banea, C., Cardie, C., Cer, D.M., Diab, M.T., Gonzalez-Agirre, A., Guo, W., Mihalcea, R., Rigau, G., Wiebe, J.: Semeval-2014 task 10: multilingual semantic textual similarity. In: Nakov, P., Zesch, T. (eds.) Proceedings of the 8th International Workshop on Semantic Evaluation, SemEval@COLING 2014, Dublin, Ireland, August 23–24, pp. 81–91. The Association for Computer Linguistics (2014). http://aclweb.org/anthology/S/S14/S14-2010.pdf
3. Bethard, S., Cer, D.M., Carpuat, M., Jurgens, D., Nakov, P., Zesch, T. (eds.): Proceedings of the 10th International Workshop on Semantic Evaluation, SemEval@NAACL-HLT 2016, San Diego, CA, USA, June 16–17, 2016. The Association for Computer Linguistics (2016). http://aclweb.org/anthology/S/S16/
4. Labadié, A., Prince, V.: The impact of corpus quality and type on topic based text segmentation evaluation. In: Proceedings of the International Multiconference on Computer Science and Information Technology, IMCSIT 2008, Wisla, Poland, 20–22 October 2008. pp. 313–319. IEEE (2008). http://dx.doi.org/10.1109/IMCSIT.2008.4747258
5. Microsoft Research: Microsoft Research Paraphrase Corpus (2010). http://research.microsoft.com/en-us/downloads/607d14d9-20cd-47e3-85bc-a2f65cd28042/
6. Microsoft Research: Microsoft Research Video Description Corpus (2010). https://www.microsoft.com/en-us/download/details.aspx?id=52422
7. Miller, G.A.: Wordnet: a lexical database for english. Commun. ACM **38**, 39–41 (1995)
8. Rus, V., Banjade, R., Lintean, M.: On Paraphrase Identification Corpora (2015). http://www.researchgate.net/publication/280690782_On_Paraphrase_Identification_Corpora/
9. Rychalska, B., Pakulska, K., Chodorowska, K., Walczak, W., Andruszkiewicz, P.: Samsung Poland NLP Team at SemEval-2016 task 1: Necessity for diversity; combining recursive autoencoders, wordnet and ensemble methods to measure semantic similarity. In: Bethard et al., [3], pp. 602–608. http://aclweb.org/anthology/S/S16/S16-1091.pdf
10. Socher, R., Huang, E.H., Pennington, J., Ng, A.Y., Manning, C.D.: Dynamic Pooling and Unfolding Recursive Autoencoders for Paraphrase Detection (2011)
11. Sultan, M.A., Bethard, S., Sumner, T.: DLS@CU: Sentence Similarity From Word Alignment and Semantic Vector (2015). http://alt.qcri.org/semeval2015/cdrom/pdf/SemEval027.pdf
12. Talvensaari, T.: Effects of aligned corpus quality and size in corpus-based CLIR. In: Macdonald, C., Ounis, I., Plachouras, V., Ruthven, I., White, R.W. (eds.) Advances in Information Retrieval, 30th European Conference on IR Research, ECIR 2008, Glasgow, UK, March 30–April 3, 2008. Proceedings. Lecture Notes in Computer Science, vol. 4956, pp. 114–125. Springer, Berlin (2008). http://dx.doi.org/10.1007/978-3-540-78646-7_13
13. Ul-Qayyum, Z., Altaf, W.: Paraphrase Identification Using Semantic Heuristic Features (2012). http://maxwellsci.com/print/rjaset/v4-4894-4904.pdf
14. Zhou, Y., Liu, P., Zong, C.: Approaches to improving corpus quality for statistical machine translation. Int. J. Comput. Proc. Oriental Lang. **23**(4), 327–348 (2011). http://dx.doi.org/10.1142/S1793840611002395

Context Sensitive Sentiment Analysis of Financial Tweets: A New Dictionary

Narges Tabari and Mirsad Hadzikadic

Abstract Sentiment analysis can make a contribution to behavioral economics and behavioral finance. It is concerned with the effect of opinions and emotions on economical or financial decisions. In sentiment analysis, or in opinion mining as they often call it, emotions or opinions of various degrees are assigned to the text (tweets in this case) under consideration. This paper describes an application of a lexicon-based domain-specific approach to a set of tweets in order to calculate sentiment analysis of the tweets. Further, we introduce a domain-specific lexicon for the financial domain and compare the results with those reported in other studies. The results show that using a context-sensitive set of positive and negative words, rather than one that includes general keywords, produces better outcomes than those achieved by humans on the same set of tweets.

Keywords Sentiment analysis · Twitter · Financial sentiment analysis · Lexicon

1 Introduction

Sentiment analysis has been a promising approach for many researchers in the past few years. This area focuses on the investigation of opinion or emotion of people on different aspects of life. With the rapid growth of textual data, such as social networks and micro-blogging applications, the need for analyzing these texts has increased as well. The ability to analyze a vast amount of information on topics as diverse as companies, products, social issues, or political events has made sentiment analysis an influential field of research, mostly because it offers a window into understanding human behavior. As an example, Bollen et al. [1] presented evidence of predicting the size of markets using social-media sentiment analysis.

N. Tabari (✉) · M. Hadzikadic
UNC Charlotte, Charlotte, USA
e-mail: nseyedit@uncc.edu

M. Hadzikadic
e-mail: mirsad@uncc.edu

R. Bembenik et al. (eds.), *Intelligent Methods and Big Data in Industrial Applications*, Studies in Big Data 40, https://doi.org/10.1007/978-3-319-77604-0_26

During the past few years with the growth in usage of micro blogging applications, such as Twitter, it has rapidly become more popular among people in various professions. Not only professionals, celebrities, companies, and politicians use Twitter regularly, but also other people such as students, employees, and costumers have been using this service widely. The popularity of Twitter helps researchers obtain proper understanding of various topics from different views of people.

Financial markets issues have become a very popular context for analyzing Twitter data. It was implied in previous research reports that, if properly modeled, Twitter data can be used to derive useful information about the markets. This is why we decided to select financial markets as the context for our research in sentiment analysis.

This paper lays the foundation for a better understanding of the relationships between financial markets and social media. In this regard, we present here a sentiment analysis of tweets in order to extract a signal for an action (buy, sell) in financial markets. For this purpose, we compared two different word lists (dictionaries) for analyzing tweets gathered about Bank of America. First, the word list was generated specifically for financial texts based on [2], while the second word list was created for this project based on the frequency of words in a large financial dataset of tweets. We then used Mechanical Turk to label our data in order to compare them to our results. Our sentiment analysis scores (with F-score = 64.9%) concluded that it is better to use word-lists based on informal texts for tweets, rather to use ones created for formal texts.

The remainder of this paper is structured as follows. A related work on tweet-level and entity-level sentiment analysis is discussed in Sect. 2. Section 3 focuses on our approach to the context sensitive analysis, while Sect. 3.2 elaborates on the results of the inquiry. The future work is covered in Sect. 4. Finally, we summarize our work in Sect. 5.

2 Related Work

2.1 General Approaches to Sentiment Analysis

Emergence and growth of large volume of social media data in recent years give rise to the need to analyze and investigate it. Understanding what people are tweeting about is an important part of this analysis. Sentiment analysis is a field of assigning sentiments (positive, negative, neutral, or other categories) to tweets, or to texts in general. Although text categorization as a research direction was introduced fairly long time ago by Salton and McGill [3], sentiment analysis of texts was introduced more recently [4].

Twitter is one of the most popular sources for text sentiment analysis. The growth of the number of tweets per day from 5000 tweets per day in 2007 to recent 500 million tweets per day demonstrates the substantially increased level of participation of people in social media. This makes Twitter as the most suitable platform for text analysis today. Another advantage of Twitter is the fact that tweeting is not reserved for people only. Often organizations and companies, through their official representatives, tweet as well. Even more importantly for the increased popularity of tweeting, celebrities are often seen as the most fervent users of the Twitter service. This gives researchers the opportunity to gain a broad perspective on public opinions in any context. Using Twitter sentiment analysis, we now have the ability to use scientific approaches to discern emotions behind people's tweets.

The main two methods for sentiment analysis include supervised and unsupervised analysis. The initial approach to text representation [3] used a bag of words method. In subsequent studies, both lexicon-based and machine learning supervised approaches relied on the bag of words method. In the machine learning approach, the bag of words is used as a classifier, whereas in the lexicon-based approach it is being used as a guide in assigning the polarity score to tweets. Although the overall polarity score of the text is calculated using various formulas, the most common method of computation is a simple summation of all polarity scores.

Supervised Machine Learning Methods. Most of the machine learning methods used in sentiment analysis are classification methods that are based on previously labeled word lists. The first step in sentiment analysis machine learning methods is to create the features to be used to learn the resulting model. This model can then ultimately differentiate between labels in the unlabeled dataset. The most popular SAMLs used in sentiment analysis are Naïve Bayes [5, 6], Support Vector Machines [7, 8], and maximum entropy [9].

One of the limitations of using supervised learning methods is that updating the training data is a very difficult job [10], especially given the rate of conversions and modifications on Twitter. Go et al. [11] used a distant supervision approach that generates an automatic training data set using emoticons included in the tweets. This approach increases the error rate of analysis, which obviously may affect the performance of classifiers [12]. Another limitation of machine learning methods is that often classifiers trained in one context perform poorly in another context [13].

Lexicon-based Methods. A lexicon-based approach is an alternative way to analyze text in order to assign emotions. It works by using the sentiment of each word in a bag-of-words approach. Emotions are assigned to each word, which enables the user of the word list, to figure out the overall sentiment of the text. Therefore, this approach dispenses with the need to procure a training data set and devise a classification technique. There are many dictionaries that have been built over time for this method, including SentiWordNet [14], MPQA subjectivity lexicon [15], the LIWC lexicon [16–18], Harvard Dictionaries, and [19]. Many of these word lists do not just assign polarity, but they also assign multiple ranges of sentiment to the text.

Although by using a lexicon-based approach we eliminate deficiencies of machine learning methods, the lexicon-based methods themselves can be restrictive by their lexicons as well. This creates inefficiencies in the process of analysis, as researchers have to use the assigned, static sentiment of words in dictionaries regardless of the context. In different research projects, new methods were introduced which manually train texts to solve this restriction [20].

Contextual Sentiment Analysis. Contextual sentiment analysis on the other hand focuses on sentiment analysis in a specific context. Most frequently used methods for contextual semantic analysis use the frequency and the co-occurrence of words. Then, mathematical approaches are applied to most frequent words to evaluate their accuracy, including: weighting the elements, smoothing the matrix, and comparing the vectors. Turney and Pantel [21] used a very simple Support Vector Machine (SVM) to calculate the value of an element in a document vector as the number of times the corresponding word occurs in the given document. Turney and Littman [22] used two different statistical measures for word association: point-wise mutual information (PMI) and latent semantic analysis (LSA), both based on co-occurrence of words. Their basic idea was that "a word is characterized by the company it keeps" (Firth 1957). In our research, we follow the most frequent words used in the chosen context.

2.2 Financial Sentiment Analysis

Financial texts, as the context of special interest in this research, are being used widely for analyzing or investigating various areas of finance such as corporate finance, financial markets, investment, banking, and asset and derivative pricing. Most of financial analyses on texts, such as news or social media, are aiming to predict either the future prices of assets and stocks or the risk of a financial crash.

A financial context sentiment analysis in [23] was implemented by applying the SentiWordNet word list in order to correlate the result to the market movement. They used the log probability of each token in the word. The log probability of all tokens in each tweet represents the probability of 'happy' and 'sad' labels for the entire tweet. Then, they counted the frequency of 'happy' and 'sad' tweets each day to calculate the sentiment percentage of all tweets per day. O'Hare et al. [6] analyzed financial blogs and showed that word-based approaches perform better than sentence-based or paragraph-based ones. Loughran and Mcdonald [2] used text analysis to show that specialized financial word lists must be created for analyzing financial texts. They developed a specialized word list for financial domains, since they found out that 73.8% of the negative word counts according to the Harvard list is attributable to words that are typically not negative in financial contexts. For example, the words "decrease" and "increase" are very ambiguous in the financial world and cannot be counted as negative or positive per se. Consequently, they created word lists that have negative/positive implications in the financial context.

In our study, we show that Loughran and McDonald's [2] word list, even though it was created for financial contexts, is not as effective when it comes to the informal texts, such as tweets.

3 Approach

This study focuses on the lexicon-based approach to context sensitive tweets. The targeted goal is to assign "Positive" or "Negative" emotions to each tweet mentioning a specific financial institution, in this case Bank of America. The data was streamed from Twitter using Twitter API. In order to simplify the analysis, we selected one context and targeted English tweets focused on Bank of America with "BofA", "Bankofamerica", or "Bank of America" keywords. This lexicon-based analysis focuses on the sentences and words that people used in each tweet. For this purpose, after selecting the data in the pre-processing step we removed from each tweet all punctuations, control characters, numbers, emoticons, and links.

3.1 Analysis

First, we created a list of tweets with 200 manually selected tweets. This list contained 100 tweets for each different absolute emotion, positive and negative. In order to calculate the sentiment score and assign polarity to each tweet, one positive point was assigned to each count of positive word in the tweet and one negative point for each negative word. Finally, the sentiment score was calculated by subtracting the positive scores from the negative scores of each tweet, resulting in the overall score for each dataset.

We decided to use Amazon's Mechanical Turk as the benchmark dataset for comparison. In Mechanical Turk each tweet was analyzed by 20 different people and assigned sentiments accordingly to each of the tweets. We used the mean of those 20 scores in order to create the overall sentiment of each tweet.

Next, we used two different dictionaries to compare with the Mechanical Turk's results. The first wordlist was from McDonald. Then, the second word list was created by us. In order to create this list, we gathered six months' worth of the filtered tweet data, from April to October 2015. We used these tweets to create a list of most frequent words used in those tweets. After eliminating the stop-words (e.g., as, is, on and which) from the list of most frequent ones, a positive or negative sentiment was then manually assigned to each of the remaining words. Finally, we created the lists of 103 positive and 97 negative words.

3.2 Comparison

As presented in Table 1, our word list achieved a better result than McDonalds' in both accuracy and f-score when referenced against the "objective" outcome of the Mechanical Turk. We used the positive and negative values in the Mechanical Turk list as our actual positive and negative sentiments. By calculating the *confusion*

Table 1 F-score and accuracy comparison of different analysis

Wordlist	Accuracy (%)	F-score (%)
Our list	65.3	64.9
McDonald	64.2	63.8

matrix in both our word list and that of McDonald's, we demonstrated improvement in both accuracy and f-score.

Therefore, we believe that we demonstrated that for context-sensitive analyses one should not use general-type word lists that have been used for many other purposes. Rather, a context-specific word list should be preferred. The improvement over McDonald's word list, which is actually created for financial purposes, is a proof to our claim that using the wordlists for formal purposed financial texts is not suitable for informal texts, such as tweets.

Furthermore, our approach was based on the words in each tweet instead of the understanding of the semantics of complete sentences/tweets, meaning that the occurrences and Part-of-Speech in sentences were not considered. Obviously, a semantic processing of each tweet would render even better results in sentiment analysis.

4 Future Work

This work is a preliminary work in context-sensitive, lexicon-based sentiment analysis. The main purpose of this work is to solve the first component of a much larger project; to investigate the effect and influences of financial markets and social media on each other. We hope to improve our word list even further using additional machine learning approaches, such as Random Forest classification. This improvement is critical to make sure that we understand the sentiment of each tweet before we can attempt to extract financial signals from tweets.

5 Conclusion

Sentiment analysis is defined as the use of various approaches to assign emotions to text. In this paper we tackled the problem of analyzing the sentiment of tweets in the context of financial markets. We collected Bank of America-related tweets and applied two different word lists, using a lexicon method on the collected data. One list was our context-sensitive word list using most frequent words in Bank of America-related tweets. The other list was that of McDonald's. Both lists were compared to the outcome of the Mechanical Turk. In the paper we demonstrated that our context-sensitive word list performed better than McDonald's in both the f-score and accuracy.

References

1. Bollen, J., Mao, H., Zeng, X.: Twitter mood predicts the stock market. J. Comput. Sci. **2**(1), 1–8 (2011)
2. Loughran, T.I.M., Mcdonald, B.: When is a Liability not a Liability ? Textual Analysis, Dictionaries, and 10-Ks Journal of Finance, forthcoming (2010)
3. Salton, G., McGill, M.: Introduction to Modern Information Retrieval, xv + 448 pp., $32.95. McGraw-Hill, New York (1983). ISBN 0-07-054484-0
4. Das, S.R., Chen, M.Y.: Yahoo! for Amazon: sentiment extraction from small talk on the web. Manag. Sci. **53**(9), 1375–1388 (2007)
5. Saif, H., He, Y., Alani, H.: Semantic sentiment analysis of twitter. In: Lecture Notes Computer Science (including Subser. Lecture Notes in Artificial Intelligence and Lecture Notes in Bioinformatics), vol. 7649 LNCS, no. PART 1, pp. 508–524 (2012)
6. O'Hare, N., Davy, M., Bermingham, A., Ferguson, P., Sheridan, P.P., Gurrin, C., Smeaton, A.F., OHare, N.: Topic-dependent sentiment analysis of financial blogs. In: International CIKM Workshop on Topic-Sentiment Analysis for Mass Opinion Measurement, pp. 9–16, 2009
7. Mohammad, S.M., Kiritchenko, S., Zhu, X.: NRC-Canada: building the state-of-the-art in sentiment analysis of tweets. In: Proceedings of the Seventh International Workshop on Semantic Evaluation. Exercises, vol. 2, no. SemEval, pp. 321–327 (2013)
8. Hamdan, H.: Experiments with DBpedia, WordNet and SentiWordNet as resources for sentiment analysis in micro-blogging. In: Seventh International Workshop on Semantic Evaluation (SemEval 2013)—Second Joint Conference on Lexical and Computational Semantics, vol. 2, no. SemEval, pp. 455–459 (2013)
9. Da Silva, N.F.F., Hruschka, E.R., Hruschka, E.R.: Tweet sentiment analysis with classifier ensembles. Decis. Support Syst. **66**, 170–179 (2014)
10. Liu, B.: Sentiment analysis and opinion mining. Synth. Lect. Hum. Lang. Technol. **5**(1), 1–167 (2012)
11. Go, A., Bhayani, R., Huang, L.: Twitter sentiment classification using distant supervision. Processing **150**(12), 1–6 (2009)
12. Speriosu, M., Sudan, N., Upadhyay, S., Baldridge, J.: Twitter polarity classification with label propagation over lexical links and the follower graph. In: Conference on Empirical Methods in Natural Language Processing, pp. 53–56 (2011)
13. Aue, A., Gamon, M.: Customizing sentiment classifiers to new domains: a case study. Proc. Recent Adv. Nat. Lang. Process. **3**(3), 16–18 (2005)
14. Baccianella, S., Esuli, A., Sebastiani, F.: SentiWordNet 3. 0: an enhanced lexical resource for sentiment analysis and opinion mining SentiWordNet. Analysis **0**, 1–12 (2010)
15. Wilson, T., Wiebe, J., Hoffman, P.: Recognizing contextual polarity in phrase level sentiment analysis. ACL **7**(5), 12–21 (2005)
16. Pennebaker, J.W., Graybeal, A.: Patterns of natural language use: disclosure, personality, and social integration. Curr. Dir. Psychol. Sci. **10**(3), 90–93 (2001)
17. Andreevskaia, A., Bergler, S.: When specialists and generalists work together: overcoming domain dependence in sentiment tagging. In: Proceedings of the ACL-08 HLT, no. June, pp. 290–298 (2008)
18. Neviarouskaya, A., Prendinger, H., Ishizuka, M.: SentiFul: generating a reliable lexicon for sentiment analysis. In: Proceedings of the 2009 3rd International Conference on Affective Computing and Intelligent Interaction Workshop, ACII (2009)
19. Hu, M., Liu, B.: Mining and summarizing customer reviews. In: Proceedings of the 2004 ACM SIGKDD International Conference on Knowledge Discovery and Data Mining, KDD 04, vol. 4, p. 168 (2004)
20. Thelwall, M., Buckley, K., Paltoglou, G., Cai, D.: Sentiment strength detection in short informal text. Am. Soc. Inf. Sci. Technol. **61**(12), 2544–2558 (2010)
21. Turney, P.D., Pantel, P.: ★★★★★ From Frequency to Meaning Vector Space Models of Semantics (讲的非常好, 但是我还只看了三分之一), vol. 37, pp. 141–188 (2010)

22. Turney, P.D., Littman, M.L.: Unsupervised learning of semantic orientation from a hundred-billion-word corpus. Tech. Rep. NRC Tech. Rep. ERB-1094, Inst. Inf. Technol., p. 11 (2002)
23. Chen, R., Lazer, M.: Sentiment Analysis of Twitter Feeds for the Prediction of Stock Market Movement, pp. 1–5 (2013)

Index

R. Bembenik et al. (eds.), *Intelligent Methods and Big Data in Industrial Applications*, Studies in Big Data 40, https://doi.org/10.1007/978-3-319-77604-0

Printed in the United States
By Bookmasters